Wiring Regulations Pocket Book

This new Routledge Pocket Book provides a user-friendly guide to the latest amendments to the 18th Edition of IET Wiring Regulations (BS 7671:2018).

This Pocket Book contains topic-based chapters that link areas of working practice with the specifics of the Regulations themselves. The requirements of the Regulations are presented in an informal, easy-to-read style that strips away confusion. Packed with useful hints and tips that highlight the most important or mandatory requirements, the book is a concise reference on all aspects of the 18th Edition of the IET Wiring Regulations.

This handy guide provides an on-the-job reference source for Electricians, Designers, Service Engineers, Inspectors, Builders and Students.

Ray Tricker (MSc, IEng, CQP-FCQI, FIET, FCMI, FIRSE) is a Senior Consultant with over 60 years continuous service in Quality, Safety, Environmental and Project Management, Communication Electronics, Railway Command, Control and Signalling Systems, Information Technology and the development of Molecular Nanotechnology.

He served with the Royal Corps of Signals (for a total of 37 years) during which time he held various managerial posts culminating in being appointed as the Chief Engineer of NATO's Communication Security Agency (ACE COMSEC).

Most of Ray's work since leaving the Services has centred on the European Railways. He has held a number of posts with the Union International des Chemins de fer (UIC) [e.g. Quality Manager of the European Train Control System (ETCS)] with the European Commission [e.g. T500 Review Team Leader, ERTMS Users Group

Project Coordinator, HEROE Project Coordinator] and currently (as well as writing over 60 books on diverse subjects for Taylor & Francis and Elsevier) he is busy assisting Small Businesses from around the world (usually on a no cost basis) to produce their own auditable Integrated Management Systems sufficient to meet the requirements of ISO 9001, ISO 14001 and OHSAS 18001, etc. He is also a UKAS Assessor and recently he was appointed as the Quality, Safety and Environmental Manager for the Project Management Consultancy responsible for overseeing the multi-billion-dollar Trinidad Rapid Rail System.

Wiring Regulations Pocket Book

Ray Tricker

LONDON AND NEW YORK

First published 2022
by Routledge
2 Park Square, Milton Park, Abingdon, Oxon OX14 4RN

and by Routledge
605 Third Avenue, New York, NY 10158

Routledge is an imprint of the Taylor & Francis Group, an informa business

British Library Cataloguing-in-Publication Data
A catalogue record for this book is available from the British Library

Library of Congress Cataloging-in-Publication Data
Names: Tricker, Ray, author.
Title: Wiring regulations pocket book / Ray Tricker.
Description: Abingdon, Oxon ; New York, NY : Routledge, 2022. |
Includes bibliographical references and index.
Identifiers: LCCN 2021019257 (print) | LCCN 2021019258 (ebook) |
ISBN 9780367760304 (hbk) | ISBN 9780367760090 (pbk) |
ISBN 9781003165170 (ebk)
Subjects: LCSH: Electric wiring–Insurance requirements–Handbooks, manuals, etc. | Electric wiring–Specifications–Handbooks, manuals, etc.
Classification: LCC TK3275 .T75 2022 (print) | LCC TK3275 (ebook) | DDC 621.319/24–dc23
LC record available at https://lccn.loc.gov/2021019257
LC ebook record available at https://lccn.loc.gov/2021019258

ISBN: 978-0-367-76030-4 (hbk)
ISBN: 978-0-367-76009-0 (pbk)
ISBN: 978-1-003-16517-0 (ebk)

DOI: 10.1201/9781003165170

Typeset in Goudy Oldstyle Std
by KnowledgeWorks Global Ltd.

Contents

Figures

Tables

Foreword

BS 7671 Requirements for Electrical Installations – IEE Wiring Regulations has been an integral part of most of my professional life as the guide to safe and acceptable practice. Yet it is a difficult task to make the requirements readily accessible to a diverse readership, whilst remaining up to date, with the evolution of legislation, technological developments, and changes in custom and practice. Commentaries and further explanations are, therefore, often helpful and are sometimes essential reading to gain sufficient understanding to carry out work associated with electrical installations.

I am, therefore, pleased to contribute in a small way to Ray's book, which sheds another useful light on the Wiring Regulations in pocketbook form and is thoughtfully condensed from Ray's Wiring Regulations in Brief – fourth edition. Significantly for me, and perhaps many other older hands, both books place these Regulations within their context in Part P of the Building Regulations and cross-references requirements. Ray's pocketbook is a handy contribution to working with the Wiring Regulations and Part P of the Building Regulations where just brief information is needed 'in the field' or on site.

I like the simple and condensed approach, not delving into the fine detail and making information easy to find by grouping it and laying it out logically. This is what most readers need much of the time. This book can save the cost of buying parts of the Building Regulations, making it a cost-effective option. Some readers may find that they do not need to buy the Wiring Regulations as well.

Modern life would be impossible without electricity. Yet it is potentially dangerous to property, humans, and many living creatures. Only the application of human ingenuity can reduce the risks to levels that are deemed acceptable. That we can live, work, and play surrounded by live electrical equipment and installations is testimony to the dedication, understanding, and skills of many people. It is also testimony to the success of the Wiring Regulations, in its many revisions and editions over the years, and books such as Ray's Wiring Regulations in Brief – fourth edition and this pocketbook, which help to make the Regulations more accessible to every user.

Nigel Moore BSc (Hons), MBA, CEng, MIET, MCIM, HNC

Preface

BS 7671:2018 '*Requirements for Electrical Installations. IET Wiring Regulation*' is the UK's main electrical safety standard aimed at ensuring that all electrical installations perform safely. First published in 1882, this Standard is now IET's 18th edition of what (in the Electrical sector) is commonly known as '*The Wiring Regs*'.

The requirements in this new Standard came into effect on 1st January 2019 and stipulates that ALL installations designed after 31st December 2018 **must** comply with this new 560 page BS 7671:2018 Standard – hence the importance for everyone in the electrical profession to ensure that they are up to speed with the new requirements – and the object of my '*Wiring Regulation*' series of books is to provide a user-friendly guide to the current requirements.

Edition 4 of '*Wiring Regulations in Brief*' (WRIB) was published on 30 Nov 2020 and has been written in compliance with BS 7671:2018 and not only includes all of its Electrical Safety Requirements but also changes to the Building Regulations and the Planning Laws that affect electrical installations.

The justification for writing a '*Wiring Regulations Pocket Book*' is (similar to my other Pocket Books) to provide a condensed working-copy of the (rather bulky) 4th Edition of WRIB. A Pocket Book that can be can be used as a quick reference to the Wiring Regulations for the busy electrician, building inspectors and/or students studying the requirements for compliance with the wiring and building regulations, etc.

Key features

- The chapters and the layout of the Pocket Book are in total harmony with WRIB-E4 and the basic structure of each chapter is effectively unchanged.

- This will enable (for example) an organisation to keep a copy of the WRIB-E4 book (which contains a complete description of BS 7671:2018) in their office library and provide their Managers, Employees and Staff to have immediate access to a cheaper, smaller, more usable version of the book in their place of work, briefcase, toolbox, glove compartment or even as an App on their mobile. A working book which can include notes and reminders, etc. made by its owner.
- Should there be insufficient detail in the Pocket Book for a particular project, then the user can have easy access to a complete description and set of regulations in the office copy of WRIB-E4 maintained in the organisation's library.
- In the beginning of 2021, COVID-19's induced changes to the planning and building control world, has resulted in new ways of working that are a blend of the old and new. At the time of writing, there has been little guidance on what 'extraordinary measures' will remain and what will go. However, the Pocket Book has been amended to reflect the situation as it emerges.
- Any changes to other National and International Standards, Regulations or Directives which occur prior to the publication have also be included in this Pocket Book.

'*Wiring Regulations in Brief*' and '*Wiring Regulations Pocket Book*' are two of the most reliable and portable guides to compliance with the Wiring Regulations. They have become essential reading for all electrical contractors and sub-contractors, site engineers, building engineers, building control officers, building surveyors, architects, construction site managers as well as DIYers.

Contents

As previously mentioned, this Pocket Book is effectively a précis WRIB E4 and thereby maintains the same layout as its predecessor as indicated below and in the Table of Contents on Page v.

Chapter 1 Introduction	This initial chapter provides a historical background to the Wiring Regulations, its contents, a description of the unique numbering system, its objectives, legal status and what it actually encompasses.

Chapter 2 Building Regulations

For many years the IET Wiring Regulations have not fully supported the requirements of Building Regulations (and vice versa). On 1st January 2019, however, it was set in law that in future, all new electrical wiring and components for Domestic Buildings (within England and Wales) must be designed and installed in accordance with the electrical safety requirements contained in Part P of the Building Regulations (which are written in compliance with BS 7671 and other National and international Standards).
Chapter 2, therefore, provides:

- details of the design, construction, installation, inspection and testing of all electrical installations and components in Domestic Buildings;
- the responsibilities for electrical safety; and
- an overview of the structure and contents of the Building Regulations and their associated 'Approved Documents'.

Chapter 3 Earthing

The third chapter reminds the reader about the different types of earthing systems and earthing arrangements. It then lists the main requirements for safety protection (direct and indirect contact), protective conductors and protective equipment before briefly touching on the test requirements for earthing.

Chapter 4 External influences

Chapter 4 provides a brief overview (together with précised extracts from the current Regulations) on how the environment in which an electrical product, service or equipment, etc. will affect it meeting the quality and safety requirements of BS 7671:2018.

Chapter 5 Safety protection

Chapter 5 lists the mandatory and fundamental requirements for safety protection contained in both BS 7671:2018 and the Building Regulations. It includes information concerning protection against electric shock, fault protection, protection against direct and indirect contact, protective conductors and protective equipment and lists the test requirements for safety protection.

Chapter 6 Electrical equipment, components, accessories and supplies

Chapter 6 provides a catalogue of all the different types of equipment, components, accessories and supplies for electrical installations identified and referred to in the Wiring Regulations (e.g. luminaires, RCDs, plugs and sockets, etc.) and then makes a list of the specific requirements that are scattered throughout the Regulations.

Chapter 7 Cables, conductors and conduits

Within the Wiring Regulations, there is frequent reference to different types of cables (e.g. single core, multicore, fixed, flexible, etc.) conductors (such as live supply, protective, bonding, etc.) and conduits, cable ducting, cable trunking and so on.

The aim of Chapter 7, therefore, is to provide a catalogue of all the different types identified and referred to in the Wiring Regulations in three main headings (namely cables, conductors and conduits) and then make a list of their essential requirements.

Chapter 8 Special installations and locations

Whilst the Wiring Regulations apply to electrical installations in all types of buildings, there are also some indoor and out-of-doors special installations and locations (such as agricultural buildings) that are subject to additional requirements due to the extra dangers they pose. This chapter considers the requirements for these special locations and installations and whilst perhaps not being a complete list, represents the most important requirements.

Chapter 9 Installation, maintenance and repair

Chapter 9 is an extremely important chapter as it provides guidance on the requirements for the safe installation, maintenance and repair of electrical equipment and installations.

Appendix 9.1 to this chapter includes example stage audit check lists for the quality, safety and environmental control of electrical equipment and electrical installations.

Chapter 10 Inspection and testing	This final chapter of the book provides guidance on the requirements for installation, maintenance, inspection, certification and repair of electrical installations. It lists the Regulations' requirements for these activities and in Appendix 10.1, provides examples of typical test equipment used to test electrical installations. Finally (for your assistance) I have provided in Appendix 10.2, a complete, bullet pointed check lists for the design, construction, inspection and testing of any new electrical installation, or new work associated with an alteration or addition to an existing installation.

These Chapters are then supported by the following Annexes:

Annex A	Symbols used in electrical installations
Annex B	List of electrical and electromechanical symbols
Annex C	SI units for existing technology
Annex D	IPX coding
Annex E	Acronyms and abbreviations
Annex F	Other books associated with the Wiring Regulations

Plus, a full Index

The following symbols (shown in the margins) will help you get the most out of this book:

An important requirement or point

A good idea, suggestion or something worth remembering

and within the text:

Note: Used to provide further amplification or information.

For your convenience (and to save you constantly having to refer backwards and forwards throughout the book for a particular Requirement), I have duplicated quite of a lot of these Requirements and shown more than once – i.e. in different chapters and/or sections of the book. I have also thought it prudent to duplicate a few of the figures and tables.

If any reader has any thoughts about the contents of this book (such as areas where perhaps they feel I have not given sufficiently coverage, omissions and/or made mistakes, etc.) then please let me know by e-mailing at ray@herne.org.uk and I will make suitable amendments in the next edition of this book.

I hope that you find this book useful!!

Ray Tricker

Appendix A – About the author (during lock down!)

Ray Tricker (MSc, IEng, CQP-FCQI, FIET, FCMI, FIRSE) is a Senior Consultant with over 50 years continuous service in Quality, Safety and Environmental Management, Project Management, Communication Electronics, Railway Command, Control and Signalling Systems, Information Technology, and the development of Molecular Nanotechnology.

He served with the Royal Corps of Signals (for a total of 37 years) during which time he held various managerial posts culminating in being appointed as the Chief Engineer of NATO's Communication Security Agency (ACE COMSEC).

Most of Ray's work since leaving the Services has centred on the European Railways. He has held a number of posts with the Union International des Chemins de fer (UIC) [e.g. Quality Manager of the European Train Control System (ETCS)] and with the European Union (EU) Commission [e.g. T500 Review Team Leader, ERTMS (European Rail Traffic Management System) Users Group Project Coordinator, HEROE Project Coordinator].

Ray was also the UKAS Assessor for the assessment of certification bodies for the harmonisation of the Trans-European, High Speed, Railway Network and in addition to being appointed as the Quality, Safety and Environmental Manager for the Project Management Consultancy responsible for overseeing the multi-billion-dollar Trinidad Rapid Rail System.

As well as writing over 60 books on diverse subjects such as, Quality, Safety and Environmental Management, Building, Wiring and Water Regulations for Taylor & Francis (under their Routledge imprints) and Elsevier (under their Butterworth-Heinemann and Newnes imprints!) he is busy assisting Small Businesses from around the world (usually on a no cost basis) to produce their own auditable Quality and/or Integrated Management Systems to meet the requirements of ISO 9001, ISO 14001 and OHSAS 18001 etc..

One day he might retire – but not before he has written his 100th book!!

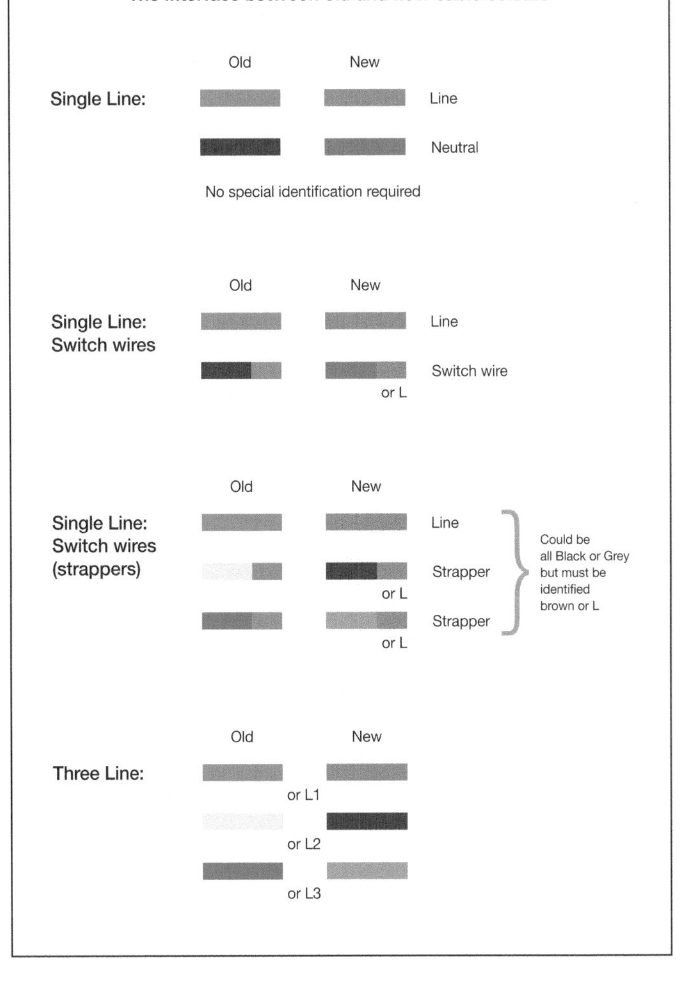
BS7671 Colour Arrangements
The interface between old and new cable colours
Old
New
Single Line:
Line
Neutral
No special identification required
Old
New
Single Line:
Switch wires
Line
Switch wire
or L
Old
New
Single Line:
Switch wires
(strappers)
Line
Strapper
or L
Strapper
or L
Could be
all Black or Grey
but must be
identified
brown or L
Old
New
Three Line:
or L1
or L2
or L3

1 Introduction

Author's Start Note

This initial chapter provides a historical background of the BS 7671 standard, what it contains; a description of its unique numbering system, its objectives, legal status and what it actually encompasses. It also discusses the effect that Wiring Regulations have on other Statuary Instruments and how this British Standard can be implemented.

1.1 Introduction

The latest edition of the IET Wiring Regulations has now grown to a massive 560-page document that defines the way in which all electrical installation work must be carried out. It does not matter whether the work is carried out by a professional electrician or an unqualified DIY enthusiast, the installation **must** always comply with the Wiring Regulations.

The current edition of the Regulations is BS 7671:2015 and is entitled '*Requirements for Electrical Installations – IET Wiring Regulations*', which is a bit of a mouthful to remember (!) and so it is normally referred to as '*The Wiring Regulations*', '*The Blue Book*', '*The 18th Edition*' (or simply) *BS 7671:2018* (whose front cover is depicted in Figure 1.1).

This British Standard is published with the full support of the BEC (i.e. the British Electrotechnical Committee, which is the UK's national body responsible for the formal standardisation within the electrotechnical sector) in partnership with the BSI (i.e. the British Standards Institution, which has the ultimate responsibility for all British Standards produced within this sector) and The Institution of Engineering and Technology (IET) – which, with more than 168,000 members worldwide

DOI: 10.1201/9781003165170-1

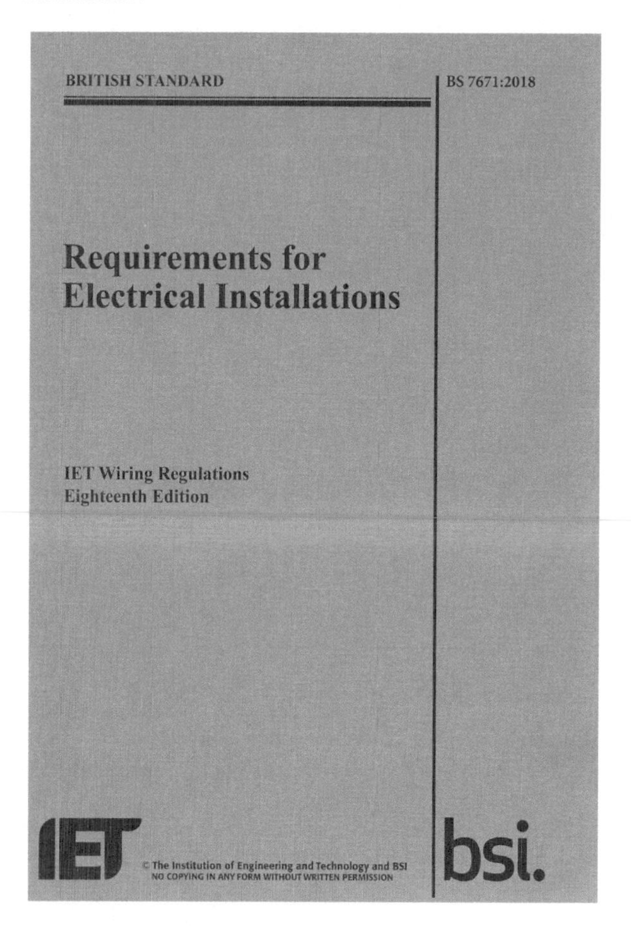

Figure 1.1 Front cover of BS 7671:2015

in 150 countries, is Europe's largest grouping of professional engineers involved in power, engineering, communications, electronics, computing, software, control, informatics and manufacturing.

The technical authority for BS 7671:2018 is the Joint IET/BSI Technical Committee (JPEL/64). BS7671:2018 came into effect on 1 January 2019, and from that date **ALL** new installations (as well as additions and alterations to existing installations) **MUST** comply with BS 7671:2018.

Please note that all references made in this book to the 'Wiring Regulations' or 'Regulation(s)' – where not otherwise specifically identified – refer to BS 7671:2018 '*Requirements for Electrical Installations*' (IET Wiring Regulations – 18th Edition).

1.2 Historical background

The first public electricity supply in the UK was at Godalming in Surrey, in November 1881 and mainly provided street lighting. At that time, there were no existing rules and regulations available to control their installations and so the electricity company just dug up the roads and laid the cables in the gutters. This particular electricity supply was discontinued in 1884.

On 12 January 1882, the steam powered Holborn Viaduct Power station officially opened and this facility supplied 110 V d.c. for both private consumption and street lighting. Once more, there was no one in authority to tell them how to lay the cables and their positioning was, therefore, dependent on the electrician responsible for that particular section of the work.

Later in 1882 *The Electric Lighting Clauses Act* (modelled on the previous 1847 *Gas Act*) was passed by Parliament and this enabled the Board of Trade to authorise the supply of electricity in any area by a local authority, company or person and to grant powers to install this electrical supply (including breaking up the streets) through the use of the 1882 '*Rules and Regulations for the prevention of Fire Risks Arising from Electric Lighting*'.

This document was the forerunner of today's Wiring Regulations.

Historically, since 1882, there has been a succession of new editions and amendments in alignment with other national and international standards. For example:

- in 1981 the 15th Edition closely corresponded to the international standard IEC 60364;
- in 1991, the 16th Edition was officially adopted by the British Standards Institute as the basis for BS 7671:1992 ['*Requirements for Electrical Installations*'];
- in 2004, the 16th Edition was officially adopted as mandatory requirements for the new Building Regulations Part P '*Electrical Safety*';
- in 2008, the 17th Edition was rewritten and included old, existing and new Comité Européen de Normalisation Electrotechnique (CENELEC), International Electrotechnical Commission (IEC) and European Normalisation (EN) Harmonised Documents. It was then republished as a British Standard (i.e. BS 7671:2008); and

- in 2018, the 18th Edition was published in order to include current changes to Fire Safety, Earth Fault Loop Impedance, Residual Current Device (RCD) protection of Socket Outlets and revised Inspection and Testing Documentations.

Since then;

- In 2020, Amendment No 1 to the 18th Edition covered minor manuscript amendments to the Section (i.e. 722) concerning Electric Vehicle Charging Installations.
- In 2022, Amendment No 2 to the 18th Edition is due to be published by the JPEL/64 committee with a number of manuscript changes and a new Part 8 on '*Functional Requirements*'. These proposed changes have been included in this edition of the book based on the current draft of this Amendment.

By aligning with all existing and new CENELEC, IEC and EN Harmonised Documents for electrical safety, in essence, BS 7671: 2008 has now virtually become a European Regulation in its own right and is widely used in many European Union (EU) countries as shown in Figure 1.2.

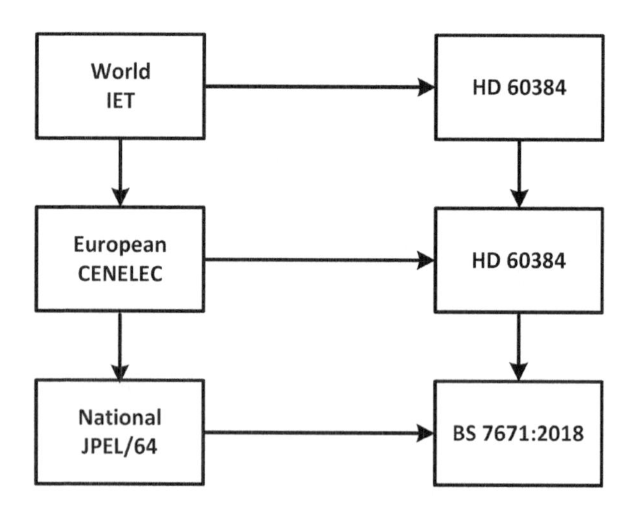

Figure 1.2 Installation Standards in world, European and national levels

1.3 What does BS 7671:2018 contain?

The Standard '*contains the rules for the design and erection of electrical installations so as to provide for safety and proper functioning for the intended use*'. It is based on the plan agreed internationally (i.e. through CENELEC) for the '*arrangement of safety rules for electrical installations*'.

The detailed structure of the seven parts making up BS 7671:2018 is shown in Table 1.1.

The seven parts of the standard are then supported by the following (17) informative Appendices shown in Table 1.2.

1.3.1 What about the standard's numbering system?

The numbering system used to identify specific requirements in BS 7671:20186 is as follows:

- the first digit signifies a Part;
- the second digit a Chapter;
- the third digit a Section.

Table 1.1 BS 7671:2018 – Structure

Part no.	*Description*
Part 1	Sets out the scope, object and fundamental principles
Part 2	Defines terms used within the Regulations
Part 3	Identifies the characteristics of an installation that will need to be taken into account in choosing and applying the requirements of the subsequent Parts of the Regulations
Part 4	Describes the basic measures available for the protection of persons, property and livestock against potential electrical hazards
Part 5	Describes precautions that need to be taken in the selection and erection of electrical installations
Part 6	Covers Inspection and Testing
Part 7	Identifies requirements for special installations or locations

Table 1.2 BS 7671:2018 – Appendices

Appendix	*Title*
1.	British Standards referred to in the Regulations
2.	Statutory Regulations and associated memoranda
3.	Time/current characteristics of overcurrent protective devices and RCDs
4.	Current-carrying capacity and voltage drop for cables
5.	Classification of external influences (extracted from HD 60364-5-51)
6.	Model forms of certification and reporting
7.	Harmonised cable core colours (for details see inside front cover of this book)
8.	Current-carrying capacity and voltage drop for busbar trunking and powertrack systems
9.	Definitions – multiple source, d.c. and other systems
10.	Protection of conductors in parallel against overcurrent
11.	~~Effect of harmonic currents~~ (moved to Appendix 4)
12.	~~Voltage drop in consumers' installations~~ (moved to Appendix 4)
13.	Methods for measuring the insulation resistance/ impedance of floors and walls to Earth or to the protective conductor system
14.	Determination of Prospective Fault Current
15.	Ring and radial final circuit arrangements
16.	Devices for protection against overvoltage
17.	Energy efficiency

For example, Section number **413** is made up as follows:

- PART 4 – PROTECTION FOR SAFETY.
    - Chapter 41 (first chapter of Part 4) – PROTECTION AGAINST ELECTRIC SHOCK.
        - Section 413 (third section of Chapter 41) – PROTECTIVE MEASURE: ELECTRICAL SEPARATION.

1.4 What are the objectives of the Wiring Regulations?

The stated intention of wiring safety codes is to '*provide technical, performance and material standards that will allow sufficient distribution of electrical energy and communication signals, at the same time protecting persons in the building from electric shock and preventing fire and explosion'.* In other words:

> **To ensure the protection of people and livestock from fire, shock or burns from any installation that complies with their requirements.**

The Regulations form the basis of safe working practice throughout the electrical industry.

1.5 What is the legal status of the original Wiring Regulations?

Although the original IET Wiring Regulations were always held in high esteem throughout Europe, they had **no** legal status. This problem was overcome in October 1992 when the IET Wiring Regulations became British Standard 7671 – thus providing them with a national/international status.

1.6 What does it cover?

The Wiring Regulations cover both electrical installations and electrical equipment.

1.6.1 Electrical installation

The Regulations apply to the design, erection and verification of electrical installations such as those for:

- agricultural and horticultural premises;
- caravans and motor caravans;
- caravan parks and similar sites;
- commercial premises;
- conducting locations with restricted movement;
- construction and demolition sites;
- electric vehicle charging;
- exhibitions, shows and stands;
- extra low voltage lighting;
- floor and ceiling heating systems;
- highway equipment and street furniture;
- locations containing a bath or shower;

- low voltage generating sets;
- marinas and similar locations;
- medical locations;
- mobile or transportable units;
- offshore units of electrical shore connections for inland navigation vessels;
- operating and maintenance gangways;
- outdoor lighting;
- prefabricated buildings;
- public premises;
- residential premises;
- rooms and cabins containing sauna heaters;
- solar photovoltaic (PV) power supply systems;
- swimming pools and other basins;
- temporary installations for structures, amusement devices and booths at fairgrounds, amusement parks and circuses including professional stage and broadcast applications.

The Regulations also include requirements for:

- consumer installations external to buildings;
- circuits supplied at voltages up to and including 1000 V a.c. or 1500 V d.c.;
- circuits that are operating at voltages greater than 1000 V;
- fixed wiring for communication and information technology, signalling; command and control, etc.;
- the addition to (or alteration of) installations and parts of existing installations;
- wiring systems and cables not specifically covered by appliance standards.

Notes:

1. The standard nominal supply voltage for domestic single-phase 50 Hz installations in the UK has been 230V a.c. (rms) since 1 January 1995. Previously it was 240 V.
2. Although the preferred frequencies are 50 Hz, 60 Hz and 400 Hz, the use of other frequencies for special purposes is not excluded.

In certain cases, the Regulations may need to be supplemented by the requirements and/or recommendations from other British, European and International Standards or by the requirements of the person ordering the work. Such cases could include (amongst others) the following:

- electrical apparatus for use in the presence of combustible dust (covered by BS EN 50281 and BS EN 61241);
- electrical installations for open-cast mines and quarries (covered by BS 69070);
- electric signs and high voltage luminous discharge tube installations (covered by BS 559 and BS EN 50107);
- electric surface heating systems (covered by BS EN 60335-2-96);
- emergency lighting (covered by BS 5266 and BS EN 1838);
- explosive atmospheres (covered by BS EN 600790);
- fire detection and alarm systems in buildings (covered by BS 5839);
- life safety and firefighting applications (covered by BS 8519 and BS 9999);
- telecommunications systems (covered by BS 6701);
- temporary electrical systems for entertainment and related purposes (covered by BS 7909).

Exclusions:

The Regulations do **not** apply to the following installations:

- aircraft equipment;
- electric fences (covered by BS EN 60335-2-76);
- electrical equipment of machines (covered by BS EN 60204);
- escalator or moving walk installations (specifically covered by relevant parts of BS 5655 and BS EN 115);
- lift installations (specifically covered by relevant parts of BS 5655 and BS EN 81-1);
- lightning protection systems (for buildings and structures covered by BS EN 62305);
- mines specifically covered by Statutory Regulations;
- mobile and fixed offshore installation equipment;
- motor vehicle equipment (except where Regulations concerning caravans or mobile units are applicable);
- on board ship equipment (covered by BS 8450, BS EN 60092-507 or BS EN 13297);
- radio interference suppression equipment (except so far as it affects safety of the electrical installation);

- railway traction, rolling stock and signalling equipment;
- systems for the distribution of electricity to the public;
- the d.c. side of cathode protection systems.

1.6.2 *Electrical equipment – Definition*

'*Electrical equipment*' means any item used for the generation, conversion, transmission, distribution or utilisation of electrical energy, such as machines, transformers, apparatus, measuring instruments, protective devices accessories and appliances, and luminaires alternators, generators and batteries.

1.7 What affect does using the Regulation have on other Statutory Instruments?

The requirements of the Wiring Regulations also have an effect on the implementation of other Statutory Instruments such as:

- The Building Act 1984;
- The Disability and Equality Act 2010;
- The Electricity at Work Regulations 1989;
- The Fire Precautions (Workplace) Regulations 1997;
- The Health and Safety at Work Act 2015;
- The Sustainable and Secure Building Act 2004.

1.7.1 *What is The Building Act 1984?*

The Building Act 1984 (as implemented by the Building Regulations 2010) is designed to ensure:

- the health, safety, welfare and convenience of persons in or about buildings;
- the conservation of fuel and power;
- the prevention of waste, undue consumption, misuse or contamination of water;
- the design and construction of buildings and the provision of services, fittings and equipment in (or in connection with) buildings, is controlled.

Note: A copy of the current Building Regulations 2010 can be downloaded from: http://www.legislation.gov.uk/uksi/2010/2214/contents/made.

In future **all** new electrical wiring or electrical components for domestic premises, or small commercial premise linked to domestic accommodation, **must** now to be designed and installed in accordance with Approved Document P (AD P) of the Building Regulations. This Approved Document is based on the fundamental principles set out in Part 7 of BS 7671:2018 which also includes the requirement for the standardisation of cable core colours of all a.c. power circuits.

In addition, **all** fixed electrical installations (i.e. wiring and appliance fixed to the building fabric – such as socket-outlets, switches, consumer units and ceiling fittings) **must** now be designed, installed, inspected, tested and certified to BS 7671:2018.

Note: Part P **only** applies to fixed electrical installations that are intended to operate at low voltage or extra-low voltage which are not controlled by the Electricity (Standards of Performance) Regulations 2015 as amended, or the Electricity at Work Regulations 1989 (as amended).

1.7.1.1 Competent persons scheme

Under Part P of the Building Regulations, all domestic installation work **must** now be inspected by Local Authority Building Control officers **unless** the work has been completed out by a '*Competent Person*' who is able to self-certify the work.

For more details about the Building Regulations visit https://www.gov.uk/building-regulations-approval or see 'Building Regulations in Brief' Edition 10 https://www.routledge.com/product/isbn/9780815368380?-source=igodigital or get a copy of the 'Building Regulations Pocket Book Edition 2' https://www.routledge.com/Building-Regulations-Pocket-Book/Tricker-Alford/p/book/9780815368380.

1.7.2 What is The Equality Act 2010?

The Equality Act replaced most of the Disability Discrimination Act (DDA) in October 2010.

From the point of view of the BS 7671:2008, The Equality Act makes it unlawful:

- for a trade organisation or qualifications body to discriminate against a disabled person;
- for service providers to make it impossible or unreasonably difficult for disabled persons to make use of that service.

More information about the Equality Act can be found on the Government Equalities Office website – http://homeoffice.gov.uk/equalities.

1.7.3 *What is the Electricity at Work Regulations 1989?*

Reducing the risk of such an accident is a legal requirement.

The Electricity at Work Regulations (EWR) 1989 (as amended) came into force on 1 April 1990 and requires precautions to be taken against the risk of death or personal injury from electricity in work activities to ensure that:

- all electrical systems have been properly constructed, maintained and are used in such a way so as not to cause danger to the user;
- maintenance of fixed electrical installations and portable appliances is carried out and regular inspections are;
- the person responsible for the building, ensures that electrical test certificates are in place to confirm that the building's installations and appliances have been appropriately tested.

Overall, the Regulations require that:

- all electrical equipment and installations are maintained in a safe condition;
- all electrical systems are constructed and maintained to prevent danger;
- all people working with electricity are competent to do the job.

Complicated tasks (i.e. equipment repairs, alterations, installation work and testing) may require a suitably qualified electrician, in which case:

- all staff are aware of an organisation's electrical safety procedure;
- all work activities are carried out so as not to give rise to danger;

- equipment and procedures are safe and suitable for the working environment;
- equipment is switched off and/or unplugged before making adjustments. *'Live working'* must be eliminated from work practices.

Electricity is recognised as a major hazard for not only can it kill (research has shown that the majority of electric shock fatalities occur at voltages up to 240 V), it can also cause fires and explosions. Even non-fatal shocks can cause severe and permanent injury.

Most electrical risks can be controlled by using suitable equipment, following safe procedures – particularly in wet surroundings, cramped spaces, work out of doors or near live parts of equipment.

Whilst the majority of the Regulations concern hardware requirements, others are more generalised. For example:

- installations shall be considered as a proper construction;
- conductors shall be insulated;
- means of cutting off the power (i.e. for electrical isolation) shall be available.

The major activity and effects of the Regulations are shown in Table 1.3.

Table 1.3 Contents of the regulations

Regulation	*Activity*	*Effect*
Regulation 4	Systems, work activities and protective equipment	Systems shall at all times be so constructed to prevent, insofar as is reasonably practicable, danger
Regulation 5	Strength and capability of electrical equipment	No electrical equipment is to be used where its strength and capability may be exceeded so as to give rise to danger
Regulation 6	Adverse or hazardous environments	Electrical equipment sited in adverse or hazardous environments must be suitable for those conditions

(continued)

Table 1.3 (Continued)

Regulation	*Activity*	*Effect*
Regulation 7	Insulation, protection and placing of conductors	Permanent safeguarding or suitable positioning of live conductors is required
Regulation 8	Earthing and other suitable precautions	Equipment must be earthed or other suitable precautions must be taken
Regulation 9	Integrity of reference conductors	Nothing is to be placed in an earthed circuit conductor which might, without suitable precautions, give rise to danger by breaking the electrical continuity or by introducing a high impedance
Regulation 10	Connections	All joints and connections in systems must be mechanically and electrical suitable for use
Regulation 11	Means for protecting from excess current	Suitable protective devices should be installed in each system to ensure all parts of the system **and** users of the system are safeguarded from the effects of fault conditions

Note: Regulations 4 to 11 in effect, therefore, place a duty on the designer, installer **and** the end user to ensure the suitability and protection of all electrical equipment.

Regulation 12	Means of cutting off the supply and for isolation	Where necessary to prevent danger, suitable means shall be available for cutting off the electrical supply to any electrical equipment and isolating that particular equipment
Regulation 13	Precautions for work on equipment made dead	Adequate precautions must be taken to prevent electrical equipment, which (in order to prevent danger) has been made dead, from becoming live – whilst any work is carried out

(Continued)

Table 1.3 (Continued)

Regulation	*Activity*	*Effect*
Regulation 14	Work on or near live conductors	No work shall be carried out on live electrical equipment unless this can be properly justified
		Which means that risk assessments are required and suitable precautions must be taken to prevent injury
Regulation 15	Working space, access and lighting	Adequate working space, means of access and a lighting shall be provided at all electrical equipment on or near where work is being completed in circumstances that may give rise to danger
Regulation 16	Persons to be competent to prevent danger and injury	No person shall engage in work which requires professional technical knowledge or professional experience, unless he has that knowledge or experience, or is under appropriate supervision

For more information about the Electricity at Work Regulations 1989, contact the Health and Safety Executive (HSE) www.hse.gov.uk/lau/lacs/19-3.htm.

Or to download a copy of the Electricity at Work Regulations 1989 (Statutory Instrument 1989 No 635), go to:

http://www.opsi.gov.uk/si/si1989/Uksi_19890635_en_1.htm.

1.7.4 The effect of the Wiring Regulations on the Fire Precautions (Workplace) Regulations

The Fire Precautions (Workplace) Regulations 1997 (as amended by the Fire Precautions (Workplace) (Amendment) Regulations 1999) stipulates that:

Fire Precautions (Workplace) Regulations

A Fire Certificate is required for:

> *'All offices, shops, railway premises and factories which have more than 20 persons employed in the building (or more than 10 person employed anywhere other than on the ground floor);*
>
> *Any hotel or boarding house providing sleeping accommodation for more than six persons (guests or staff)* that is above the first floor or below the ground floor;

When a Fire Certificate is issued the owner or occupier is required to provide and maintain:

- a means of escape;
- means of fighting fire;
- means of providing warning in case of fire;
- other means for ensuring that the means of escape can be safely and effectively used at all material times.

For further information about the Fire Precautions (Workplace) (Amendment) Regulations 1999 (Statutory Instrument 1999 No 1877), visit:

https://www.legislation.gov.uk/uksi/1999/1877/contents/made

1.7.5 The effect of the Wiring Regulations on the Health and Safety at Work Act 1974?

Any company with more than five employees is legally obliged to possess a comprehensive health and safety policy.

Installations that conform to BS 7671 (as amended) are regarded by the HSE (Health & Safety Executive) as likely to achieve conformity with the relevant parts of the Electricity at Work Regulations 1989. In certain instances where the Regulations have been used, they may also be accompanied by Codes of Practice approved under Section 16 of the Health and Safety at Work Act 1974.

Although some existing installations may have been designed and installed to conform to the standards set by earlier editions of the Wiring Regulations, this does **not** necessarily mean that they will fail to achieve conformity with the relevant parts of the Electricity at Work Regulations 1989.

For further information about the Health and Safety at Work Act 1974, visit:

http://www.legislation.gov.uk/ukpga/1974/37/contents

1.8 How are the IET Wiring Regulations implemented?

Although the IET Wiring Regulations rely (primarily) on British Standards for their implementation they do, however, include the policy decisions made in a number of Statutory Instruments and a relative EU Harmonised Directives.

1.8.1 Statutory instruments

As shown in Table 1.4, Great Britain and Northern Ireland's various classes of electrical installations are required to comply with Statutory Regulations.

Table 1.4 Statutory instruments affecting electrical installations

Type of electrical installation	*Statutory instrument*
Buildings generally (subject to certain exemptions)	Building Regulations 2010 (as amended) (for England and Wales) • SI 2010 No 2214 (as amended by SI 2011/1515) Building (Scotland) Amendment Regulations 2011(as amended) • Scottish SI 2011 No 120 Building Regulations (Northern Ireland) 2000 (as amended) • Statutory Rule 2000 No 398

(continued)

Table 1.4 (Continued)

Type of electrical installation	*Statutory instrument*
Cinematograph installations	Cinematograph (Safety) Regulations 1955 (as amended under the Cinematograph Act, 1909, and/or Cinematograph Act, 1952) • SI 1982 No 1856
Distributors' installations generally (subject to certain exemptions)	Electricity Safety, Quality and Continuity Regulations 2002 • SI 2002 No 2665 • SI 2006 No 1521
High voltage luminous tube	Conditions of licence: • In England and Wales. – The Local Government (Miscellaneous provisions) Act 1982 • In Scotland, The Civic Government (Scotland) Act 1982
Machinery	The Supply of Machinery (Safety) Regulations 1992 as amended • SI 1992 No 3073 • SI 1994 No 2063
Theatres and other places licensed for public entertainment, music, dancing, etc.	Conditions of licence under: • In England and Wales. – The Local Government (Miscellaneous provisions) Act 1982 • In Scotland, The Civic Government (Scotland) Act 1982
Work activity Places of work Non-domestic installations	Conditions of licence under: • The Electricity at Work Regulations 1989 as amended

The full texts of **all** Statutory Instruments that have been published since 1987 are now available at https://www.legislation.gov.uk/

1.8.2 *British Standards, International Standards and Harmonised Documents*

Along with British and International Standards, the Wiring Regulations also take account of the technical substance of agreements reached in CENELEC.

For your convenience, the current (i.e. at the time of writing this book) Standards and Directives relevant to the BS 7671:2018 have been listed in Appendix 1 to this Standard.

Author's End Note

In Chapter 2 we shall now take a look at the structure and contents of the Building Regulations. The responsibilities for electrical safety. The need to oversee and control extensions (material alterations and material changes of use). The electricity distributor's responsibilities (earthing, electrical installation work, types of wiring and wiring systems, etc.) and, inspection, testing and self-certification.

2 Building regulations

Author's Start Note

For many years the Wiring Regulations have not fully supported the requirements of the Building Regulations UNTIL (owing to the growing number of electrical accidents occurring in domestic buildings) the Government decided that in future:

> ***All new electrical wiring and components (within England and Wales) must be designed and installed in accordance with the electrical safety requirements contained in Part P of the Building Regulations.***

This chapter, *therefore, provides:*

- *an overview of the structure and contents of the Building Regulations and their 'Approved Documents';*
- *responsibilities for electrical safety;*
- *the need to oversee and control extensions, alterations and changes of use);*
- *electricity distributor's responsibilities;*
- *inspection and testing;*
- *responsibilities of a competent firm or person;*
- *how a building can be self-certified; and*
- *Figure 2.1 is used as a reminder to electricians of their responsibilities.*

DOI: 10.1201/9781003165170-2

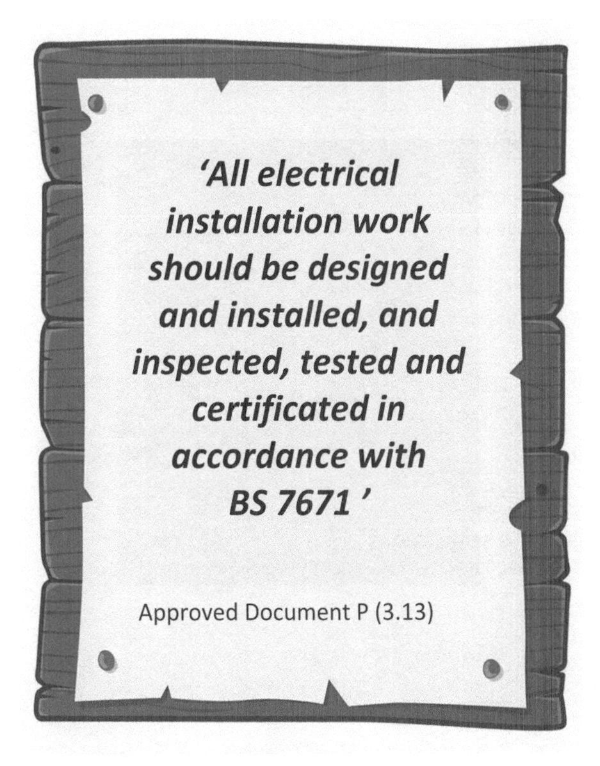

Figure 2.1 Mandatory requirements for domestic and non-domestic buildings

2.1 The Building Act 1984

The Building Act 1984 imposes a set of requirements for:

- buildings under construction and during use;
- conservation of fuel and power;
- preventing and detecting crime;
- preventing waste, undue consumption or contamination of water;
- protection and enhancement of the environment;
- securing the health, safety, welfare and convenience of persons regarding buildings;
- services, fittings and use of installed equipment;
- sustainable development.

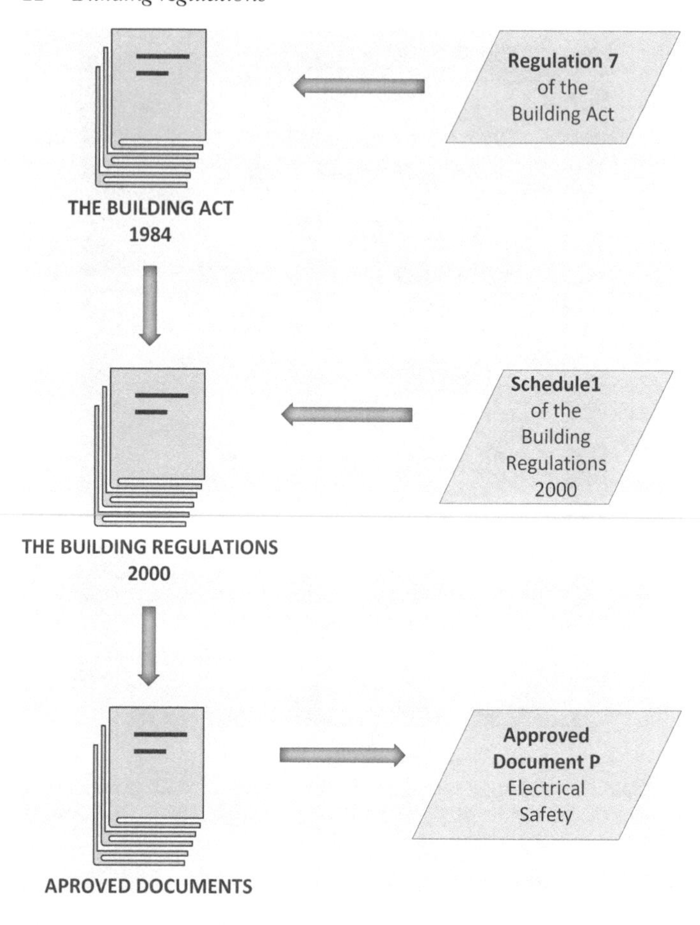

Figure 2.2 Implementing the Building Act

The Building Act 1984 consists of a number of Regulations, Schedules and Approved Documents (as depicted in Figure 2.2) but it does **not** apply to Scotland or to Northern Ireland who under devolved government have their own legislation – see Section 2.3.1 for details.

2.1.1 *What does the Building Act 1984 contain?*

The Building Act 1984 is made up of five parts (see Table 2.1).

Table 2.1 The structure of the Building Act

Building Act	*Description*
Part 1	The Building Regulations
Part 2	Supervision of building work, etc. other than by a Local Authority
Part 3	Other provisions about buildings
Part 4	General
Part 5	Supplementary

2.1.2 What about the rest of the United Kingdom?

The Building Act 1984 **only** applies to England and Wales. Separate Acts and Regulations apply in Scotland or Northern Ireland. These are shown in Table 2.2.

In Scotland, the building requirements are very similar to England and Wales, except that a building warrant is required before work can start.

Note: Table 2.3 provides a summary of the titles of the sections of Building Regulations that are applicable in the United Kingdom.

A more detailed summary of the requirements of the Building Act 1984, Building Regulations 2010 and their associated Approved Documents etc., is contained in the latest Edition (i.e. 10th) of '*Building Regulations in Brief*' by Ray Tricker & Samantha Alford. (Routledge).

Table 2.2 Building legislation within the United Kingdom

Nation	*Act*	*Regulations*	*Implementation*
England and Wales	Building Act 1984	Building Regulations 2010	Approved Documents
Scotland	Building (Scotland) Act 2003	Building (Scotland) 2004 (as amended)	Technical Handbooks
Northern Ireland	Building Regulations (Northern Ireland) Order 1979 (as amended)	Building Regulations (Northern Ireland) 2012 (as amended)	Technical Booklets

Table 2.3 Building Regulations for the United Kingdom

England and Wales		*Scotland*		*Northern Ireland*
Part A	Structure	Section 1 – Structure	Technical Handbooks 2019	Part D – Structure
Part B	Fire safety	Section 2 – Fire	Technical Handbooks 2019	Part E – Fire safety
Part C	Site preparation and resistance to contaminants and water	Section 3 – Environment	Technical Handbooks 2019	Part C – Site preparation and resistance to contaminants and moisture
Part D	Toxic substances	Section 3 – Environment	Technical Handbooks 2019	Part B – Materials and workmanship
Part E	Resistance to the passage of sound	Section 5 – Noise	Technical Handbooks 2019	Part G – Resistance to the passage of sound
Part F	Ventilation	Section 3 – Environment	Technical Handbooks 2019	Part K – Ventilation
Part G	Sanitation, hot water safety and water efficiency	Section 3 – Environment	Technical Handbooks 2019	Part P – Sanitary appliances and unvented hot water storage systems and reducing the risk of scalding
Part H	Drainage and waste disposal	Section 3 – Environment	Technical Handbooks 2019	Part N – Drainage
Part J	Combustion appliances and fuel storage	Section 3 – Environment Section 4 – Safety	Technical Handbooks 2019	Part L – Combustion appliances and fuel storage systems
Part K	Protection from falling, collision and impact	Section 4 – Safety	Technical Handbooks 2017	Part H – Stairs, ramps, guarding and protection from impact

(Continued)

Table 2.3 (Continued)

England and Wales		*Scotland*		*Northern Ireland*
Part L	Conservation of fuel and power	Section 6 – Energy	Technical Handbooks 2019	Part F – Conservation of fuel and power
Part M	Access and facilities for disabled people	Section 4 – Safety	Technical Handbooks 2019	Part R – Access to and use of buildings
Part P	Electrical safety	Section 4 – Safety	Technical Handbooks 2019	Booklet E – Fire safety
Part Q	Security in dwellings	Section 4 – Safety	Technical Handbooks 2019	Booklet R
Part R	High speed electronic communications networks	Section 4 – Safety	Technical Handbooks 2019	Technical Booklet M – Access and facilities for disabled people

2.1.3 *Who polices the Building Act?*

Local authorities are responsible for ensuring that any building work conforms to the requirements of the associated Building Regulations. They have the authority to:

- make you complete alterations so that your work complies with the Building Regulations;
- make you take down and remove or rebuild anything that contravenes a regulations; or
- employ a third party (and then send you the bill!) to take down and rebuild non-conforming buildings or parts of buildings.

2.2 What are the Supplementary Regulations?

Part 5 of the Building Act contains seven Supplementary Regulations which provide scheduled details of how the Building Regulations are to be controlled by local authorities. These schedules (full copies of which are contained in BS 7671:2018) are as follows:

Schedule 1	Building Regulations
Schedule 2	Relaxation of Building Regulations for existing work
Schedule 3	Inner London
Schedule 4	Provisions consequential upon Public Body's notice
Schedule 5	Transitional provisions
Schedule 6	Consequential amendments
Schedule 7	Repeals

Note: Schedule 1 is the most important schedule from the point of view of the electrical trade because it shows how the Building Regulations will be administered by local authorities, the approved methods of construction and the approved types of materials that are to be used in connection with building work.

2.2.1 *What requirements are listed in Schedule 1 of the Supplementary Regulations?*

The Building Regulations 2010 state the minimum requirements for implementing the Building Act 1984 and are designed to secure the health, safety and welfare of people in and around buildings as well as the conservation of fuel and energy in England and Wales.

The level of safety and acceptable standards are set out as guidance in **Approved Documents** (as shown below) and compliance with these Approved Documents is normally considered as evidence that the Building Regulations themselves have been fulfilled.

2.2.2 *How is my building work evaluated for conformance with the Building Regulations?*

The Local Authority will make regular checks that all building work being completed, is in conformance with the agreed approved plan and the Building Regulations themselves. These checks will include:

- tests of any service, fitting or equipment that is being provided in connection with a building;
- tests of any material, component or combination of components that is being used in the construction of a building;
- tests of the soil or sub-soil of the site of the building.

The Local Authority is authorised to ask the person responsible for the building work to complete some of these tests on their behalf!

2.3 Approved Documents

Approved Documents describe how the requirements of Schedule 1 and Regulation 7 of the Building Act 1984 can be met.

Each Approved Document reproduces the actual requirements contained in the Building Regulations relevant to the subject area (e.g. Approved Document P deals with Electrical Safety). This is then followed by practical and technical guidance (together with examples) showing how the requirements can be met in some of the more common building situations.

The current sets of Approved Documents are in 15 Parts, A to R (less 'I', 'O' and 'N') and consist of:

Approved Document A (Structure);
Approved Document B (Fire Safety);
Approved Document C (Site preparation and resistance to contaminants and moisture);
Approved Document D (Toxic substances);
Approved Document E (Resistance to the passage of sound);
Approved Document F (Ventilation);
Approved Document G (Sanitation, hot water safety and water efficiency);
Approved Document H (Drainage and waste disposal);
Approved Document J (Combustion appliances and fuel storage systems);
Approved Document K (Protection from falling, collision and impact);
Approved Document L (Conservation of fuel and power);
Approved Document M (Access to and use of Buildings);
Approved Document P (Electrical safety);
Approved Document Q (Security in Dwellings);
Approved Document R (High speed electronic communications networks);
Regulation 7 (Materials and workmanship).

Free pdf copies of **all** of these Approved Documents are available from https://www.gov.uk/government/collections/approved-documents

2.4 Electrical safety

2.4.1 The main requirements from the Approved Documents for electrical insulations

Although:

- Part E (Resistance to sound);
- Part J (Combustion appliances and fuel storage systems); and
- Part K (Protection from falling, collision and impact);

have a number of requirements concerning electrical safety and electrical installations (see below for details), the main requirements are contained in Part P (Electrical safety) together with:

- Part M (Access to and use of buildings);
- Part L (Conservation of fuel and power); and
- Part B (Fire safety).

As shown in Figure 2.3.

2.4.2 Part P – Electrical safety – Dwellings

For many years, the UK has managed to maintain relatively high electrical safety standards with the support of voluntary controls based on BS 7671. However, with a growing number of electrical accidents occurring in the home, the government has been forced to consider the legal requirement for safety in electrical installation work in dwellings.

The aim of Part P, therefore, is to increase the safety of householders by improving the design, installation, inspection and testing of electrical installations in dwellings when they (i.e. the installations) are being newly built, installed, extended or altered.

Since 1 January 2005, therefore:

- all new electrical wiring or electrical components used for domestic premises (as well as small commercial premises linked to domestic accommodation) must be designed and installed in accordance with Approved Document P of the Building Regulations;
- all fixed electrical installations (i.e. wiring and appliances fixed to the building fabric such as socket-outlets, switches, consumer units and ceiling fittings) must be designed, installed, inspected, tested and certified to BS 7671:2018.

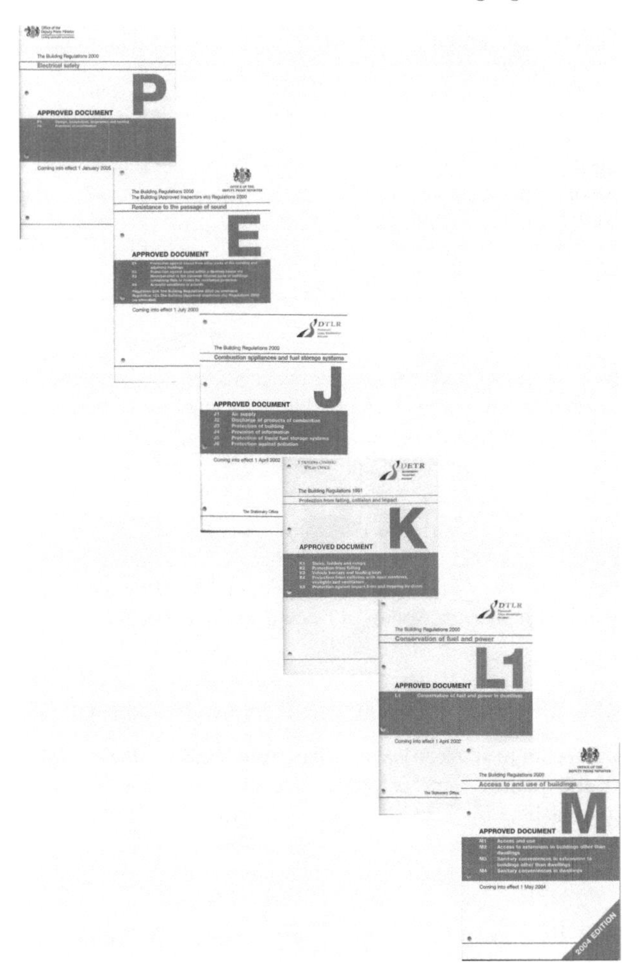

Figure 2.3 The Building Regulations

Part P also introduced new requirements for cable core colours for a.c. power circuits. With effect from 31 March 2006, all new installations or alterations to existing installations **must** use the new (harmonised) colour cables that are shown inside the front cover of this book and listed in Table 2.4. (Cable identification colours for extra-low voltage and d.c. power circuits are available in Appendix 7 of BS 7671).

Table 2.4 Identification of conductors in a.c. power and lighting circuits

Conductor	*Colour*
Protective conductor	Green and yellow
Neutral	Blue
Phase of single-phase circuit	Brown
Phase 1 of 3-phase circuit	Brown
Phase 2 of 3-phase circuit	Black
Phase 3 of 3-phase circuit	Grey

Note: Currently, electricians carrying out work in England and Wales **must** comply with Part P of the Building Regulations. In Scotland it is the Building Standards system whilst Northern Ireland has no equivalent statutory requirement.

The Government is currently in the process of introducing a mandatory scheme that all domestic installations shall be checked at regular intervals (as well as when they are sold and/or purchased) to make sure that they fully comply with Approved Document P **and** BS 7671:2018.

2.4.2.1 *What types of building does Approved Document P cover?*

Part P applies to **all** electrical installations in (**and around**) buildings or parts of buildings. Table 2.5 provides details of the types of works that are notifiable to the Local Authority and which need to be completed by a company registered as a Competent Firm.

2.4.2.2 *Where can I get more information about the requirements of Part P?*

Further guidance concerning the requirements of Part P (Electrical safety) is available from the:

- IET (Institution of Engineering Technology) at http://electrical.theiet.org/building-regulations/part-p/index.cfm
- The NICEIC (National Inspection Council for Electrical Installation Contracting) at www.niceic.org.uk
- the ECA (Electrical Contractors' Association) at www.eca.co.uk

Table 2.5 Notifiable work

Locations	*Type of work*	
Where work is being completed	*Extensions and modifications to circuits*	*New circuits*
Bathrooms	Yes	Yes
Bedrooms	Yes	Yes
Ceiling (overhead) heating	Yes	Yes
Communal areas of flats	Yes	Yes
Computer cabling	No	No
Conservatories	No	Yes
Dining rooms	No	Yes
Garages (integral)	No	Yes
Garages (detached)	No	Yes*
Garden – lighting	Yes	Yes
Garden – power	Yes	Yes
Greenhouses	Yes	Yes
Halls	No	Yes
Hot air Saunas	Yes	Yes
Kitchen	Yes	Yes
Kitchen diners	Yes	Yes
Landings	No	Yes
Lounge	No	Yes
Paddling pools	Yes	Yes
Remote buildings	Yes	Yes
Sheds	Yes	Yes
Shower rooms	Yes	Yes
Small scale generators	Yes	Yes
Solar power systems	Yes	Yes
Stairways	No	Yes
Studies	No	Yes
Swimming pools	Yes	Yes
Telephone cabling	No	Yes
TV Rooms	No	Yes
Under floor heating	Yes	Yes
Workshops (remote)	Yes	Yes

Note:
* If the installation requires outdoor wiring.

2.4.3 *Part M – Access to and use of buildings*

In addition to the requirements of the Disability and the Equality Act 2019, reasonable provision and precautions need to be taken to ensure that people, regardless of their disability, age or gender to safely gain access to and to make use of the facilities of the buildings.

The current edition of Part M no longer primarily concentrates on wheelchair users, but includes people using walking aids, people with impaired sight (and other mobility and sensory problems), mothers with prams as well as people with luggage, etc.

2.4.4 *Part L1 – Conservation of fuel and power*

Energy efficiency measures shall be provided which:

- provide lighting systems that utilise energy-efficient lamps with manual switching or automatic switching controls;
- provide the building occupiers instructions and/or diagrams of how the heating and hot water services can be efficiently operated and maintained.

Note: Part L consists of four separate sub-Parts. L1A, L1B (for domestic buildings) and L2A and L2B (for non-domestic buildings).

2.4.5 *Part B – Fire safety*

The building needs to be designed and constructed so that:

- in the event of fire, stability will be maintained for a reasonable period;
- there are adequate fire warning facilities;
- there are sufficient means of escape in case of fire from the building to a place of safety outside the building;
- that the unseen spread of fire and smoke within concealed spaces in its structure and fabric is inhibited;
- a wall common to two or more buildings is designed and constructed so that it capable of resisting the spread of fire between those buildings.

To prevent the spread of fire **within** the building, the internal linings shall:

- resist the amount of flame over their surfaces; and
- if ignited, have a reasonable heat release rate or fire growth rate.

Following the Glenville Towers disaster, Part B has been completely revised and now contains detailed advice on fire alarms, ventilation, measures to control the spread of fire and access for firefighters.

2.5 Who is responsible for electrical safety?

Basically, there are three people who are responsible for the electrical safety of (and within) buildings. These are:

The owner – needs to determine whether the works carried out are either minor or notifiable work.

The designer – needs to ensure that all electrical work is designed, constructed, inspected and tested in accordance with the Requirements of BS 7671:2018.

The builder/developer – needs to ensure that they have electricians who can self-certify their work or who are qualified/experienced enough to enable them to sign off under the Electrical Installation Certification form.

2.5.1 What are the statutory requirements for electrical installations?

All electrical installations need to:

- be designed and installed so that they:
  - protect against mechanical and thermal damage;
  - will not present an electrical shock and/or fire hazard;
- be tested and inspected to meet relevant equipment/installation standards;
- provide sufficient information so that users can operate, maintain or alter an electrical installation with reasonable safety;
- comply with the requirements of the Building Regulations and BS 7671:2018.

2.5.1.1 What does all this mean?

With a few exceptions, **any** electrical work undertaken in a home must be reported to the Local Authority Building Control for inspection if it includes the addition of a new electrical circuit, or involves work in a, kitchen, bathroom or a garden area.

The **ONLY** exception is when the installer has been approved by a Competent Persons organisation such as those shown in Table 2.6 which details the type of installation and its associated competent person self-certification scheme.

Table 2.6 Authorised competent person self-certification schemes for installers

Type of installation	*Schemes*
In dwellings – installation of fixed low or extra-low voltage electrical installations	BESCA, Blue Flame Certification, Certsure, NAPIT, OFTEC, Stroma
In dwellings – fixed low or extra-low voltage electrical installations as part of other work being carried out by the registered person	APHC, BESCA, Blue Flame Certification, Certsure, APIT, Stroma
Buildings other than dwellings – installation of lighting or electrical heating systems	BESCA, Blue Flame Certification, Certsure, NAPIT, Stroma

2.5.2 *Requirements for electrical installations*

All electrical installations:

- shall provide adequate protection against the risks of electric shock, burn or fire injuries;
- should be designed and installed to provide mechanical and thermal protection.

Electrical installations shall be inspected and tested during installation, at the end of installation and before they are taken into service to verify that they:

- are safe to use, maintain and alter;
- meet the relevant equipment and installation standards;
- meet the requirements of the Building Regulations;
- comply with Part P and BS 7671:2018.

2.5.3 *TrustMark*

TrustMark is the Government Endorsed Quality Scheme covering any work that consumers choose to have carried out in or around their home (Figure 2.4).

The scheme ensures that when a consumer uses a TrustMark Registered Business, they know they are engaging an organisation that has been thoroughly vetted to meet required standards, and has made a considerable commitment to good customer service, technical competence and trading practices.

Figure 2.4 The TrustMark initiative

Source: Logo produced courtesy of TrustMark.

Note: The TrustMark replaced the Quality Mark scheme which closed on 31 December 2004 because too few firms joined. For more information about TrustMark, see their website at www.trustmark.org.uk.

2.6 Certification

2.6.1 What is a competent firm?

For the purposes of Part P, the Government has defined *Competent Firms* as electrical contractors:

- who work in compliance with the requirements to BS 7671:2018;
- whose standard of electrical work has been assessed by a third party;
- who are registered under the NICEIC Approved Contractor scheme and the Electrotechnical Assessment Scheme (EAS).

2.6.2 What is a competent person responsible for?

When a competent person undertakes installation work, that person is responsible for:

- ensuring compliance with the latest edition of BS 7671:2018 and all relevant Building Regulations;
- providing the person ordering the work with a signed Building Regulations self-certification certificate and a completed Electrical Installation Certificate;
- providing the relevant Building Control Body with an information copy of the certificate.

2.6.3 *Who is entitled to self-certify an installation?*

Part P affects **every** electrical contractor carrying out fixed installation and/or alteration work in homes. **Only** registered installers are entitled to self-certify the electrical work and they **must** be registered as a competent person under one of the schemes shown in Table 2.6.

2.6.4 *What if the work completed by a contractor or an installer?*

If the work is of a notifiable nature, then the installer(s) must be registered with one of the schemes shown in Table 2.6.

Figure 2.5 provides a quick guide to the requirements.

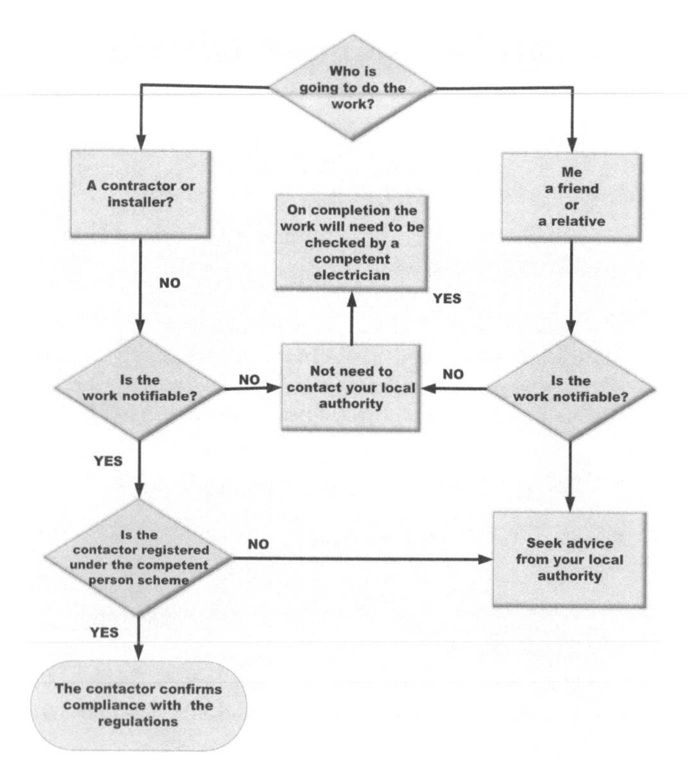

Figure 2.5 How to meet the new rules

2.6.5 What if the work is completed by a friend?

You do **not** need to tell your Local Authority's Building Control Department about non-notifiable work such as:

- repairs, replacements and maintenance work;
- extra power points or lighting points or other alterations to existing circuits (unless that are in a kitchen, bathroom, or is outdoors).

You **do**, however, need to tell them about most other work.

2.6.6 When do I have to inform the Local Authority Building Control Body?

All proposals to carry out electrical installation work **must** be notified to the Local Authority's Building Control Body before work begins, **unless** the proposed installation work is undertaken by a person who is a competent person registered under a government approved Part P Self Certification Scheme or the work is agreed non-notifiable work, such as:

- adding a telephone, extra-low voltage wiring and equipment for communications, information technology, signalling, command, control and other similar purposes;
- adding lighting points to an existing circuit;
- adding prefabricated equipment sets with integral plug and socket connections;
- adding socket outlets and fused spurs to an existing ring or radial circuit;
- connecting an electric gate or garage door to an existing isolator;
- fitting and replacing cookers and electric showers;
- installing equipment (e.g. security lighting);
- installing fixed equipment where the final connection is via a 13 A plug and socket;
- installing mechanical protection to existing fixed protective devices;
- installing or upgrading main or supplementary equipotential;
- installing prefabricated, 'modular' systems such as kitchen lighting systems;
- re-fixing or replacing the enclosures of existing installations or components;
- replacing fixed electrical installations;

- replacement, repair and maintenance jobs;
- replacing the cable of a single circuit cable (damaged, for example, by fire, rodent or impact).

ALL DIY electrical work (unless completed by a qualified professional) will still need to be checked, certified and tested by a competent electrician.

Any work that involves adding a new circuit to a dwelling will need to be either notified to the Building Control Body (who will then inspect the work) or needs to be carried out by a competent person who is registered under a government-approved Part P Self-Certification Scheme.

Work involving any of the following will also have to be notified:

- consumer unit replacements;
- electric floor or ceiling heating systems;
- extra-low voltage lighting;
- garden lighting and/or power installations;
- installation of a socket outlet on an external wall;
- installation of outdoor lighting and/or power installations in the garden or that involves crossing the garden;
- installation of new central heating control wiring;
- solar photovoltaic (PV) power supply systems;
- small-scale generators such as micro-CHP (Combined Heat & Power Generation) units.

Note: Where a person who is **not** registered to self-certify, intends to carry out the electrical installation, then a Building Regulation (i.e. a Building Notice or Full Plans) application will need to be submitted and the Building Control Body will arrange to have the electrical installation inspected at first-fix stage and tested upon completion.

However, the electrical work will **still** need to be certified under BS 7671:2018 by a suitably competent person who will be responsible for the design, installation, inspection and testing of the system (on completion).

The main things to remember are:

- **is** the work notifiable or non-notifiable?
- does the person undertaking the work need to be registered as a competent person?
- what records (if any) need to be kept of the installation?

2.7 What inspections and tests will have to be completed and recorded?

As shown in Table 2.7 there are four types of electrical installation certificates and one Building Regulation compliance certificate that have to be completed.

Table 2.7 Types of installation

Type of inspection	*When is it used?*	*What should it contain?*	*Remarks*
Electrical Installation Certificate	For the initial certification of a new installation, alteration or addition to an existing installation	A schedule of inspections and test results as required by Part 6 (of BS 7671:2018) <u>Together with:</u> A certificate (from Appendix 6 of BS 7671:2018) providing guidance for recipients	The original Certificate shall be given to the person ordering the work and a duplicate retained by the contractor
Minor Electrical Installation Works Certificate	For additions and alterations to an installation such as an extra socket-outlet or lighting point to an existing circuit, the relocation of a light switch, etc.	Relevant provisions of Part 6 of BS 7671:2018	This Certificate may also be used for the replacement of equipment such as accessories or luminaires, but **not** for the replacement of distribution boards or the provision of a new circuit
Electrical Installation Report	For the inspection of an existing electrical installation	A schedule of inspections and test results as required by Part 6 of BS 7671:2018	For safety reasons, the electrical installation will need to be inspected at appropriate intervals by a competent person

(Continued)

Table 2.7 (Continued)

Type of inspection	*When is it used?*	*What should it contain?*	*Remarks*
Building Regulations Compliance Certificate	Confirmation that the work carried out complies with the Building Regulations (specifically Part P)	The basic details of the installation, the location, completion date and name of the installer	A purchaser's solicitor may request this document when you come to sell your property In the future, this *may* be required as part of your Home Information Pack

Copies of these various Certificates and Forms are contained in Appendix 6 of BS 7671:2018.

Figure 2.6 indicates how to choose what type of electrical inspection is required.

2.7.1 Design

Electrical installations should be designed and installed (suitably enclosed and appropriately separated) so that they:

- are safe to use, maintain and alter;
- comply with the requirements of BS 7671:2018;
- comply with Part P (and any other relevant parts of the Building Regulations);
- comply with relevant equipment and installation standards;
- do not present an electric shock or fire hazard to people;
- provide adequate protection against mechanical and thermal damage;
- provide adequate protection for persons against the risks of electric shock, burn or fire injuries.

Appendix A of Part P to the Building Regulations provides details of the types of electrical services normally found in dwellings; some of the ways that they can be connected and the complexity of wiring and protective systems that can be used to supply them.

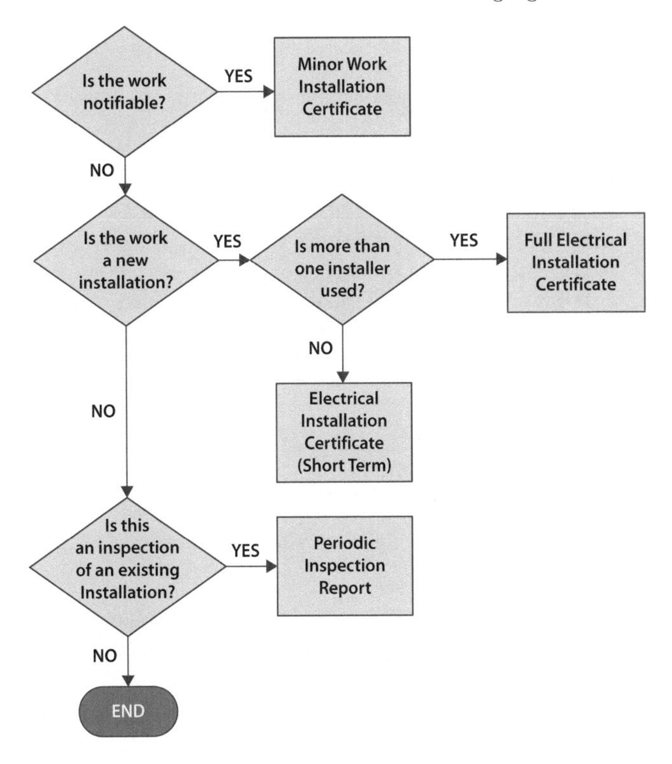

Figure 2.6 Choosing the correct Inspection Certificate

2.7.2 *Earthing*

Distributors are required to provide an earthing facility for all new connections and to ensure:

- all electrical installations are properly earthed;
- all lighting circuits include a circuit protective conductor;
- all socket-outlets that have a rating of 32A or less and which may be used to supply portable equipment for use outdoors must be protected by a Residual Current Device (RCD).

Note: The most usual type of earthing is an electricity distributor's earthing terminal, which is provided for this purpose, near the electricity meter.

In addition, Distributors should note and ensure that:

- new or replacement, non-metallic light fittings, switches or other components do not require earthing unless **new** circuit protective (earthing) conductors are provided;
- socket-outlets that are capable of accepting unearthed (2-pin) plugs must **not** be used to supply equipment that needs to be earthed;
- where electrical installation work is classified as an extension, a material alteration or a material change of use, the work **must** consider and include that the earthing and bonding systems are satisfactory and meet the requirements (see Figure 2.7 which shows some of the Earth and bonding conductors that might be part of an electrical installation).

As a safety precaution, **all** accessible consumer units should be fitted with a child-proof cover or installed in a lockable cupboard.

2.7.3 *Extensions, material alterations and material changes of use*

If electrical installation work is classified as an extension, a material alteration or a material change of use, then the following work needs to be considered:

- is the mains supply equipment suitable?
- what additions and alterations are required?
- what earthing and bonding systems are required and are these available?
- what type of protective measures required?

Figure 2.8 shows details of some of the types of electrical services that will normally be found in a typical new (or upgraded) dwelling.

2.7.4 *Electricity distributors' responsibilities*

The Electricity Distributor is responsible for:

- ensuring that their installation is mechanically protected and can be safely maintained;
- evaluating and agreeing proposals for new installations or significant alterations to existing ones;
- installing the cut-out and meter;
- taking into consideration the possible risk of flooding.

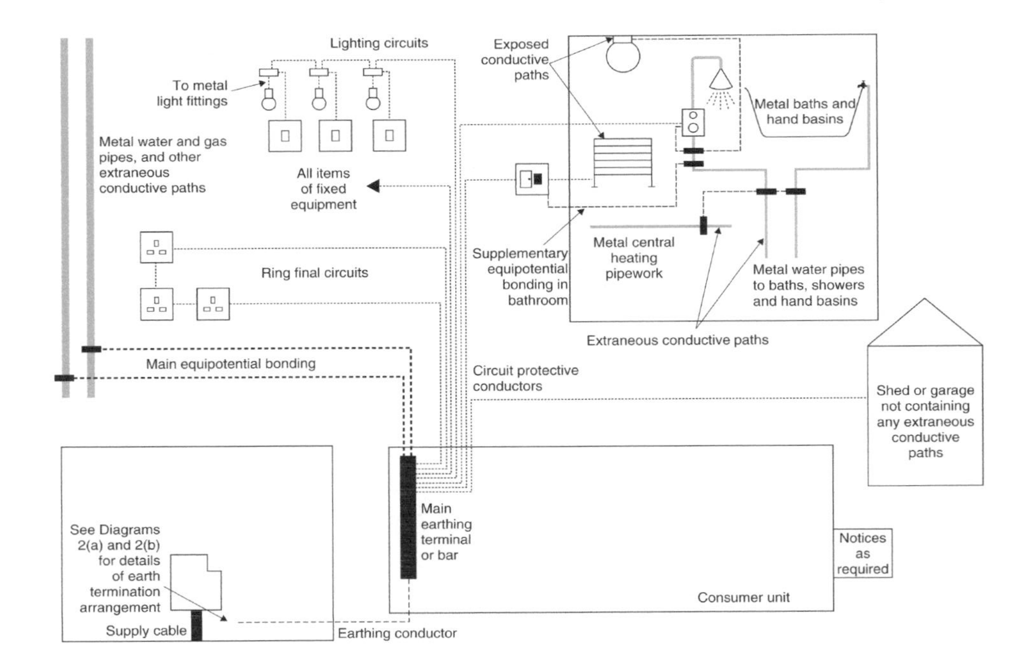

Figure 2.7 Typical Earth and bonding conductors that might be part of the electrical installation consumer units

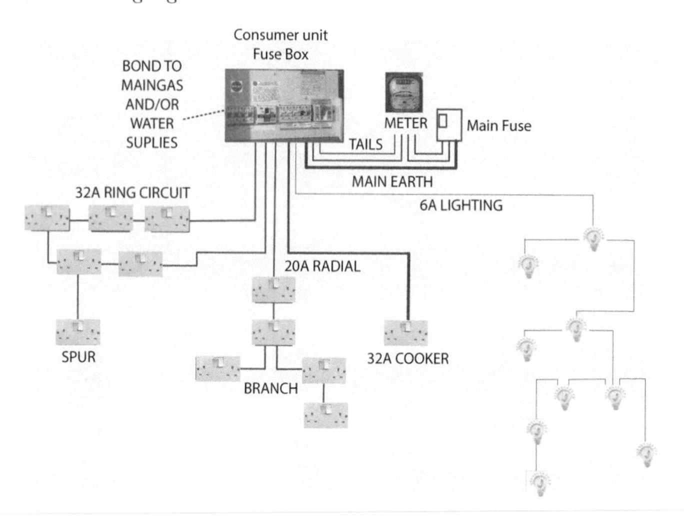

Figure 2.8 Typical fixed installations that might be encountered in new (or upgraded) existing dwellings

Distributors are required to:

- maintain the supply within defined tolerance limits;
- provide an earthing facility for new connections;
- provide certain technical and safety information to consumers to enable them to design their installations.

Distributors and metre operators must ensure that equipment that is designed to remain on the consumers' premises:

- is safe;
- is suitable;
- shows the polarity of the conductors.

Distributors may disconnect consumers' installations if they:

- do not comply with BS 7671:2018;
- are a source of danger; or
- cause interference with other installations.

Detailed Guidance on these Regulations is available at https://www.gov.uk/building-regulations-approval

2.7.5 Electrical installation work

All electrical installation work:

- shall be carried out professionally;
- shall comply with the Electricity at Work Regulations 1989 (as amended);
- may only be carried out by competent person.

2.7.5.1 Types of wiring or wiring system

The wiring system chosen for electrical installations will depend on its intended location and designed use. For example:

- cables concealed in floors need to have an earthed metal covering and be enclosed in steel conduit, or given some other form of mechanical protection;
- cables to an outside building (if run underground) should:
    - give protection against electric shock;
    - provide protection against mechanical damage to a cable;
    - include underground coloured plastic warning tape installed directly above the service line (at a depth of 6–8 inches).

2.7.5.2 Equipotential bonding conductors

- Main equipotential bonding conductors are required to water service pipes, gas installation pipes, oil supply pipes plus certain other 'earthy' metalwork that may be present on the premises.
- The installation of supplementary equipotential bonding conductors is required for installations and locations where there is an increased risk of electric shock (e.g. bathrooms and shower rooms).
- The minimum size of supplementary equipotential bonding conductors (without mechanical protection) is 4 mm^2.

2.7.6 Inspections and tests

Electrical installations need to be inspected and tested during, at the end of installation and before they are taken into service to verify that they:

- are safe to use;
- comply with BS 7671:2018;

- meet the relevant equipment and installation standards;
- meet the requirements of the Building Regulations.

All electrical work should be inspected (during installation as well as on completion) to verify that the components have:

- been selected, tested and installed in accordance with BS 7671:2018;
- been evaluated against external influences (e.g. moisture);
- not been visibly damaged or are defective;
- been tested to ensure satisfactory performance.

2.7.7 Alarm systems

2.7.7.1 Emergency alarms

Emergency alarm pull cords should be:

- coloured red;
- located as close to a wall as possible;
- have two red 50 mm diameter bangles.

Front plates should contrast visually with their backgrounds.

The colours red and green should **not** be used in combination as indicators of 'ON' and 'OFF' for switches and controls:

2.7.7.2 Emergency assistance alarms

Emergency assistance alarm systems should have:

- visual and audible indicators to confirm that an emergency call has been received;
- a reset control reachable from a wheelchair, WC, or from a shower/changing seat; and
- a signal that is distinguishable visually and audibly from the fire alarm.

2.7.7.3 Fire alarms

Fire alarms should emit an audio and visual signal to warn occupants with hearing or visual impairments.

2.7.8 Controls and switches

All controls and switches should be easy to operate, visible and free from obstruction and:

- should be located between 750 mm and 1200 mm above the floor;
- should not require the simultaneous use of both hands to operate (unless absolutely necessary for safety reasons);
- switched socket outlets should indicate whether they are 'ON';
- mains and circuit isolator switches should clearly indicate whether they 'ON' or 'OFF';

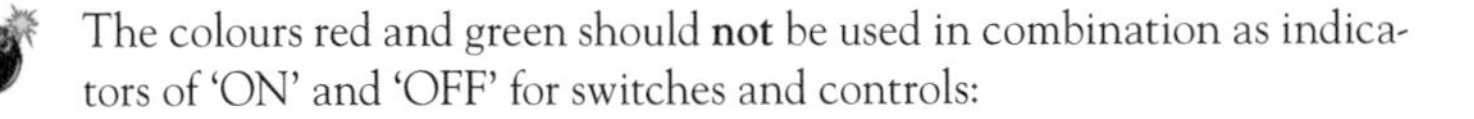

The colours red and green should **not** be used in combination as indicators of 'ON' and 'OFF' for switches and controls:

- front plates should contrast visually with their backgrounds;
- light switches with large push pads should, where possible, be used in preference to pull cords;
- individual switches on panels and on multiple socket outlets should be well separated;
- the operation of all switches, outlets and controls should not require the simultaneous use of both hands (unless necessary for safety reasons);
- controls that need close vision (e.g. thermostats) should be located between 1200 mm and 1400 mm above the floor.

2.7.9 Heat emitters

- Heat emitters should either be screened or have their exposed surfaces kept at a temperature below 43°C.

In toilets and bathrooms, heat emitters (if located) should **not** restrict:

- wheelchair manoeuvring space;
- the space beside a WC used to transfer from the wheelchair to the toilet.

2.7.10 Lighting circuits

All lighting circuits shall include a circuit protective conductor.

Light switches should be:

- aligned horizontally with door handles and have large push pads (in preference to pull cords);
- within 900 to 1100 mm from the entrance door opening;
- located between 750 mm and 1200 mm above the floor.

2.7.10.1 Fixed lighting

In locations, where lighting is expected to have most use, fixed lighting with a luminous efficacy greater than 40 lumens per circuit-watt should be available.

Note: Table 2.8 is an indication of recommended number of locations (excluding garages, lofts and outhouses) that need to be equipped with efficient lighting.

2.7.10.1.1 EXTERNAL LIGHTING FIXED TO THE BUILDING

External lighting (including lighting in porches, but not in garages and carports) should:

- automatically extinguish when there is enough daylight and when not required at night;
- have sockets that can only be used with lamps greater than 40 lumens per circuit Watt.

2.7.11 Power operated doors

If a door requires a force greater than 20 N to open or shut, then it shall be provided with a power operated door opening and closing system.

Table 2.8 Lighting requirements

Number of rooms created (Hall, stairs and landing(s) count as one room as does a conservatory)	*Recommended minimum number of locations (fixed lighting)*
1–3	1
4–6	2
7–9	3
10–12	4

Once open, all single leaf doors (or one leaf of a double leaf door) used as an accessible entrance should be wide enough to allow unrestricted passage for a variety of users, including wheelchair users, people carrying luggage, people with assistance dogs, and parents with pushchairs and small children. Table 2.9 provides details of the minimum opening widths of these doors for new and well as existing building undergoing renovation (see Figure 2.9).

Table 2.9 Minimum opening widths of doors

Direction and width of approach	*New buildings (mm)*	*Existing buildings (mm)*
Straight-on (without a turn or oblique approach)	800	750
At right angles access route at least 1500 mm wide	800	750
At right angles to an access route ate least 1200 mm wide	825	775
External doors to buildings used by the general public	1000	775

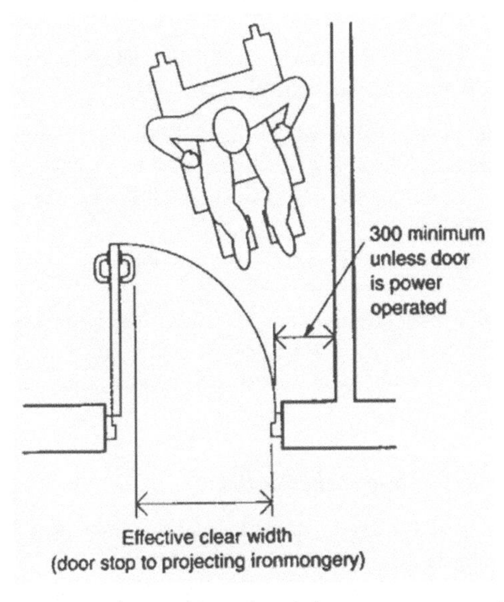

Figure 2.9 Effective clear width and visibility requirements of doors

Power operated entrance doors should either:

- be a manually controlled sliding, swinging or folding action door; or
- be automatically controlled by a motion sensor; or
- have a proximity sensor such as contact mat.

Power operated entrance doors should, when and where necessary for health or safety:

- be provided with a manual or automatic opening device;
- open towards people approaching the doors;
- provide visual and audible warnings that they are operating (or about to operate);
- incorporate automatic sensors to ensure that they open early enough; (and stay open long enough to permit safe entry and exit);
- incorporate a safety stop that is activated if the doors begin to close when a person is passing through;
- have a readily identifiable and accessible stop switch;
- have safety features to prevent injury to people who are struck or trapped;
- revert to manual control in the open position in the event of a power failure;
- when open, should not project into any adjacent access route;
- ensure that its manual controls:
    - are located between 750 mm and 1000 mm above floor level;
    - are operable with a closed fist;
    - are set back 1400 mm from the leading edge of the door when fully open;
    - contrast visually with the background.

Note: Revolving doors are **not** considered 'accessible' as they create particular difficulties for people who are visually impaired, people with assistance dogs or mobility problems and for parents with children and/or pushchairs.

2.7.12 Switches and socket outlets

- Switches and socket outlets for lighting and other equipment should be easily reachable.
- Switches and socket outlets for lighting should be installed between 450 mm and 1200 mm from the finished floor level (see Figure 2.10).

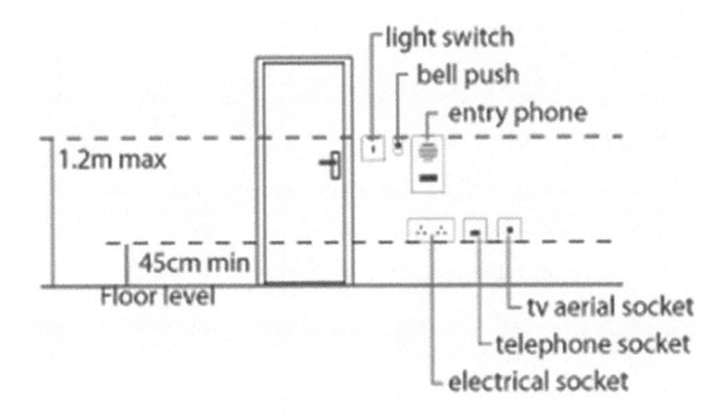

Figure 2.10 Heights of switches, sockets, etc.

2.7.12.1 Portable equipment for use outdoors

- All socket-outlets which have a rating of 32A or less which may be used outdoors as a supply for portable equipment must be protected by an RCD.

2.7.12.2 Socket outlets

- Non-fused older types of socket-outlet plugs must **not** be connected to a ring circuit.
- Socket-outlets that will accept unearthed (2-pin) plugs, must **not** be used supply equipment that needs to be earthed.
- Socket outlets should comply with the requirements of Part M (*Access to and use of buildings*).

2.7.12.2.1 SWITCHED SOCKET OUTLETS

All switched socket outlets should be wall-mounted and:

- be located no nearer than 350 mm from room corners;
- front plates should contrast visually with their backgrounds; and
- indicate whether they are ON.

Mains and circuit isolator switches should clearly indicate whether they 'ON' or 'OFF'.

Individual switches on panels and on multiple socket outlets should be well separated.

2.7.12.2.2 WALL SOCKETS

Wall sockets shall meet the following requirements (see Table 2.10).

2.7.13 *Telephone points and TV sockets*

Telephone points and TV sockets should be located between 400 mm and 1000 mm above the floor.

2.7.14 *Other considerations*

2.7.14.1 Cellars or basements

Liquefied Petroleum Gas (LPG) storage vessels and LPG fired appliances fitted with automatic ignition devices or pilot lights must **not** be installed in cellars or basements.

Table 2.10 Building Regulations requirements for wall sockets

Type of wall	*Requirement*
Timber framed	Power points may be set in the linings provided there is a similar thickness of cladding behind the socket box Power points should **not** be placed back to back across the wall
Solid masonry	Deep sockets and chases should **not** be used in separating walls Stagger the position of sockets on opposite sides of the separating wall
Cavity masonry	Stagger the position of sockets on opposite sides of the separating wall Deep sockets and chases should **not** be used in a separating wall or placed back to back
Framed walls with absorbent material	Sockets should • be positioned on opposite sides of a separating wall • not be connected back to back • be staggered a minimum of 150 mm edge to edge

2.7.14.2 Lecture/conference facilities

Artificial lighting should be designed to:

- give good colour rendering of all surfaces;
- be compatible with other electronic and radio frequency installations.

2.7.14.3 Swimming pools and saunas

Swimming pools and saunas are subject to special requirements specified in Part 7 of BS 7671: 2018.

Author's End Note

Having seen how the Building Regulations have an impact on the installation and maintenance of electrical installations, Chapter 3 looks at the different types of earthing systems that are available for both domestic and nondomestic building.

3 Earthing

Author's Start Note

This chapter *reminds the reader about the different types of earthing systems and earthing arrangements. It then lists the main requirements from the various Approved Documents and Regulations for safety protection (direct and indirect contact), protective conductors and protective equipment before briefly touching on the test requirements for earthing.*

More detailed information concerning BS 7671:2018's 'earthing arrangements and protective conductors' is contained in Chapter 54 *of the standard.*

From an electrical point of view, the world is effectively a huge conductor at zero potential and is used as a reference point which is called '*Earth*' (in the UK) or '*Ground*' in the USA. People and animals are normally in contact with Earth and so if another part, which is open to touch, becomes charged at a different voltage from Earth, then a shock hazard will exist.

One lightning bolt (see Figure 3.1) has enough electricity to service 200,000 homes!!!

BS 7671:2018's requirements for bonding and earthing (as depicted in Figure 3.2) have been amended (in accordance with the CENELEC harmonisation documents) and now include:

- the requirement to protect low voltage installations against temporary overvoltages (caused by Earth faults) in high and low voltage systems;

DOI: 10.1201/9781003165170-3

Figure 3.1 Typical lightning strike

Source: Picture courtesy of Ireland on UNsplash

- the measurement of Earth fault loop impedance (taking into consideration the increased resistance of conductors with an increased temperature);
- methods for measuring the insulation resistance and impedance of floors and walls to Earth or to the protective conductor system.

It has been agreed that using a gas, water or any other metal service pipe as a means of earthing for an electrical installation is **not** permitted unless some form of equipotential bonding has been connected to these pipes.

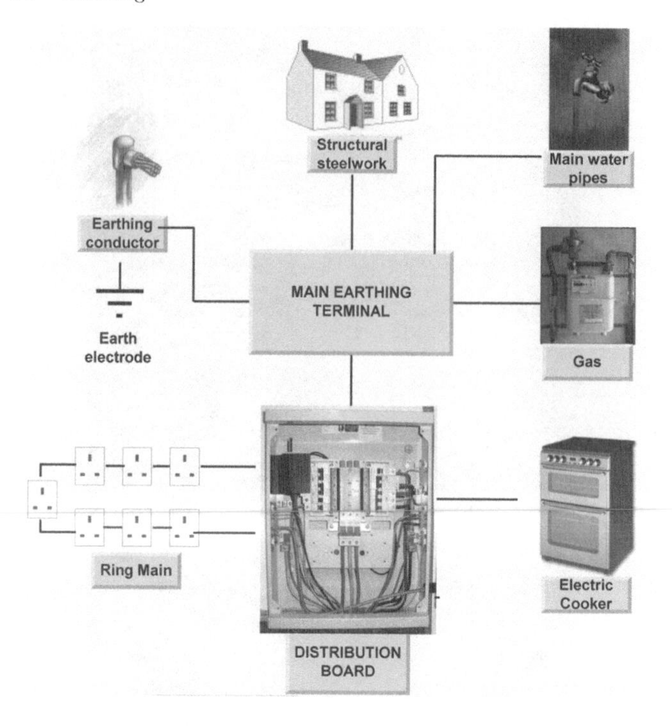

Figure 3.2 Bonding and earthing

Source: Courtesy Herne

3.1 What is Earth?

In electrical terms, 'Earth' is defined as:

> ***The conductive mass of the Earth, whose electric potential at any point is conventionally taken as zero.***

3.2 But what is meant by 'earthing' and how is it used?

'Earthing' is the process of transferring the immediate discharge of electricity directly to the main Earth terminal of that installation, by means of low resistance electrical cables or wires.

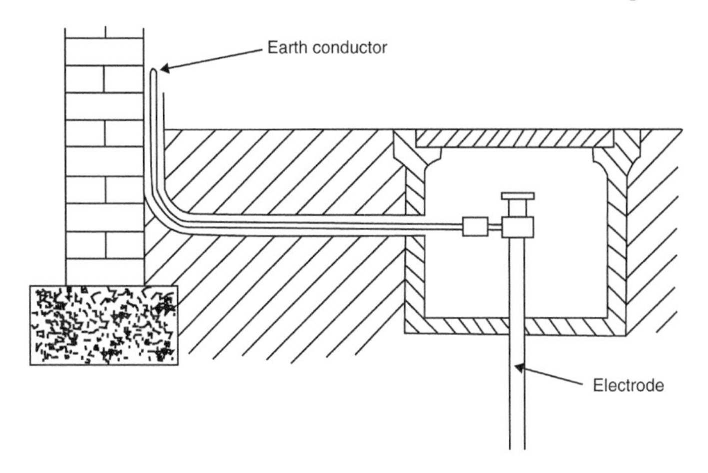

Figure 3.3 Earth conductor and electrode

An **'earthing system'**, on the other hand, defines the electrical potential of the conductors relative to the Earth's conductive surface.

An **'Earth electrode'** is that part of the system directly in contact with Earth. This can be just a metal (usually copper) rod or stake driven into the Earth (as shown in Figure 3.3) or a connection to a buried metal service, pipe or a complex system of buried rods and wires.

The resistance of the electrode-to-Earth connection will determine its quality and this can be improved by:

- increasing the surface area of the electrode that is in contact with Earth;
- increasing the depth to which the electrode is driven;
- using several connected ground rods;
- increasing the moisture of the soil;
- improving the conductive mineral content of the soil; and
- increasing the land area covered by the ground system.

A Protective Earth (PE) connection ensures that all exposed conductive surfaces are at the same electrical potential as Earth to avoid risk of an electrical shock if a person, or an animal, touches an equipment (or device) in which an insulation fault has occurred.

A Functional Earth (FE) connection, as well as providing protection against electrical shock' can also carry a current during the normal operation of a device – a facility that is often required by devices such as surge suppression and electromagnetic-compatibility filters, some types of antennas as well as a number of measuring instruments.

3.3 Advantages of earthing

The main advantages of earthing are that:

- an electrical system is tied to the general mass of Earth and cannot 'float' at another potential;
- it saves people liable to get electric shocks;
- it prevents the risk of fire caused by current leakage;
- it stops excessive current running through the circuit;
- it reduces the cost of having to provide protective conductors and Earth electrodes, etc.

By connecting Earth to metalwork (that is not intended to carry current), a path is provided for a fault current which can be detected by a protective conductor and, if necessary, broken. The path for this fault current is shown in Figure 3.4.

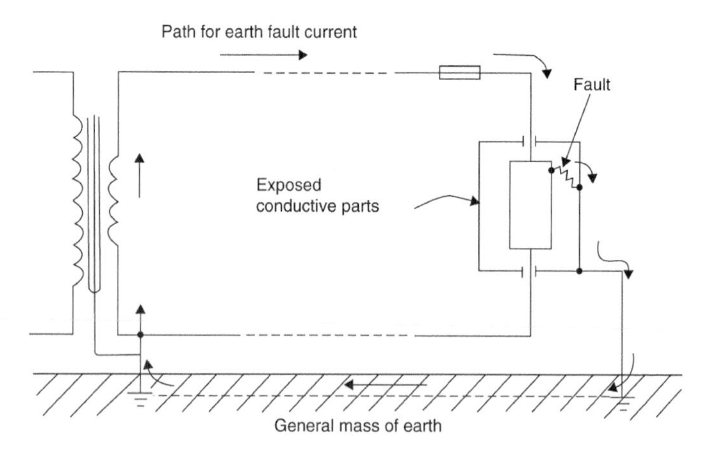

Figure 3.4 Path for Earth fault current (shown by arrows)

3.4 Precautions that should be taken for earthing

Earthing arrangements should ensure that:

- they are sufficiently robust (or have additional mechanical protection) to external influences;
- the impedance from the consumer's main earthing terminal to the earthed point of the supply, meets the protective and functional requirements of the installation;
- all Earth fault currents and protective conductor currents that occur are carried without danger (particularly from thermal, thermomechanical and electromechanical stresses).

If a number of installations have separate earthing arrangements, then any protective conductor that is common to one of these installations:

- shall either be capable of carrying the maximum fault current likely to flow through them; or
- shall Earth one installation and be insulated from the earthing arrangements of the other installation(s).

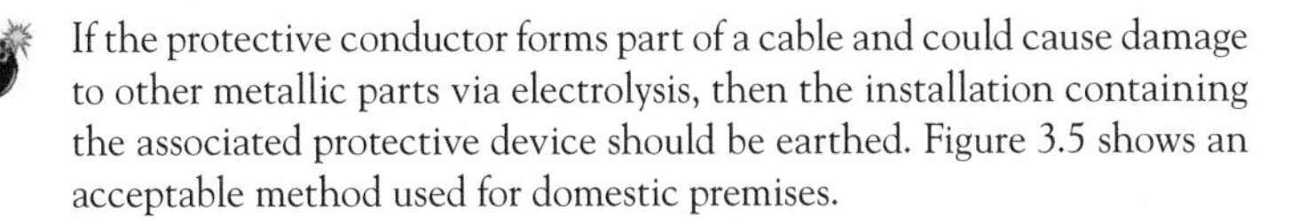

If the protective conductor forms part of a cable and could cause damage to other metallic parts via electrolysis, then the installation containing the associated protective device should be earthed. Figure 3.5 shows an acceptable method used for domestic premises.

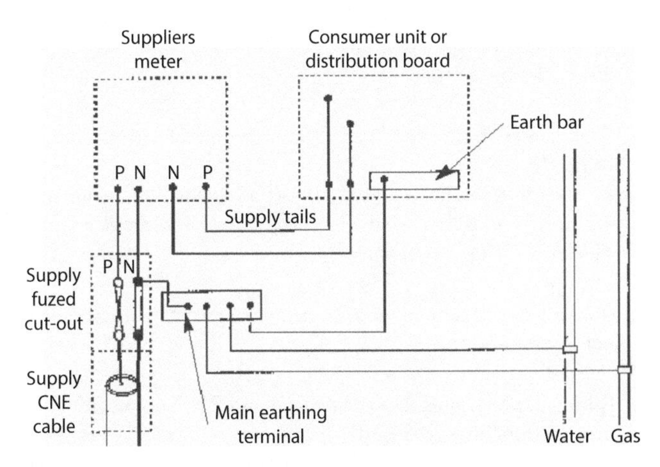

Figure 3.5 Domestic earthing arrangement Earth electrodes

3.5 What types of earthing systems are there?

BS 7671:2018 defines an electrical system as consisting '*of a single source of electrical energy and an installation*' and the type of system depends on the link between the source and the exposed conductive parts of the installation, to Earth.

Note: In this context, an '*exposed conductive part*' means any part of an equipment which can be touched and which is not (i.e. currently) a live part, but which '*may*' become live under fault conditions.

3.5.1 System classifications

T = Earth, N = Neutral, S = Separate, C = Combined, I = Isolated

In order to identify the different systems, a unique four-lettered code is used whereby:

The first letter indicates the type of Earth supply, so that:

- **T = Earth** (from the French word Terre) indicates that one or more points of the supply are directly earthed (for example, the earthed neutral at the transformer);
- **I = Isolated,** indicates either that the supply system is not earthed (at all) or that the earthing includes a deliberately inserted impedance, in order to limit fault current.

The second letter provides details of the actual earthing arrangements in the installation, so that:

- T indicates that all exposed conductive metalwork is connected to Earth;
- N (Neutral), indicates that all exposed conductive metalwork is connected to an earthed supply conductor provided by the Electricity Supply Company.

The third and fourth letters show the arrangement of the earthed supply conductor system so that:

- S (Separate) ensures that the neutral and Earth conductor systems are quite unconnected; and
- C (Combined) ensures that the neutral and Earth conductor systems are combined into a single conductor.

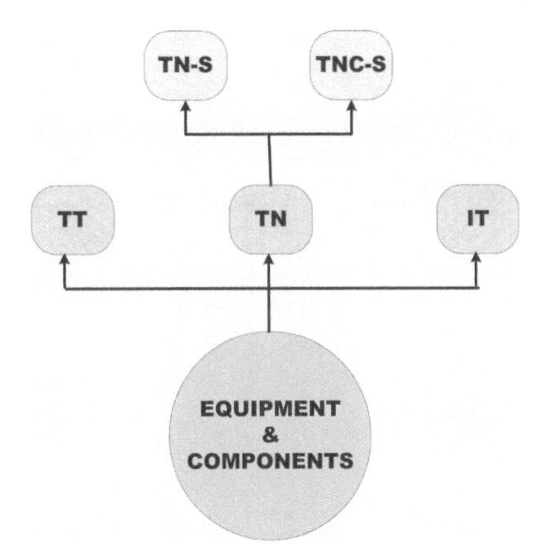

Figure 3.6 Three phased earthing systems

3.5.2 Three phased earthing systems

The **three** most common three phased systems are TN (TN-S, and TN-C-S), TT and IT as shown in Figure 3.6.

3.5.2.1 TT system

This type of installation is usually found in rural locations where the system is not provided with an Earth terminal by the Electricity Supply Company and the installation is fed from an overhead supply. It has one point of the energy source directly earthed and the exposed-conductive parts of the consumer's installation are provided with a local connection to Earth, **independent** of any Earth connection at the generator (see Figure 3.7).

3.5.2.2 TN-S system

The TN-S is the most common earthing system in the UK and one where the electricity supply company provides an Earth terminal (usually the armour and/or sheath of the underground supply cable) at the incoming mains position. Separate Protective Earth (PE) and Neutral (N) conductors are only connected together near the power source and remain separated throughout the system (see Figure 3.8).

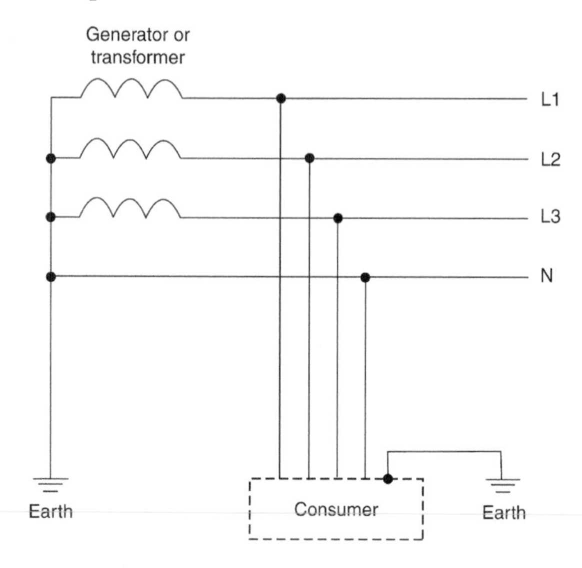

Figure 3.7 TT system

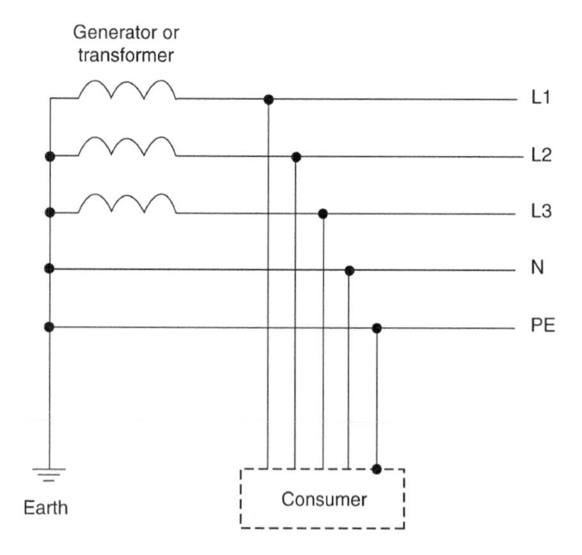

Figure 3.8 TN-S system

Note: The main advantage of a TN-S system concerns Electromagnetic Compatibility as the consumer has a low-noise connection to Earth and, therefore, does not suffer from the voltage that appears on the neutral conductor because of return currents and the impedance of that conductor. This is particularly important for some types of telecommunication and measurement.

In TN-S systems an RCD can be used for additional protection.

3.5.2.3 TN-C-S system

In the UK, the TN-C-S system is also known as Protective Multiple Earthing (PME) as it connects the combined neutral and Earth, to real Earth (at many locations) – thereby reducing the risk of broken neutrals.

To achieve this requirement, a TN-C-S earthing system [also known as Protective Multiple Earthing (PME)] connects the combined neutral and Earth, to real Earth (at many locations) – thereby reducing the risk of broken neutrals – but separate PE and N conductors to fixed indoor wiring and flexible power cords (see Figure 3.9).

The use of TN-C-S is **not** recommended for locations such as petrol stations, etc. where there are lots of buried metalwork and explosive gasses.

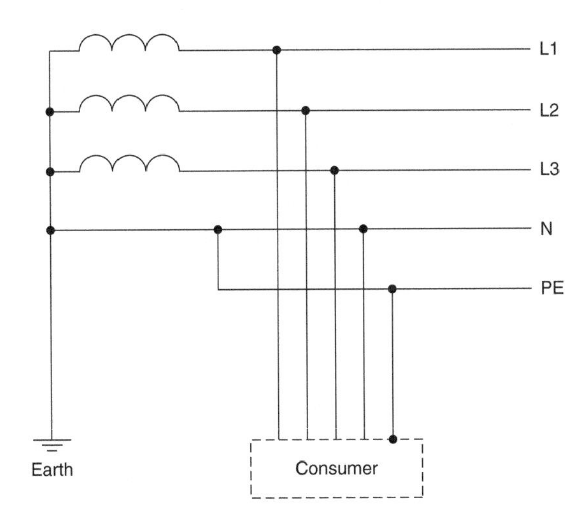

Figure 3.9 TN-C-S System

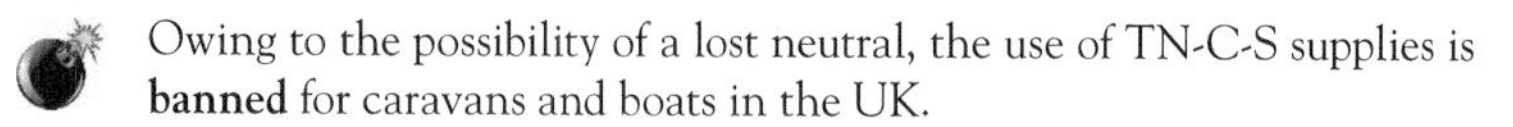

Owing to the possibility of a lost neutral, the use of TN-C-S supplies is **banned** for caravans and boats in the UK.

3.5.2.4 *IT system*

In an IT system, the main earthing terminal is connected via an earthing conductor to an Earth electrode (see Figure 3.10).

In an IT system:

- all live parts are insulated from Earth (or connected to Earth through a sufficiently high impedance connector);
- exposed-conductive-parts are earthed individually, in groups, or collectively.

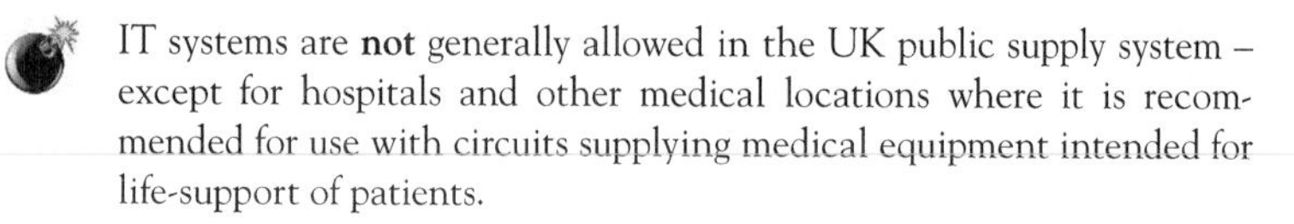

IT systems are **not** generally allowed in the UK public supply system – except for hospitals and other medical locations where it is recommended for use with circuits supplying medical equipment intended for life-support of patients.

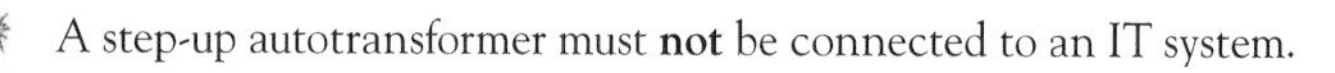

A step-up autotransformer must **not** be connected to an IT system.

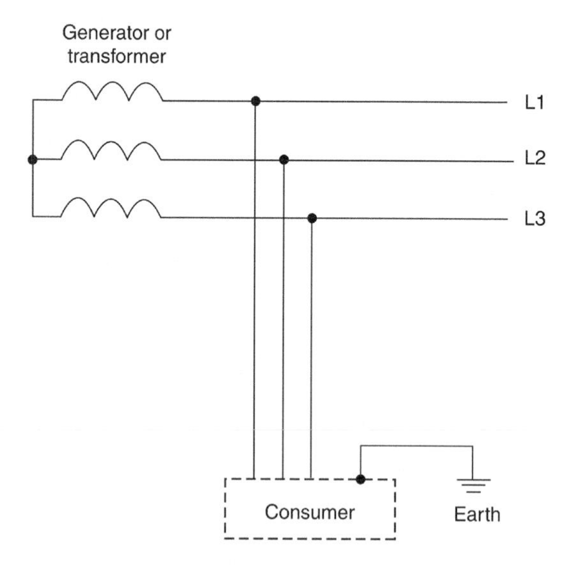

Figure 3.10 IT system

3.6 Earthing points

The resistance area around an Earth electrode depends on the size of the electrode and the type of soil and this is particularly important with regard to the voltage at the surface of the ground. For example (as shown in Figure 3.11) if you consider a 2 m rod with its top at ground level, under fault conditions at least 80–90% of the voltage at the electrode will have dropped in the first 2.5–3 m from the electrode.

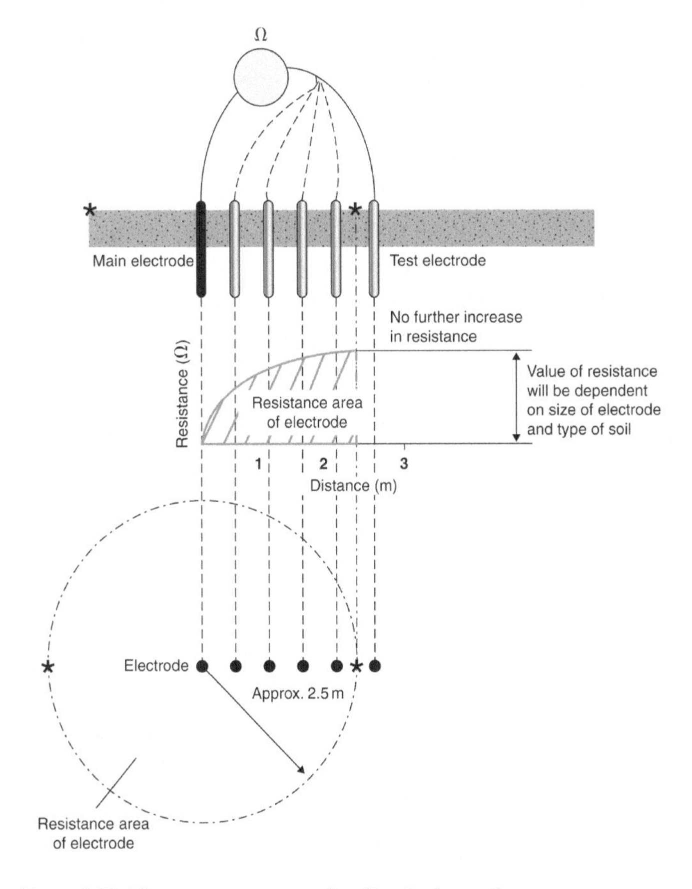

Figure 3.11 The resistance area of an Earth electrode

Source: Courtesy Brian Scaddan

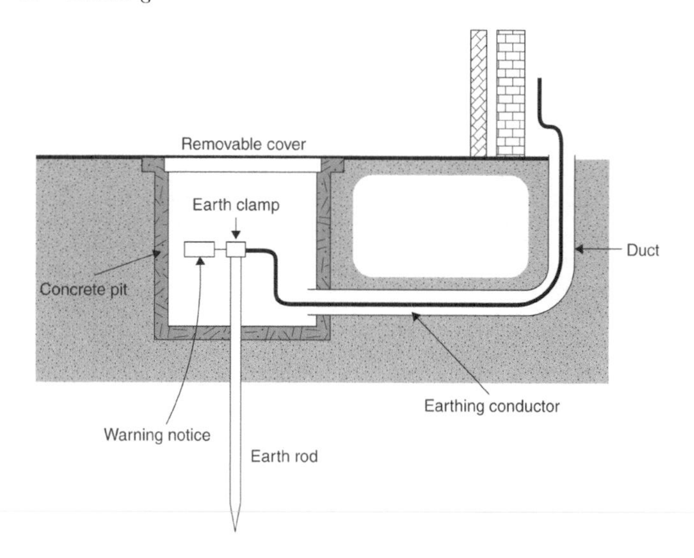

Figure 3.12 An Earth electrode protected by a pit below ground level

Source: Courtesy Brian Scaddan

Earthing can be particularly dangerous where livestock is concerned. For example, a grazing cow might have its forelegs inside the resistive area, whilst its hindlegs are outside. Bearing in mind that a potential difference of 25 V can be lethal to an animal, measures have to be taken to reduce this risk. One method is to house the Earth electrode in a pit that is below ground level (as shown in Figure 3.12).

3.7 Earthing terminals

The Main Earthing Terminal (MET) in a building is where the main Earth, the main equipotential bonding conductors and the connection to the circuit protective conductors for installation's circuits meets and this can be either a bar, plate or even a copper internal 'ring' conductor.

Protective bonding conductors of each installation connect all other extraneous-conductive-parts of the installation to the main earthing terminal. These include:

- central heating and air conditioning systems;
- exposed metallic structural parts of the building;

- gas installation pipes;
- water installation pipes;
- other installation pipework and ducting.

Joints in an earthing conductor shall **only** be capable of being disconnected by means of a specialist tool!

The means of connecting the different electrical systems is shown in Table 3.1.

The main earthing terminal is connected to Earth as shown below and in Figure 3.13.

The Earth electrode should be positioned as close as possible to the main earthing terminal.

The main earthing terminal connects to the following earthing conductors:

- the circuit protective conductors;
- the protective bonding conductors;
- functional earthing conductors (if required); and
- lightning protection system bonding conductor (if any).

Table 3.1 Connections of main earthing terminals

System	*Means of connection*
TN	All exposed-conductive-parts of the installation shall be connected by a protective conductor to the main earthing terminal of that installation (which shall, in turn, be connected to the earthed point of the power supply system)
TN-S	The main earthing terminal of the installation shall be connected to the earthed point of the source of energy
TN-C-S	Where protective multiple earthing is provided; the main earthing terminal of the installation shall be connected, by the distributor, to the neutral of the source of energy
TT	Exposed conductive-parts of the installation shall be protected by a single protective device that is connected (via the main earthing terminal) to a common Earth electrode
IT	The main earthing terminal shall be connected via an earthing conductor to an Earth electrode

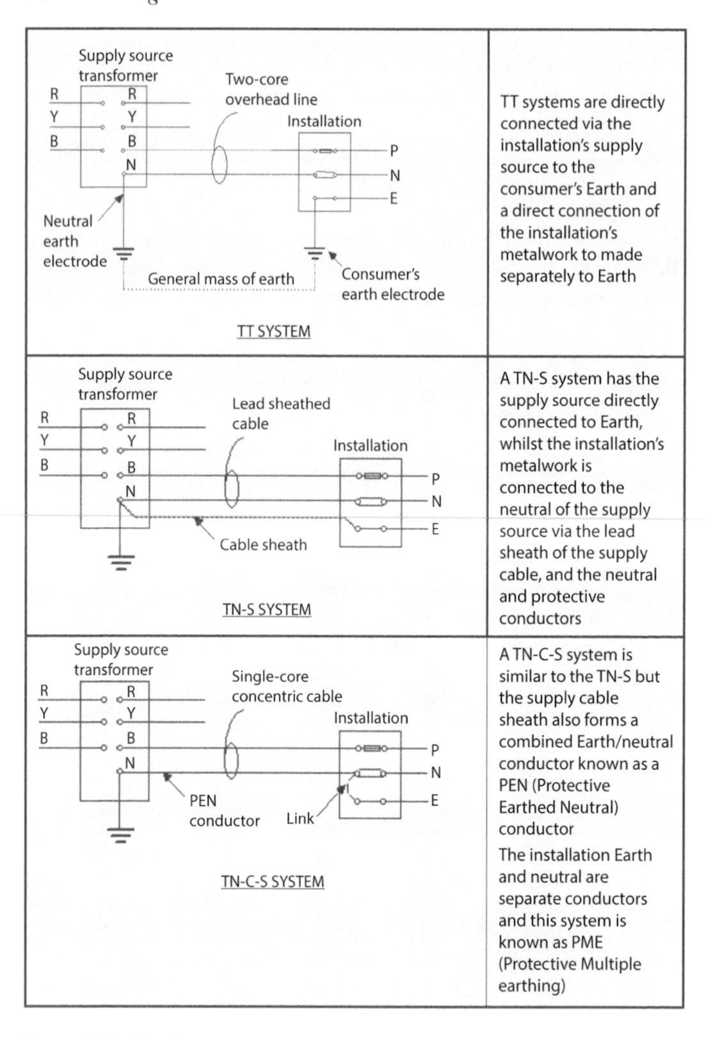

Figure 3.13 Earth provision

3.8 Conductor arrangement and system earthing

In physics and electrical engineering, a conductor is an object or type of material that allows the flow of charge (electrical current) in one or more directions. Materials made of metal are common electrical

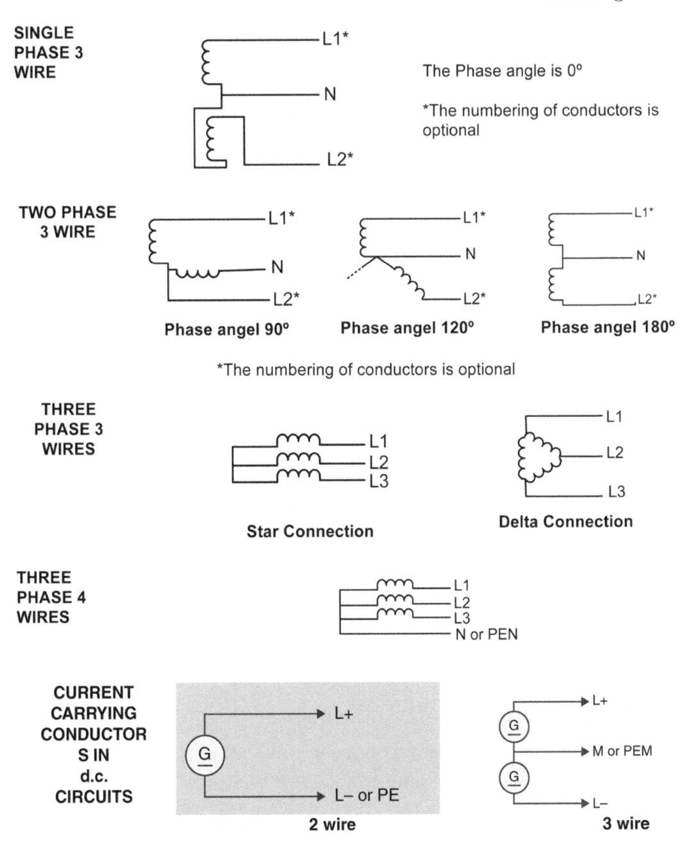

Figure 3.14 Conductor arrangements and system earthing

conductors and these (see Figure 3.14) are the most common carrying conductors currently used.

3.9 Current carrying conductors

The standard method of joining an electrical supply system to Earth is to make a direct connection at the supply transformer so that the neutral conductor (often the star point of a three-phase supply – see Figure 3.15) is connected to Earth using an Earth electrode or the metal sheath and/or armouring of a buried cable.

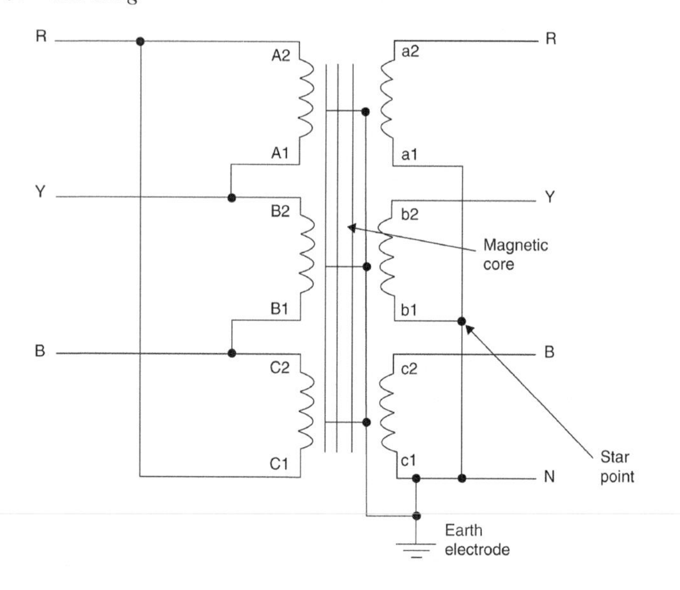

Figure 3.15 Three-phase delta/star transformer showing earthing arrangements

3.10 Protective conductors

A protective conductor is a system of conductors joining all exposed conductive parts together and connecting them to the main earthing terminal. The circuit protective conductor can take many forms, such as:

- a single core cable;
- a conductor in a cable;
- an insulated or bare conductor inside in a common enclosure with live conductors;
- a fixed bar or insulated conductor;
- a metal covering (e.g. the sheath, screen or armouring of a cable)
- a metal conduit, trunking, ducting or the metal sheath and/or armour of a cable.

A gas pipe, an oil pipe, flexible or pliable conduit, support wires or other flexible metallic parts, or constructional parts that are subject to mechanical stress in normal service, shall **NOT** be selected as a protective conductor.

3.11 Installations with separate earthing arrangements

Where a number of installations have separate earthing arrangements, protective conductors common to any of these installations shall either:

- be capable of carrying the maximum fault current likely to flow through them; or
- be earthed within one installation only and insulated from the earthing arrangements of any other installation.

Where electrical monitoring of earthing is used, dedicated devices (such as operating sensors, coils, etc.) shall **not** be connected in series with the protective conductor.

3.11.1 Equipment having a protective conductor current exceeding 10 mA

In these cases, the conductor must be connected to the supply either:

- permanently via the wiring of the installation; or
- via a flexible cable with a plug and socket-outlet; or
- via a protective conductor with an Earth monitoring system.

If this is going to be a permanent connection, it must also be by means of a flexible cable.

3.11.2 Switching devices

A switching device shall **not** be inserted in a protective conductor unless:

- the switch is between the neutral point and the means of earthing; and
- the switch is a linked switch which can disconnect and connect the earthing conductor, at virtually the same time as the related live conductors (see BS 7671:2018 Chapter 46 for more technical details concerning of isolation devices).

3.12 Earth electrodes

An Earth electrode is the conductive part of an electrical installation that connects the main earthing terminal to Earth.

The following types of Earth electrode may be used for electrical installations:

- Earth rods or pipes;
- Earth tapes or wires;
- Earth plates;
- lead sheaths and other metal cable covers;
- other suitable underground metalwork;
- structural metalwork embedded in foundations capable of withstanding corrosion;
- welded metal reinforced concrete embedded in the Earth.

The following types of Earth electrode may **not** be used for electrical installations:

- metallic pipes used for gases or flammable liquids;
- the metallic pipe from a water facility; or
- metallic objects immersed in water.

Further information on Earth electrodes can be found in BS 7430 and [if a Lightening Protection System (LPS) is in place] BS 62305-1.

3.13 Earthing conductors

The earthing conductor (commonly called the 'earthing lead') joins the installation's earthing terminal to either the Consumer's Earth electrode or to the Earth terminal provided by the Electricity Supply Company (see Figure 3.16).

Buried earthing rods must be:

- protected against corrosion;
- driven into virgin soil (as opposed to backfilled or previously disturbed) so as to make an effective contact with the surrounding material and should have a cross-sectional area not less than that shown in Table 3.2.

Note: For further information about earthing conductors, see the latest edition of BS 7430 (*Code of practice for protective earthing of electrical installations*).

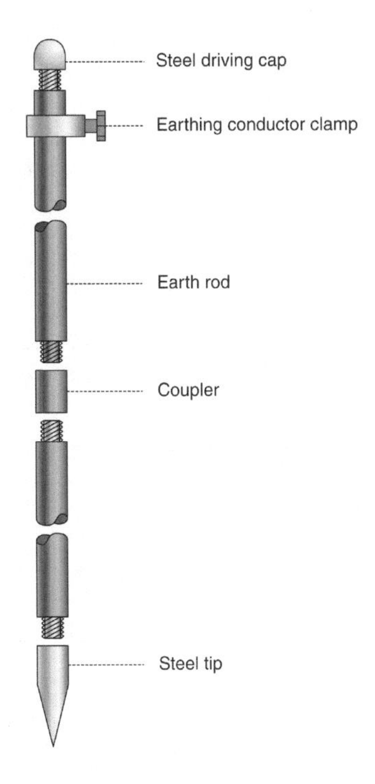

Figure 3.16 An example of an earthing conductor

Source: Courtesy Brian Scaddan

Table 3.2 Minimum cross-sectional area of buried earthing connector

Protected against corrosion	*Protected against mechanical damage*	*Not protected against mechanical damage*
Protected by a sheath	2.5 mm^2 copper, 10 mm^2 steel	16 mm^2 copper, 16 mm^2 steel
Not protected	25 mm^2 copper, 50 mm^2 steel	

3.14 Protective measures

To avoid unauthorised changes being made, protective measures (see Figure 3.17) may only be applied where the installation is under the supervision of skilled or instructed persons as stated in BS EN 61140. Namely:

Note: BS EN 61140 Requirement

Hazardous live parts shall not be accessible and accessible conductive parts shall not be hazardous live, when in use without a fault or in a single fault condition.

To meet this requirement, the following protective measures are required:

- a combination of basic protection and an independent fault protection facility;
- an enhanced protective condition which provides both basic and fault protection; plus
- one or more of the following:

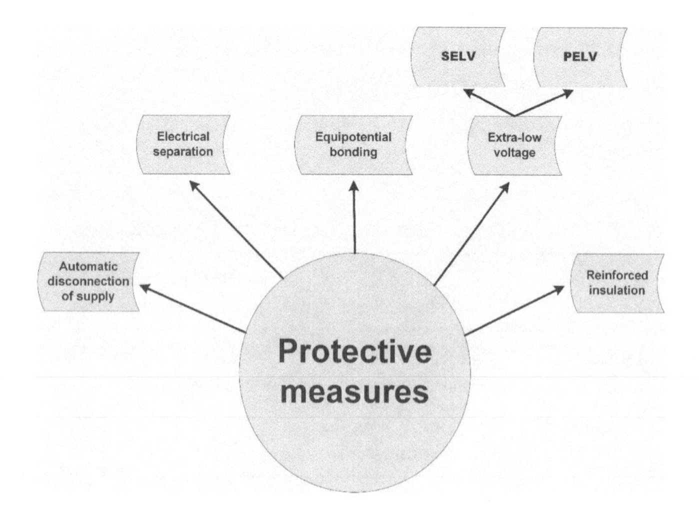

Figure 3.17 Protective measures

3.14.1 *Automatic disconnection of supply*

Automatic disconnection of supply provided by protective earthing, protective equipotential bonding and automatic disconnection in case of a fault.

3.14.2 *Electrical separation for the supply to one item of current-using equipment*

Electrical separation that provides:

- protection to live parts via enclosures or barriers; or
- fault protection by the separation of one circuit from other circuits and from Earth.

3.14.3 *Equipotential bonding*

Main protective bonding conductors that connect extraneous-conductive-parts to the main earthing terminal – such as:

- central heating and air conditioning systems;
- exposed metallic structural parts of the building;
- gas installation pipes;
- water installation pipes;
- other installation pipework and ducting.

3.14.4 *Extra-low voltage (SELV and PELV)*

A SELV (Safety Extra Low Voltage) secondary circuit that is designed and protected so that its voltages do not exceed a safe value (e.g. a modern cordless hand tool, which derives its power via a transformer, converter or equivalent isolation device).

A PELV (Protective Extra Low Voltage) circuit whose voltage cannot exceed the Extra Low Voltage (ELV) (e.g. a computer with a Class I power supply).

3.14.5 *Reinforced insulation*

Reinforced insulation, whose improved basic insulation provides the same degree of protection against electrical shock as double insulation.

3.15 Protective devices

Unless back-up protection is provided, the rated breaking capacity of a protective device shall **not** be less than the maximum prospective short-circuit or Earth fault current at the point of installation.

3.15.1 *Protective earthing*

All exposed-conductive-parts:

- **must** be connected to a protective conductor;
- **must** be connected to the same earthing system, either individually, in groups or collectively.

3.15.2 *Protective multiple earthing*

PME is effectively a high integrity TN-C-S system and is the most common form of earthing provided by the Electricity Supply Companies at new installations. Basically, PME utilises a single conductor for the neutral and earthing functions within their network and provides a PME Earth terminal at the customer's installation (see Figure 3.18).

The great virtue of the PME system is that neutral is bonded to Earth so that a phase to Earth fault is automatically a phase to neutral fault. The Earth-fault loop impedance will then be low, resulting in a high value of fault current which will operate the protective device quickly.

The Electricity Supply Regulations forbid the use of PME supplies to feed caravans and caravan sites.

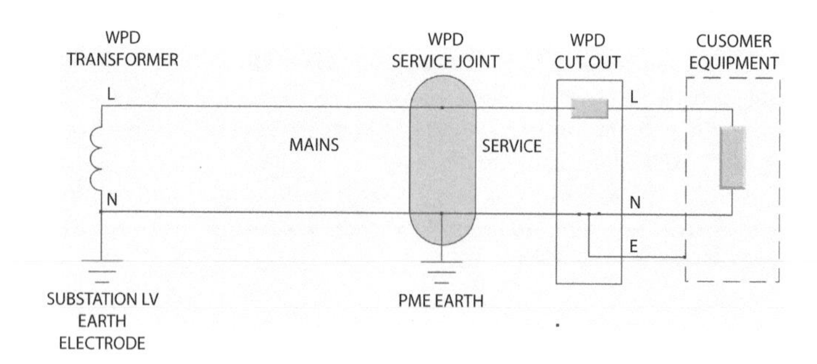

Figure 3.18 Protective multiple earthing

Source: Courtesy HEC

3.15.3 *Protective and neutral (PEN) conductors*

The neutral conductor is a protective conductor (i.e. a PEN conductor – see Figure 3.19) that is connected to a number of Earth electrodes in the installation.

1. PEN conductors may **only** be used within an installation where the installation is supplied by a privately owned transformer or converter.
2. In Great Britain, Regulation 8(4) of the '*Electricity Safety, Quality and Continuity Regulations 2002*' prohibits the use of PEN conductors in consumers' installations.
3. PEN conductors shall **not** be used in medical locations and medical buildings downstream of the main distribution board.

3.15.4 *Protective switches*

Switches, circuit-breakers (except where linked), or fuses shall **only** be inserted in an earthed neutral.

Any linked switch or linked circuit-breaker that is used with an earthed neutral conductor shall be capable of breaking **all** of the related line conductors.

Where an installation is supplied from more than one source of energy with different earthing means, a switch may be inserted in the connection between the neutral point and the means of earthing, **provided** that the device:

- is a linked switch arranged to disconnect and connect the earthing conductor for the appropriate source, at substantially the same time as related live conductors; or
- an interlinked switching device inserted in related live conductors.

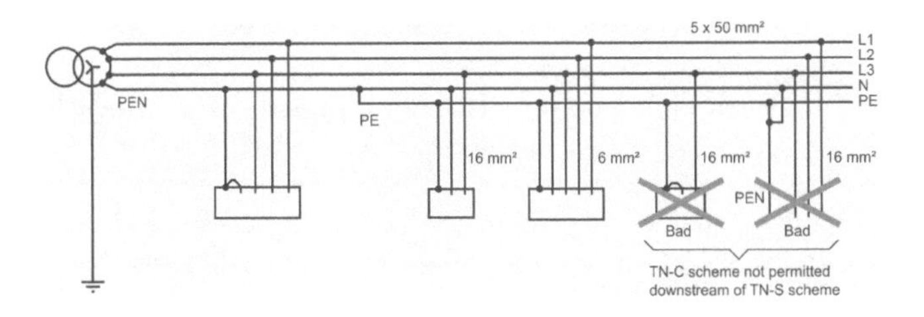

Figure 3.19 PEN conductors in consumers' installations

Source: Courtesy HEC

3.16 Earth fault

The following 'Earth fault' section provides the requirements for the safety of a low voltage installation in the event of:

- a fault between the high voltage system and Earth;
- the loss of the supply neutral in the low voltage system;
- a short-circuit in the low voltage installation;
- accidental earthing of a line conductor of a low voltage IT system.

3.16.1 *Earth fault protection*

Earth fault protection may be omitted for:

- unearthed street furniture supplied from an overhead line provided that it is out of arms reach;
- exposed conductive parts which (because of their size) cannot be gripped or come into major contact with the human body.

In all other cases:

- fault protection **must** happen by automatic disconnection of the power supply by means of an overcurrent protective device, in each line conductor, or by an RCD.

Note: if an RCD is used, the product of the residual operating current (in amperes) and the Earth fault loop resistance (in ohms) shall not exceed 50 V.

- all exposed-conductive-parts of the reduced low voltage system **must** be connected to Earth;
- the Earth fault loop impedance at all points of use (including socket-outlets), should have a disconnection time less than 5 s;
- live parts of the separated circuit shall not be connected at any point to another circuit or to Earth or to a protective conductor;
- flexible cables – particularly where they are liable to mechanical damage – **shall** be practical throughout their entire length;
- exposed-conductive-parts of a separated circuit must be connected to either the protective conductor, exposed-conductive-parts of other circuits, or to Earth.

3.16.2 *Earth fault loop impendence*

Earth fault loop impedance (Zs) is the impedance of the intended path of an Earth fault current (i.e. the Earth fault loop) starting and ending at the point of the fault to Earth.

As shown in Figure 3.20, the Earth fault loop starts at the point of the fault and comprises:

- the Circuit Protective Conductor (CPC);
- the consumer's MET and earthing conductor;
- the Earth return path (for TT and IT systems) or the metallic return path (for TN systems);
- the path through the earthed neutral point of the transformer;
- the transformer winding;
- the line (phase) conductor from the transformer to the point of fault.

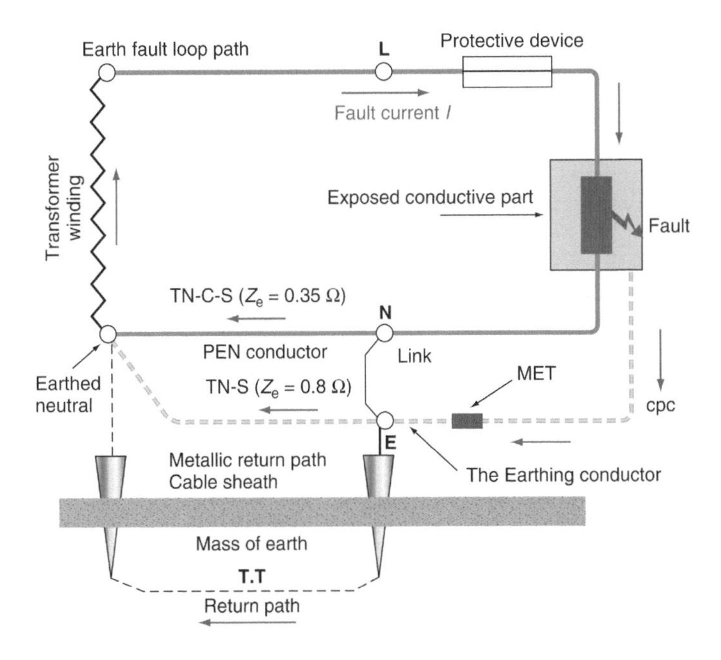

Figure 3.20 Example of a TT Earth fault loop impedance path

Source: Based on a diagram by Pathos Electrician

It is recommended that radial wiring patterns are used to avoid 'Earth loops' that may cause electromagnetic interference – particularly in medical locations!

3.17 Insulation monitoring devices for IT systems

An IMD (Insulation Monitoring Device such as an RCD) is designed to indicate when an Earth fault is detected and to allow that fault to be located and eliminated, as soon as possible, in order to restore normal operating conditions.

The IMD shall:

- act as a switching device that disconnects all live poles and which is only energized in the event of an emergency;
- be permanently connected to an IT system in order to continuously monitor the insulation resistance of the complete system;
- be connected between Earth and a live conductor of the monitored equipment;
- have its the Earth or functional Earth terminal connected to the main Earth terminal of the installation.

In some d.c. IT two-conductor installations, a passive IMD that does not inject current into the system may be used, **provided** that:

- the insulation of all live distributed conductors is monitored; and
- all exposed-conductive-parts of the installation are interconnected; and
- circuit conductors are selected and installed so as to reduce the risk of an Earth fault to a minimum.

An IMD is **not** intended to provide protection against electric shock.

3.18 Residual current devices

Currents, of around 30 mA, through the human body are potentially sufficient to cause a cardiac arrest or serious harm if it persists for more than a small fraction of a second!

An RCD (see Figure 3.21) is designed to prevent you from getting a fatal electric shock if you touch something live, such as a bare wire.

RCDs can also provide some protection against electrical fires!

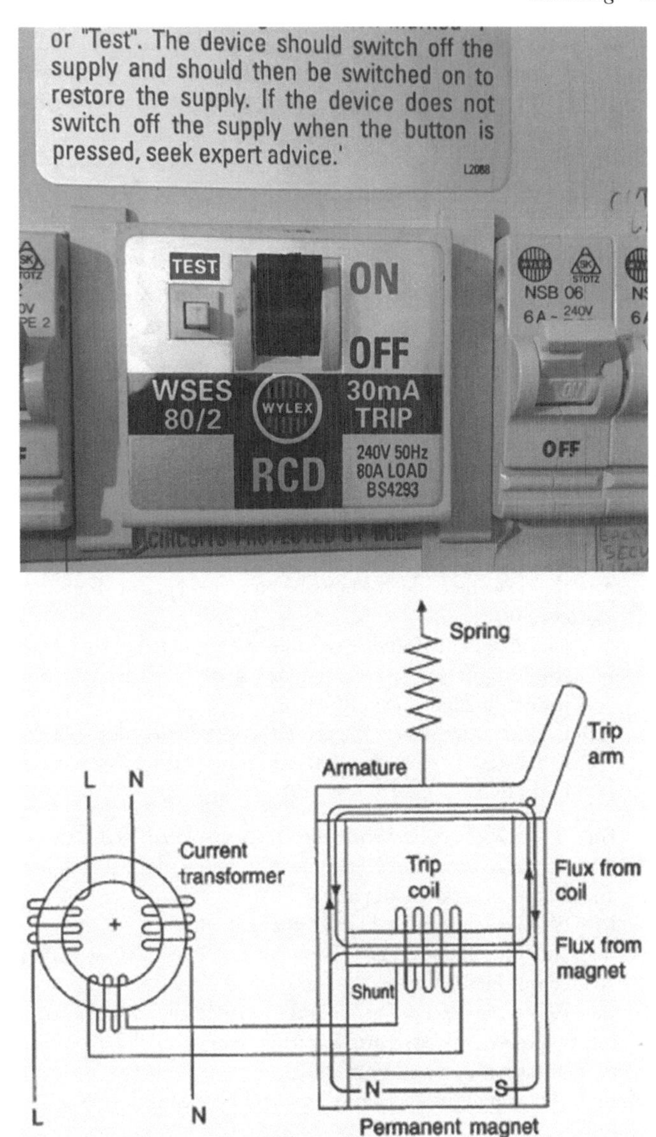

Figure 3.21 Residual current device

Source: Courtesy Riddiford

When a fault current flows there is a difference between the load and return currents, which generates a resultant flux in the toroid which induces a current in the detecting winding – which opens the main contacts of the RCD.

An RCD can be opened and closed manually to switch normal load currents, and will open automatically when an Earth fault current flows which is more than 50 per cent of the rated tripping current.

3.18.1 *The use of RCDs in different types of supplies*

In a TN, TT or IT system, one or more RCDs with a rated residual operating current of not more than 30 mA shall be installed to protect every circuit.

- In a TN system:
  - the neutral conductor must be protected against short-circuit current;
  - wiring systems (other than mineral insulated cables, busbar trunking or powertrack systems) shall be protected against insulation faults by an RCD;
  - if a protective device in part of the installation does not completely satisfy safety requirements, that part may be protected by an RCD.
- For a TN-S system where the neutral is not isolated, RCDs are positioned so as to avoid incorrect operation owing to the possibility of any parallel neutral-Earth path.
- In a TN-C system, an RCD shall **not** be used.
- If an RCD is used in a TN-C-S system, a PEN conductor shall **not** be used on the load side.
- In a TT system, except for mineral insulated cables, busbar trunking or powertrack systems, a wiring system shall be protected against insulation faults by an RCD.
- In an IT system, protection is provided by means of one or more RCDs.

If a medical IT system is used, additional protection by means of an RCD need not be used.

3.18.2 Power supply

If an RCD is powered from an independent auxiliary source that does not operate automatically in the case of failure of the auxiliary source, it shall **only** be used if:

- fault protection is maintained even in the case of failure of the auxiliary source; or
- the device is part of an installation that is inspected and tested by a competent person.

Where RCDs are also used for protection against fire, the conditions for protection by automatic disconnection of the supply shall be verified.

If an RCD is required for additional protection, the effectiveness of automatic disconnection of supply shall be verified using suitable test equipment according to BS EN 61557-6.

3.18.2.1 Warning notices: Periodic inspection and testing

Where an installation incorporates an RCD, a notice (similar to that shown in Figure 3.22) shall be fixed in a prominent position at or near the origin of the installation and shall read as follows:

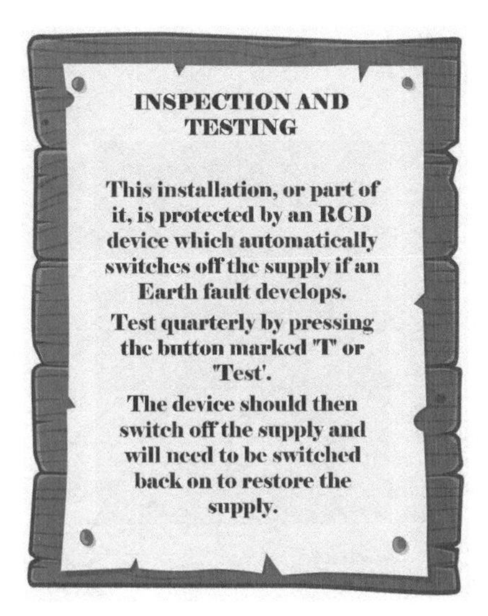

Figure 3.22 Warning notice – RCD protection

3.19 Residual current monitors

Residual Current Monitors (RCMs) are used to monitor earthed systems in TN and TT systems for fault currents or residual currents. They detect deteriorations of the insulation level at an early stage and in a reliable way and are (as long as the fault persists) used to send an audible and/or visual signal to the user long before reaching the shutdown threshold of the RCD.

For further information about RCMs, see BS EN 62020.

An RCM is **not** intended to provide protection against electric shock.

3.20 Test and inspections

Before any electrical installation is energised, a series of tests (as shown in Figure 3.23) will need to be completed to ensure that no addition or alteration (temporary or permanent) has been made to an existing installation **unless** it has been verified that the earthing and bonding arrangements used as a protective measure, are adequate.

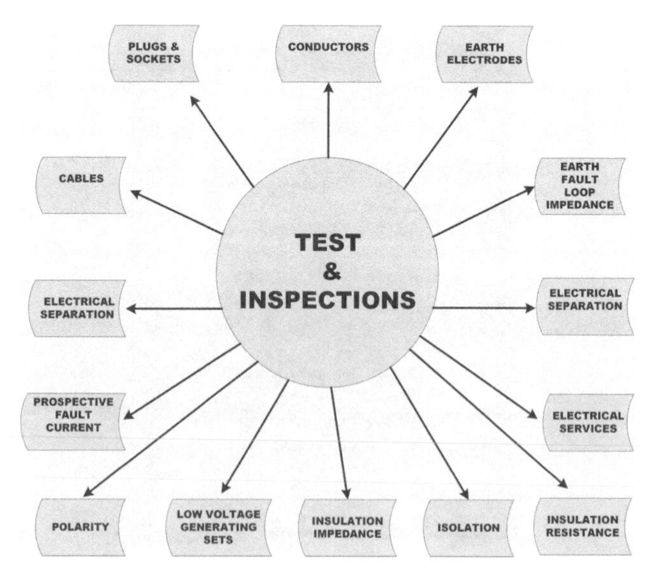

Figure 3.23 Test and inspections

For more details concerning test requirements and maintenance inspections see Chapter 9.

3.20.1 *Cables*

A cable passing through a joist within a floor or ceiling construction or through a ceiling support (e.g. under floorboards), shall:

- include in an earthed metallic covering; or
- be enclosed in an earthed conduit; or
- be enclosed in earthed trunking or ducting; or
- be mechanically protected against damage sufficient to prevent penetration of the cable by nails, screws etc.; or
- be at least 50 mm measured vertically from the top, or bottom as appropriate, of the joist or batten.

A cable concealed in a wall or partition at a depth of less than 50 mm from a surface of the wall or partition shall

- be installed in an area within 150 mm from the top of the wall or partition or within 150 mm of an angle formed by two adjoining walls or partitions.

If the cables of an installation are concealed in a wall or partition (the internal construction of which includes metallic parts, other than metallic fixings such as nails, screws and the like, then (in addition to the above requirements) it shall:

- be mechanically protected sufficiently to avoid damage to the cable during construction of the wall or partition and during installation of the cable; or
- be provided with additional protection by means of an RCD.

Notes:

1. Where the installation is not intended to be under the supervision of a skilled person, consideration may also be given to providing additional protection by means of an RCD.
2. If a cable is not put in a conduit or duct that is directly buried in the ground, then it **must** include a suitable protective conductor such as earthed armour or metal sheath or both, together with a warning tape.

3.20.2 Conductors

The continuity of conductors and connections to unprotected or superfluous conductive parts must be confirmed by measuring the resistance of:

- protective conductors;
- protective bonding conductors;
- live conductors (in the case of ring circuits).

3.20.3 Earth electrodes

Where the earthing system incorporates an Earth electrode as part of the installation, the electrode resistance to Earth shall be measured.

3.20.4 Earth fault loop impedance

If protective measures are used that need to know the Earth fault loop impedance of the installation, then the relevant impedances shall be measured.

Note: Further information on the measurement of Earth fault loop impedance can be found in Appendix 14 to BS 7671:2018.

3.20.5 Electrical separation

Electrical separation of individual circuits is designed to prevent shock currents energising exposed conductive parts in the basic circuit insulation and is a protective measure in which:

- basic protection is provided by simple insulation of live parts by barriers or enclosures; and
- fault protection is provided by ensuring that the electrical circuit is separated from other circuits and from Earth.

Electrical separation may only supply one item of current-using equipment from one unearthed source with simple separation. If there are two or more items from the same electrical source, then a warning notice (see example in Figure 3.24) must be fixed – in a prominent position – beside all points of access.

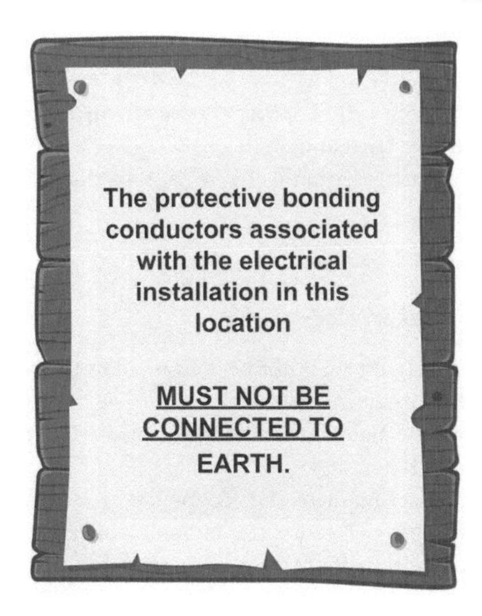

Figure 3.24 Warning notice – protective bonding conductors

Also, in accordance with BS 951, a permanent label, with the words shown in Figure 3.25, shall be permanently fixed, in a visible position, at or near:

Figure 3.25 Warning notice – earthing and bonding

- the point of connection of every earthing conductor to an Earth electrode;
- the point of connection of every bonding conductor to an extraneous-conductive-part; and
- the main Earth terminal (if it is separated from the main switchgear).

3.20.6 Electrical services

A voltage Band I circuit shall **not** be contained in the same wiring system as a Band II circuit unless their cores are separated by an earthed metal screen of equivalent current-carrying capacity to that of the largest core of a Band II circuit.

The neutral (star) point of the secondary windings of three-phase transformers and generators (or the midpoint of the secondary windings of single-phase transformers and generators) **shall** be connected to Earth.

3.20.7 Insulation resistance

Insulation resistance shall be measured:

- between live conductors; and
- between live conductors and the protective conductor connected to the earthing arrangement.

using the test voltages shown in Table 3.3.

Table 3.3 Minimum values of insulation resistance

Circuit nominal voltage (V)	*Test voltage d.c. (V)*	*Minimum insulation resistance (MΩ)*
SELV and PELV	250	0.5
Up to and including 500 V with the exception of the above systems	500	1.0
Above 500 V	1000	1.0

More stringent requirements are applicable for the wiring of fire alarm systems in buildings (see BS 5839-1).

Note: If the circuit includes electronic devices which are likely to influence the results, or be damaged, then <u>only</u> measurements between the live conductors connected together and the earthing arrangement shall be made.

3.20.8 *Isolation*

For safety reasons, it is important that **every** circuit is capable of being isolated from each source of electric energy **and** from each live supply conductor.

Notes:

1. If the 'isolation' is by means of a lock or removable handle, then the key or handle shall be non-interchangeable with any other used for a similar purpose within the premises.
2. If a switch is provided for this purpose:
   - it shall be capable of cutting off the full load current of the relevant part of the installation; but
   - if used as a device for switching off for mechanical maintenance, the switch need not necessarily interrupt the neutral conductor.

3.20.9 *Insulation resistance/impedance of floors and walls*

In a non-conducting location, at least three measurements shall be made in the same location. One must be within 1 m of any accessible extraneous-conductive-part in the location, whilst the other two measurements can be made further away.

Note: Further information on measurement of the insulation resistance and impedance of floors and walls can be found in Appendix 13 of BS 7671:2018.

3.20.10 *Low voltage generating sets*

When a generator is being used as a switched alternative to a TN system, the earthed point of the public electricity distribution system cannot be relied on for protection of the system and an alternative means of earthing shall be provided.

3.20.11 *Polarity*

Where necessary, a polarity test of the supply shall be made at the origin of the installation to verify that:

- every fuse, single pole control and protective device is connected to the line conductor – only, and (except for E 14 and E27 lamp holders according to BS EN 60238);
- wiring has been correctly connected to the socket outlets and similar accessories; and
- all wiring has been correctly terminated throughout the installation.

3.20.12 *Prospective fault current*

The prospective short-circuit current and prospective Earth fault current should be capable of being measured at the origin, as well as at other relevant points in the installation.

3.20.13 *Protection by electrical separation*

The separation of the live parts from those of other circuits and from Earth shall be confirmed by measuring the insulation resistance are in accordance with Table 3.3 (see para 3.20.7).

3.21 Special locations and installations

Similar to Domestic and non-domestic buildings, there are a number of other locations (see Figure 3.26) where Wiring Regulations are applicable.

3.21.1 *Agricultural and horticultural premises*

In agricultural and horticultural premises, a TN-C system shall **not** be used.

In locations intended for livestock, normal protective measures used in other locations (like placing obstacles out of reach shall not be used) instead, supplementary bonding shall connect all exposed-conductive-parts and extraneous conductive-parts that could possibly be touched by livestock.

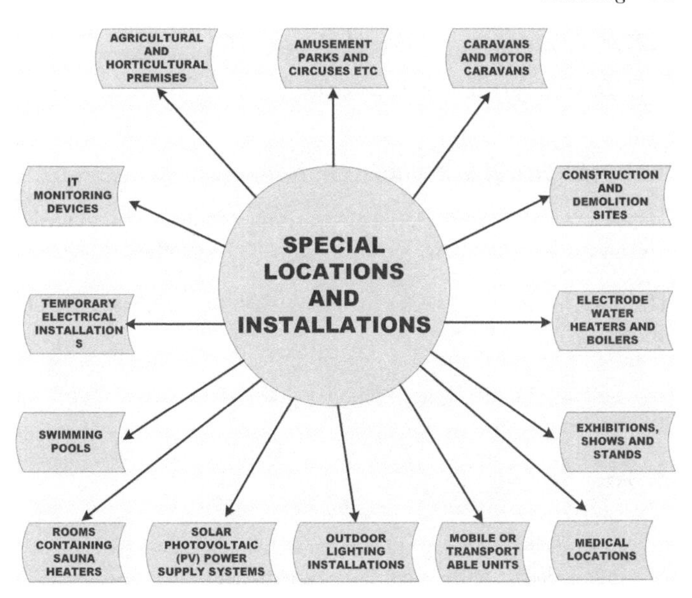

Figure 3.26 Special locations and installations

In circuits (whatever the type of earthing system is used) shall be protected by RCDs as shown in Table 3.4.

Precautions must also be taken to ensure that:

- persons or livestock cannot unintentionally touch live parts;
- all live parts are inside barriers or enclosures and suitable warning signs will be made available;
- barriers and/or enclosures are secured by a bolt or a key.

Table 3.4 RCDs used in agricultural and horticultural premises

Type of circuit	*Rated current*	*RCD operating time*
Final circuit	<32 A	<40 ms
Final circuit	>32 A	<100 ms
Other circuits		<300 ms

3.21.2 Amusement parks and circuses etc.

A PME earthing facility shall **not** be used.

3.21.3 Electrical installations in caravans, motor caravans

The use of a PME earthing facility for earthing a caravan is **prohibited** by The Electricity Safety, Quality and Continuity Regulations (ESQCR).

The nominal supply voltages for caravans shall (in accordance with BS EN 60038) be as follows:

- a.c. supplies, 230 V single phase or 400 V three-phase;
- d.c. supplies, 48 V.

BS 7671:2018 also states that:

- structural metal parts within the caravan shall be connected via bonding conductors to the main earthing terminal within the caravan;
- electrical separation shall not be used except for the shaver socket.

3.21.4 Electrical installations at construction and demolition sites

A PME earthing facility shall **not** be used for earthing an installation at a construction or demolition site unless all extraneous-conductive parts are reliably connected to the main earthing terminal.

If a functional Earth is required for certain equipment (for example, measuring and control equipment):

- supplementary equipotential bonding shall be provided;
- the unearthed source shall have simple form of separation.

3.21.5 Electrode water heaters and boilers

When an electrode water heater or boiler is directly connected:

- to a supply exceeding low voltage, the installation shall include an RCD;
- to a three-phase low voltage supply, it shall be connected to the neutral of the supply as well as to the earthing conductor;

- is not piped to a water supply or is in physical contact with any earthed metal, a fuse in the line conductor may be substituted for the circuit-breaker.

Note: If the supply to the electrode water heater or boiler is single-phase and one electrode is connected to a neutral conductor earthed by the distributor, then the shell of the water heater or boiler must be connected to the neutral of the supply **as well as** to the earthing conductor.

3.21.6 *Exhibitions, shows and stands*

The following requirements are intended for temporary electrical installations at exhibitions, shows, stands and mobile (or portable) displays and equipment:

- any cable supplying temporary structures shall be protected by an RCD;
- all accessible structural metallic parts shall be connected through the main protective bonding conductors to the main earthing terminal within the unit;
- all accessible socket-outlet circuits (other than those for emergency lighting) shall be protected by and RCD;
- all live parts shall be covered by insulation, and located inside barriers and shelters;
- PME earthing shall not be used unless the installation is continuously monitored and a suitable means of Earth has been confirmed before the connection;
- the protective measures of non-conducting location and Earth-free local equipotential bonding are not permitted.

3.21.7 *Medical locations*

HTM 06-01(a Government guidance document entitled *Electrical services supply and distribution*) provides the legal requirements, design applications, operation and maintenance of the electrical infrastructure within healthcare premises.

Although the requirements of BS 7671:2018, Section 710 mainly refer to patient healthcare facilities (such as hospitals, private clinics, medical and dental practices, healthcare centres and dedicated medical rooms in the workplace) they also equally apply to electrical installations in locations designed for medical research and (where applicable) to veterinary clinics. They do **not**, however, apply to the medical electrical equipment itself, as this is fully covered in ISO 13485!

In patient healthcare facilities, the risk to patients (owing to reduced body resistance) is enhanced and it is extremely important that the following rules and regulations are strictly adhered to.

3.21.7.1 Electromagnetic disturbances

In the following medical locations special considerations have to be made with respect to the possibility of Electromagnetic Interference (EMI) and Electromagnetic Compatibility (EMC):

- current-using equipment in a Group 1 medical location;
- in a Group 2 medical location where, applied parts are intended to be used;
- in Group 2 medical locations using SELV and/or PELV circuits not exceeding 25 V. a.c. rms or 60 V ripple-free d.c. and protection by basic insulation of live parts or by barriers or enclosures has been provided;
- In Group 2 medical locations, where PELV is used, exposed-conductive-parts of equipment (e.g. operating theatre luminaires) shall be connected to the circuit protective conductor.

3.21.7.2 Inspection and testing

3.21.7.2.1 INITIAL VERIFICATION

In addition to the requirements of Chapter 64 of the Wiring Regulations and HTM 06-01 (Part A) the following tests **shall** be carried out, prior to commissioning, after alteration or repairs **and** before re-commissioning:

- Confirmation of the correct functioning of the RCM and the IMD;
- measurements of leakage current from the IT transformers of the output circuit and enclosure under no-load conditions;
- measurements to verify that the resistance of the supplementary equipotential bonding is within stipulated limits.

3.21.7.2.2 PERIODIC INSPECTION AND TESTING

In addition to the requirements of BS 7671:2018 (Chapter 62) periodic inspection and testing should be carried out in accordance with

Health Technical Memorandum (HTM) 06-01 (Part B) and local Health Authority requirements as follows and at the given intervals:

- **Annually** – complete functional tests of all IMDs associated with the medical IT system including insulation failure, transformer high temperature, overload, discontinuity and the acoustic/visual alarms linked to them.
- **Annually** – measurements to verify that the resistance of the supplementary equipotential bonding is within the stipulated limits.
- **Every 3 years** – complete measurements of leakage current of the output circuit and of the enclosure of the medical IT transformers under a no-load condition.

The dates and results of each verification **shall** be recorded on an electrical installation report.

3.21.7.2.3 PEN CONDUCTORS

PEN conductors shall **not** be used in medical locations and medical buildings downstream of the main distribution board.

3.21.7.3 Medical facility Earth faults

In the event of a first fault to Earth, a total loss of supply in Group 2 locations shall be prevented.

3.21.7.3.1 RCDS

Chapter 710 of BS 7671:2018 lists the following requirements and recommendations for the use of RCDs in Medical locations:

- care shall be taken to ensure that simultaneous use of large numbers of equipment connected to the same circuit cannot cause unwanted tripping of the RCD;
- only type A (according to BS EN 61008 and BS EN 61009) or type B (according to IEC 62423) RCDs shall be used in Group 1 and Group 2 medical locations;
- type a.c. RCDs shall **not** be used;
- in TN systems, additional protection by RCDs in Group 1 and Group 2 final circuits shall be according to their rated current;

- RCDs shall be used in TN-S systems and the insulation level of all live conductors shall be monitored;
- in Group 1 and Group 2 medical locations that include a TT system, RCDs shall be used.

Where a medical IT subsystem is used in a Group 1 location, additional protection via an RDC is **not** required.

- in Group 2 medical locations (except for a medical IT system), RCDs shall only be used on circuits for:
    - the supply of movements of fixed operating tables; or
    - X-ray units; or
    - large equipment with a rated power greater than 5 kVA.

They should **not** be used for:

- final circuits supplying medical electrical equipment and systems intended for life support;
- surgical applications; and
- 'other' electrical equipment located or that may be moved into the 'patient environment'.

Note: For each circuit that is protected by an RCD, the possibility of the RCDs unwanted tripping due to excessive protective conductor currents produced by equipment in normal operation shall be considered.

- in Group 2 medical locations, where PELV is used, exposed-conductive-parts of equipment (e.g. operating theatre luminaires) shall be connected to the circuit protective conductor.

3.21.7.4 Socket outlets

It is a **mandatory requirement** that socket-outlet circuits in the medical IT system Group 2 for medical locations:

- intended to supply medical electrical equipment shall be unswitched;
- shall be coloured blue and clearly and permanently marked '*Medical equipment only*'.

In addition, at each patient's place of treatment (e.g. bedheads):

- each socket-outlet shall be supplied by an individually protected circuit; or
- several socket-outlets shall be separately supplied by a minimum of two circuits.

3.21.7.5 Equipotential bonding busbar

The equipotential bonding busbar shall be located in or near the medical location using a protective conductor.

3.21.7.6 Supplementary equipotential bonding

In Group 1 and Group 2 medical locations, supplementary equipotential bonding shall be installed for the parts which are located in the '*patient environment*' as follows:

- Group 1: one per patient location;
- Group 2: a minimum of 25% of the total number of individual medical IT socket outlets provided per patient location.

3.21.7.7 Supplies

In medical locations at least two different sources of supply shall be provided, one for the electrical supply system and one for safety services.

Automatic changeover facilities shall comply with BS EN 60947-6-1 and the distribution system shall be designed and installed so that in the case of a mains failure there is an automatic within:

- 0.5 s for luminaires in an operating theatre, light sources for essential Medical Electrical (ME) (e.g. endoscopes, and monitors, etc.) and life support ME equipment;
- 15 s for safety lighting and other services such as firefighters lifts vigilance systems for smoke extraction, paging communication systems and fire detection, etc.;
- more than 15 s for the maintenance of healthcare installation like sterilising equipment etc.

3.21.7.8 Transformers

Transformers **shall** be installed in close proximity to a medical location and with the following additional requirements:

- they shall comply with the requirements of BS EN 61439;
- the leakage current of the output winding to Earth and the leakage current of the enclosure shall not exceed 0.5 mA;
- at least one single-phase transformer per room (or functional group of rooms) shall be used to form the IT systems for mobile and fixed equipment;
- if several transformers are required to supply equipment in one room, they shall not be connected in parallel;
- if the supply of three-phase loads via an IT system is also required, a separate three-phase transformer shall be provided for this purpose.

Capacitors **shall not** be used in transformers for medical IT systems.

3.21.8 Mobile or transportable units

3.21.8.1 Protective measures

> **Author's Hint**
>
> *For the purpose of this particular requirement, the term 'unit' can mean a vehicle and/or mobile transportable structure in which all or part of an electric structure is contained.*

The following protective measures are not permitted:

- obstacles and placing out of reach;
- non-conducting location; and
- Earth-free local equipotential bonding, is not recommended.

In mobile (or transportable units) additional protection by an RCD shall be provided for every socket-outlet intended to supply current-using equipment outside the unit, with the exception of socket-outlets which are supplied from circuits protected by:

- SELV;
- PELV; or
- electrical separation.

3.21.8.1.1 PROTECTIVE EQUIPOTENTIAL BONDING

The following accessible parts of a transportable unit will be connected (through the main protective bonding conductors) to the main Earth terminal;

- central heating and air conditioning systems;
- exposed metallic structural parts of the building;
- gas installation pipes;
- water installation pipes;
- other installation pipework and ducting.

3.21.8.1.2 TN SYSTEM

A PME system shall **not** be used as a means of earthing, except:

- where the installation is continuously under the supervision of a skilled or instructed person; and
- the suitability and effectiveness of the means of earthing has been confirmed.

3.21.8.1.3 IT SYSTEM

An IT system can be provided by either:

- an isolating transformer or a low voltage generating set; or
- an installation fault location system; or
- a transformer providing simple separation via an RCD; and
- an Earth electrode that has been installed so that it provides automatic disconnection of the supply in case of failure in the transformer.

3.21.9 *Outdoor lighting installations*

BS 7671:2018 Section 714 concerns all outdoor lighting installations comprising one (or more) luminaires, a wiring system and accessories and the relevant highway power supplies and street furniture.

3.21.9.1 *Protective measure: Automatic disconnection of supply*

Where automatic disconnection of supply is used, all live parts of electrical equipment shall incorporate basic protection either by insulation, barriers or enclosures.

3.21.9.1.1 PROVISIONS FOR BASIC PROTECTION

- enclosed live parts shall **only** be accessible with a key or a tool;
- a door giving access to electrical equipment and located less than 2.50 m above ground level shall be locked with a key or require a tool for access;
- access to the light source for a luminaire which are positioned less than 2.80 m above ground level, shall only be possible using a tool to remove a barrier or an enclosure.

3.21.9.1.2 ADDITIONAL PROTECTION

Lighting in places such as telephone kiosks, bus shelters and advertising panels, etc. shall be provided with an RCD for additional protection.

3.21.10 *Solar photovoltaic (PV) power supply systems*

Section 712 of BS 7671:2018 concerns PV power supply systems (including those with a.c. modules) and the following rules apply:

- earthing of one of the live conductors of the d.c. side of a solar photovoltaic is permitted, if there is at least simple separation between the a.c. side and the d.c. side;
- Earth connections on the d.c. side should be electrically connected so as to avoid corrosion (see BS EN 13636 and BS EN 15112);
- PV string cables, array cables and d.c. main cables must minimise the risk of Earth faults and short-circuits.

The protective measures of non-conducting location and Earth-free local equipotential bonding are **not** permitted on the d.c. side.

3.21.11 *Rooms and cabins containing sauna heaters*

The protective measures of non-conducting location Earth-free local equipotential bonding are **not** permitted.

3.21.12 *Swimming pools and other basins*

The following requirements apply to the basins of swimming pools, the basins of fountains and the basins of paddling pools as well as to the surrounding zones of these basins.

In these areas, the risk of electric shock is increased by a reduction in body resistance and contact of the body with Earth potential.

3.21.12.1 Extra-low voltage provided by SELV or PELV

Where SELV is used (regardless of the nominal voltage) normal protection shall be provided by basic insulation, barriers or enclosures.

3.21.12.2 Supplementary protective equipotential bonding

Extraneous-conductive-parts in Zones 0, 1 and 2 shall be connected by supplementary protective bonding conductors to the protective conductors of any exposed-conductive-parts of equipment situated in these zones.

It is permitted to install an electric heating unit embedded in the floor, provided that:

- it is protected by SELV (or if its supply circuit is protected by an RCD); or
- it includes an earthed metallic sheath connected to the supplementary protective equipotential bonding; **and**
- its supply circuit is protected by an RCD; or
- it is covered by an embedded earthed metallic grid connected to the supplementary protective equipotential.

The protective measures of non-conducting location and Earth-free local equipotential bonding are **not** permitted.

3.21.12.3 Underwater luminaires for swimming pools

Underwater luminaires or luminaires in contact with the water shall be fixed and shall comply with BS EN 60598-2-18.

Special requirements may be necessary for swimming pools for medical purposes.

3.21.13 Temporary electrical installations for structures, amusement devices and booths at fairgrounds, amusement parks and circuses

3.21.13.1 Restrictions

- Protective measures such as non-conducting location and Earth-free local equipotential bonding are **not** permitted for installations of this type.
- A PME earthing facility shall **not** be used for any electrical installation covered by this section or for the supply to caravans or similar constructions.
- If a TN system is used, a PEN conductor shall **not** be used downstream of the origin of the temporary electrical installation.

3.21.13.2 Generators

Where a generator supplies a temporary installation, that is part of a TN, TT or IT system, then an Earth electrode shall be capable of withstanding damage and take into account the possibility of corrosion.

3.21.13.3 Automatic disconnection of supply

If an RCD is used as part of the supplies to a.c. motors, then it should be in accordance with BS EN 60947-2 (i.e. of the time-delayed type).

As additional protection, all final circuits for:

- lighting;
- mobile equipment connected by means of a flexible cable with a current-carrying capacity up to 32 A; and
- socket-outlets rated up to 32 A,

shall be protected by RCDs.

3.21.13.4 Supplementary protective equipotential bonding

In livestock locations, supplementary bonding shall connect all exposed conductive parts and extraneous-conductive-parts that can be touched by livestock.

3.21.13.5 Water heaters having immersed and uninsulated heating elements

All metal parts of the heater or boiler which are in contact with the water (other than current-carrying parts) shall be connected to the metal water pipe which supplies water to the heater or boiler – provided that that water pipe is connected to the main earthing terminal independently of the circuit protective conductor.

3.21.14 Monitoring devices in an IT system

In IT systems, continuous insulation monitoring devices shall be provided which give an auditable and visual indication in the event of a first fault and this indication shall continue as long as the fault persists.

The following monitoring devices and protective devices may be used:

- Residual Current Devices (RCDs);
- Residual Current Monitoring devices (RCMs);
- Insulation Monitoring Devices (IMDs);
- Insulation fault location systems;
- Overcurrent protective devices.

Consideration should also be given to the possibility that, if a line conductor of an IT system is earthed accidentally, then the insulation (or components) that are normally rated for the voltage between line and neutral conductors, can be temporarily stressed with the line-to-line voltage.

In locations where there is a strong risk of fire due to the nature of processed or stored materials, the wiring system of an IT system (except for mineral insulated cables, busbar trunking systems or powertrack systems) should be protected against insulation faults, by an IMD with audible and visual signals.

Author's End Note

Having looked at how important earthing is to electrical installations, the next thing to consider, before going any further, is to see if there are any 'external influences' that will have an effect on these products once they are installed.

In the following chapter, we will see how equipment and their interconnections can be simultaneously exposed to a large number of external influences (such as temperature, humidity, weather, pollutants, fire, electromagnetic and mechanical etc.) and why a combination of environmental factors has become so important and needs to be considered during manufacture, operation as well as maintenance and repair of electrical installations.

4 External influences

Author's Start Note

Whilst Chapter 51 of BS 7671:2018 still contains the requirements for external influences (see below), the 18th Edition of the Wiring Regulations has now been further developed in accordance with new ISO standards and the BS EN 60721 and BS EN 61000 series on environmental conditions.

This chapter provides a brief overview (together with précised extracts from the current Regulations) on how the environment in which an electrical product, service or equipment, etc. will affect it meeting the quality and safety requirements of BS 7671:2018.

Note: A list of **all** external influences and their characteristics are included Appendix 5 of BS 7671:2018 and so the following notes are merely offered as guidance and reminders.

BS 7671:2018 Requirement

512.2 External influences

- *Equipment shall be of a design appropriate to the situation in which it is to be used and shall take account of the conditions likely to be, encountered.*
- *If the equipment does not have the characteristics relevant to the external influences of its location, it may nevertheless be used provided the installation has been equipped with additional protection that will not adversely affect the operation of the equipment it has been designed to protect.*

DOI: 10.1201/9781003165170-4

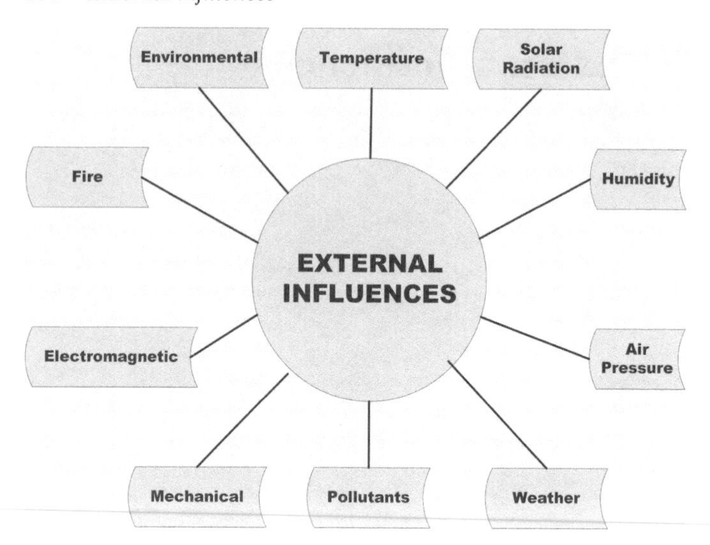

Figure 4.1 External influences

The conditions affecting electrical equipment mainly consist of the environment (ambient and created) the location where the equipment will be operating from – and how it will be used.

For simplicity this can be broken down into two basis categories:

Conditions	The environmental conditions that have been identified as having an effect on equipment (Table 4.1)
Situations	The main uses of electronic equipment (Table 4.2)

4.1 Environmental conditions

There are eight basic conditions that have a direct effect on electrical equipment and electrical installations. These are:

Table 4.1 Basic conditions

Climatic	*Externally generated influences*
• altitude ambient temperature • ambient temperature • atmospheric pressure • condensation • precipitation (i.e. rain, snow and hail) • relative atmospheric humidity • solar radiation • wind	• air movement • dust • temperature • precipitation (e.g. water spray) • pressure changes (e.g. tunnels)
Mechanical	*Ergonomic aspects*
• shock (sinusoidal and random) • vibration	• achieving maximum task effectiveness • protecting the health of the engineer and end user • the comfort of the operator and end user
Electrical	*Chemical*
• earthing and bonding • electromagnetic environment (EMC and EMI) • power supplies • susceptibility and generation • transients (spikes and surges)	• corrosion • dangerous substances • pollution and contamination • resistance to solvents
Biological	*General*
• animals • humans (vandalism) • vegetation	• components • design of equipment • earthquakes • flammability and fire hazardous areas • maintainability • safety • reliability • waste

Table 4.2 Environmental conditions

Operational	*Storage*	*Transportation*
• when installed and operational • when installed and not in us	• when in storage	• when being transported

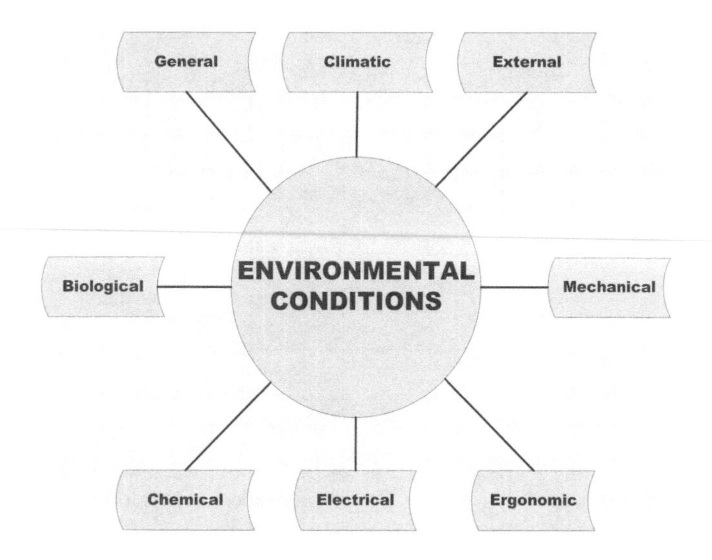

Figure 4.2 Environmental conditions

4.2 Equipment situations

Obviously not all equipment will be fully operational, all of the time, and so various equipment '*situations*' also have to be considered.

4.2.1 Requirements from the regulations for environmental factors and influences

Electrical equipment shall be selected to withstand the stresses, environmental conditions and characteristics of its location as shown in Table 4.3.

Table 4.3 Requirements from the Regulations for environmental factors and influences

Design	The design of electrical equipment shall • take into account the environmental conditions it will be subjected to • be suitable against all external influences • have a suitable level of immunity against electromagnetic disturbances and emissions Equipment that is located within an area susceptible to the risk of fire or explosion shall be adequately constructed and protected
Wiring system	The type of wiring system will depend on the following • The location • The structure supporting the wiring • The accessibility of wiring to persons and livestock • Voltage • Electromagnetic stresses likely to occur • Electromagnetic Interference (EMI) • Other external influences (such as mechanical, thermal and those associated with fire)
Enclosures	 Enclosures containing electrical connections shall provide protection against mechanical and external influence Sealing arrangements for wiring penetrations shall • resist the products of combustion • be protected from water penetration • be compatible with the wiring system's material that it is in contact with • permit thermal movement of the wiring system without a reduction in sealing quality • be mechanically stable, sufficient to withstand and damage to the wiring support system due to fire

4.3 Air pressure and altitude

Air pressure (frequently referred to as atmospheric pressure) is '*the force exerted on a surface of a unit area caused by the Earth's gravitational attraction on the air vertically above that area*'.

Earth's atmosphere has a series of layers, each with its own specific traits based on how the temperature in that layer, changes with altitude and the layer's temperature gradient. Moving upward from ground level, these layers are named the troposphere, stratosphere, mesosphere, thermosphere and above that, comes the exosphere.

Air pressure varies with altitude and location. For instance, at the equator where the trade winds of both hemispheres converge, there is a low-pressure zone (known as the ITCZ or International Conveyance Zone), which is characterised by high humidity.

It is not widely appreciated that the location of equipment, especially with respect to its altitude above sea level, can affect the working of that equipment. But it is not just the height above sea level that has the most effect! Even air pressure variations at ground level have to be considered.

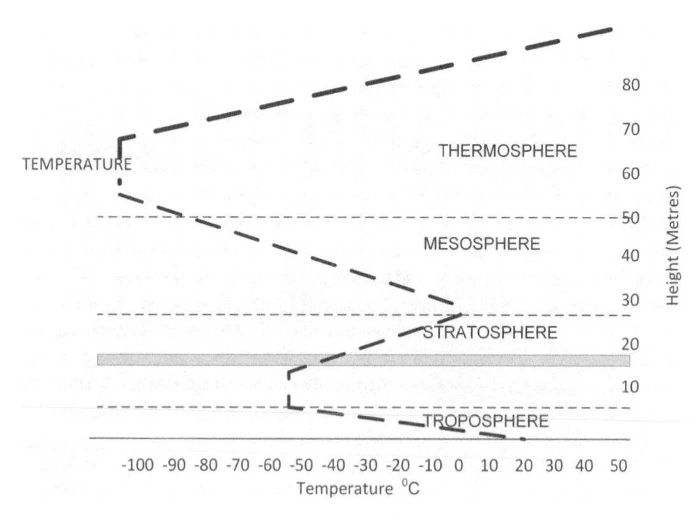

Figure 4.3 Atmospheric structure

4.3.1 *Typical requirements – Air pressure and altitude*

The following are the most common environmental requirements concerning air pressure and altitude.

- **Low air pressure** – At altitudes above sea level, low air pressure can cause:
    - change of physical or chemical properties;
    - decreased efficiency of heat dissipation, that will affect equipment cooling;
    - erratic breakdown or malfunction of equipment from arcing or corona;
    - leakage of gases or fluids from gasket sealed containers;
    - ruptures of pressurised containers;
    - temperature changes effecting the equipment (e.g. volatilisation of plasticisers evaporation of lubricants, etc.).
- **High air pressure – High air pressure occurring in natural depressions and mines can have a mechanical effect on sealed containers.**
- *Electrical equipment* – must be capable of working to an altitude between 120 and 2000 m above sea level – which corresponds to an air pressure range from 110.4 to 74.8 kPa.

4.4 Ambient temperature

Of all the elements that have an effect on man and equipment, none is more vital than temperature which is a particularly important from an environmental point of view and its accurate measurement and definition requires careful consideration.

The ambient temperature at any given time is the temperature of the air measured under standardised conditions. Temperature figures with respect to climate are generally '*shade temperatures*' (i.e. the temperature of the air measured from a location that excludes the influence of the direct rays of the sun) and it is usual for the temperature to be much higher in the direct sunshine. Many mountain areas have air temperatures in the region of zero in winter but the presence of bright sunshine will produce a feeling of warmth and permits the wearing of light clothing.

Figure 4.4 Enjoying the environment

Source: Courtesy of HEC Ltd

Seasonal fluctuations in temperature do not pass below ground deeper than 60–80 ft. Below that depth, borings and mine-shafts show that the temperature increases (downwards) depending on the geographical position, location and depth. And on average, a rise of about 1°C may be taken for each 65 ft of descent.

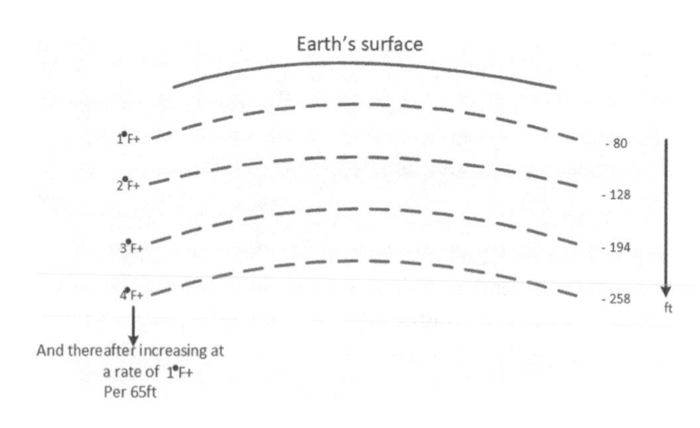

Figure 4.5 Temperature changes below the Earth's crust

4.5 Weather and precipitation

Water is most often seen either as liquid (water), solid (ice) or gas (steam) and the main points to remember are shown the following sub sections:

4.5.1 *Water*

Water is a major cause of failure in every application of electrical and electronic components especially as humidity possesses a certain amount of electrical conductivity which increases the possibility of corrosion of metals. Similarly, the ingress of water followed by freezing within electronic equipment can result in malfunction.

Salt water has an electrochemical effect on metallic materials (i.e. corrosion) which can damage and degrade the performance of equipment.

Ice – Water in the form of ice can cause problems in the cooling of equipment or freezing and thawing, which will result in cracks occurring, breaking cases, etc.

Weathering (or exfoliation) where the surface of equipment is warmed by the sun during the day, it can cool at night possibly damaging, or warping, it's outer surface.

Freeze thaw – when water freezes it turns to ice, expanding by about one twelfth of its volume. If this water is in the joint or a crack in a casing then the space will become enlarged and the casing on either side will be forced apart. When the ice eventually thaws more water will penetrate into the crack and the cycle repeats itself with the crack constantly enlarging.

Chemical weathering – water can pick up small quantities of sulphur dioxide from the atmosphere which will form a weak solution of acid that which can attack certain equipment housings.

Erosion – when the wind blows over dry ground it collects grit and 'throws' it vigorously against the surfaces nearby which will gradually wear away the surface of the item it comes into contact with.

4.5.2 *Requirements from the regulations – Weather and precipitation*

As shown in Table 4.4, the following are the most common environmental requirements concerning weather and precipitation:

Table 4.4 Requirements from the Regulations – Weather and precipitation

Operation	All equipment should be capable of operating during rain, snow, hail and be unaffected by ice, salt and water
Rain	All equipment should be capable of operating in rain and be capable of preventing the penetration of rainfall at a minimum rate of 13 cm/hour and an accompanying wind rate of 25 m/sec
Ice	Wiring systems for solar photovoltaic shall withstand the expected external influences such as ice formation
Snow and hail	Consideration needs to be given to the effect of all forms of snow and/or hail
Water	Wiring systems must be selected and erected so that no damage is caused by condensation or ingress of water during their installation, use and/or maintenance
Salt water	Equipment should be capable of operating in (or be protected from) heavy salt spray at seacoast areas and from salted roadways
	Note: See BS 7671:2018 Chapter 709 for more detailed requirements relating to marinas and similar locations
Weather protection	Equipment should be capable of operating adjacent to the sea shore or on mountain ranges and to function equally well as the same equipment housed in arid deserts

4.6 Humidity

The atmosphere is normally described as '*a shallow skin or envelope of gases surrounding the surface of the Earth which is made up of nitrogen, oxygen and a number of other gases which are present in very small quantities*'. The water vapour content of the atmosphere is subject to extremely wide fluctuations and the amount of water present in the air is referred to as '*humidity*'.

Temperature and the relative humidity of air (in varying combinations) are climatic factors which act upon electrical equipment and installations during storage, transportation or operation. Humidity and the electrolytic damage resulting from moisture, mostly affects plug points, soldered joints (in particular dry joints), bare conductors, relay contacts and switches. Humidity also promotes metal corrosion (see Section 4.8 – Pollutants and Contaminants) owing to its electrical conductivity.

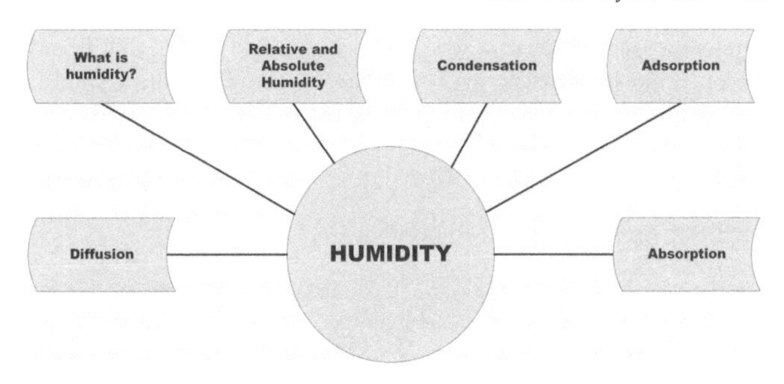

Figure 4.6 The effects of humidity

Note: Humidity (in the context of this book) has been taken to cover Relative Humidity, Absolute Humidity, Condensation, Adsorption, Absorption and Diffusion and details of these 'sub sets' are provided in the following paragraphs.

4.6.1 *Relative and Absolute Humidity and their effect on equipment performance*

The performance of virtually all electrical equipment is influenced and limited by its internal temperature which, in turn, is dependent on the external ambient conditions and on the heat generated within the device itself. Fortunately, most electrical and electronic components (especially resistors) will normally remain dry when under load owing to the amount of internal/external heat dissipation.

4.6.1.1 Externally mounted equipment

Equipment and components that are mounted in external cabinets, run the risk of coming into contact with water or water vapour (e.g. drifting snow, fog, dew, rain, spray water or water from hoses) and the equipment must, therefore, be adequately protected from such humidity in order to prevent the ingress of vapour into the system within the casing.

4.6.1.2 Housed equipment

In most locations (e.g. cabinets, equipment rooms, workshops and laboratories) although temperatures above 30°C may often occur, they are normally combined with a lower relative humidity than that found in the open air. In other rooms (e.g. offices), however, where several heat sources are present, temperatures and relative humidities can differ dramatically across the room.

4.6.1.3 Condensation

Condensation occurs when the surface temperature of an item is lower than that of the dew point (i.e. the temperature with a relative humidity of 100% at which condensation occurs) and which can change electrical characteristics (e.g. decrease surface resistance) between the absolute point at which atmospheric vapour condenses into droplets (i.e. the dew point), absolute humidity and vapour pressure.

4.6.1.4 Absorption

The quantity of water that can be absorbed by a material depends largely on the water content of the ambient air and the speed of penetration of the water molecules generally increases with the temperature.

4.6.1.5 Adsorption

Adsorption is the amount of humidity that may adhere to the surface of a material and depends on the type of material, the surface structure and the vapour pressure. This layer of water (no matter how small) can cause electrical short circuits and material distortion, etc.

4.6.1.6 Diffusion

Water vapour can penetrate encapsulations of organic material (e.g. into a capacitor or semiconductor) by way of the sealing compound and into the casing. This factor is frequently overlooked and can become a problem, especially as the moisture absorbed by an insulating material can cause a variation in a number of electrical characteristics.

4.6.1.7 Protection

The effects of humidity mainly depend on temperature, temperature changes and impurities in the air. As shown in Table 4.5, there are three basic methods of protecting the active parts of equipment and components from humidity.

Table 4.5 Protective methods – Humidity

Effect of humidity	*Protective methods*
Heating the surrounding air so that the relative humidity cannot reach high values	This method normally requires a separate heat source. The disadvantage of this method is the reliability of the circuit being dependent on the efficiency of the heating
Hermetically sealing components or assemblies using hydroscopic materials	Hermetic sealing is an extremely difficult process as the smallest crack or split can allow moisture to penetrate the component particularly in the area of connecting wire entry points
Ventilation and the use of moisture-absorbing materials	Most water-retaining materials and paint, etc. are suitable for the temporary absorption of excessive high air humidity in the casing

4.6.2 *Requirements from the regulations – Typical requirements against humidity*

The following comments shown in Table 4.6 are the most common environmental requirements concerning Electromagnetic compatibility (EMC):

Table 4.6 Environmental requirements concerning Electromagnetic compatibility

Equipment in cubicles and cases	The design of equipment should take into account temperature rises within cubicles and equipment cases in order to ensure that the components do not exceed their specified temperature ranges
Equipment interoperability	Equipment that is operated adjacent to the sea shore (and, therefore, subject to extreme humidity) must be able to function equally well as the same equipment housed in the low humidity of (for example) the desert
External humidity levels	Equipment should be designed and manufactured to meet external humidity levels, over the complete range of ambient temperature values anticipated

(Continued)

Table 4.6 (Continued)

Condensation	Operationally caused infrequent and slight moisture condensation should not lead to malfunction or failure of the equipment
Indoor installations	In all indoor installations, provision must be made for limiting the humidity of the ambient air to a maximum of 75% at −5°C
Product configuration	All proposed and in date equipment, components or other articles must be tested in their production configuration without the use of any additional external devices that have been added expressly for the purpose of passing humidity testing
Peripheral units	Peripheral units (e.g. measuring transducers, etc.) or equipment employed in a decentralised configuration (i.e. where ambient temperature ranges are exceeded) the actual temperature occurring at the location of the equipment concerned should be utilised when designing equipment
Wiring System	A wiring system shall be selected so that no damage is caused by condensation or ingress of water during installation, use and maintenance • Condensation which might form in a wiring system or where water might collect shall be made, is swiftly eliminated • Wiring systems that could be subjected to waves shall be protected from mechanical damage • Where corrosive or polluting substances (including water) could cause corrosion and/ or deterioration, the parts of the wiring system that are likely to be affected shall be suitably protected (e.g. by protective tapes, paints or grease) and/or manufactured from a material resistant to such substances • If a wiring system is routed below services that are liable to cause condensation (such as water, a steam or gas services), precautions shall be taken to protect the wiring system from harmful effects Special consideration needs to be given to wiring systems that are liable to frequent splashing, immersion or submersion

4.7 Solar radiation

Of all the factors that control the weather, the sun is by far the most powerful and practically everything that occurs on the Earth is controlled, directly or indirectly, by it.

Less than one-millionth of the energy emitted from the sun's surface travels the 90-odd million miles to reach this planet. The sun's energy crosses those miles in the form of short electromagnetic radio waves, identical in nature to those used in broadcasting, which pass through the atmosphere and are absorbed by the Earth's surface. These waves warm the Earth's surface and are then re-radiated back to space.

The wavelength of the energy emitted by the Earth is very much longer than that emitted by the sun (because the Earth is much cooler than the sun) and these longer waves are not able to pass through the atmosphere as freely as short waves. For this reason, a large proportion of the energy emitted by the Earth is absorbed by the water vapour and water droplets in the lower atmosphere which in turn is re-radiated back to Earth. Thus, the Earth plays the part of a receiving station absorbing short electro-magnetic waves and converting them into longer electro-magnetic waves, while the atmosphere acts as a trap containing most of the longer electro-magnetic waves before they are lost to space.

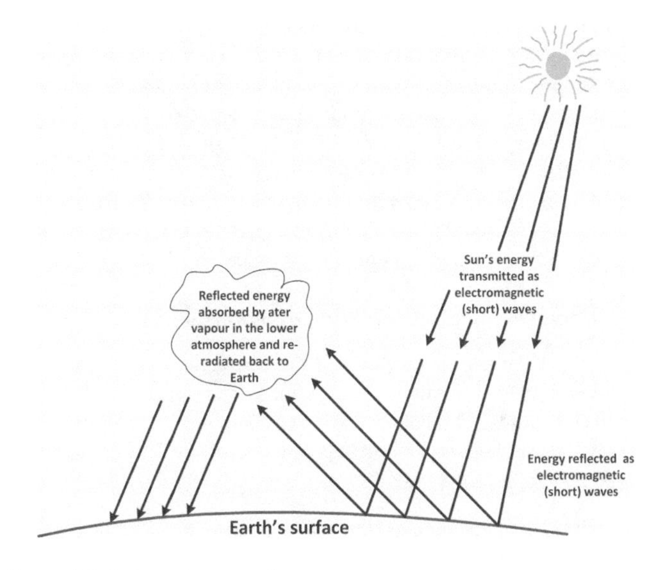

Figure 4.7 Solar radiation – Energy

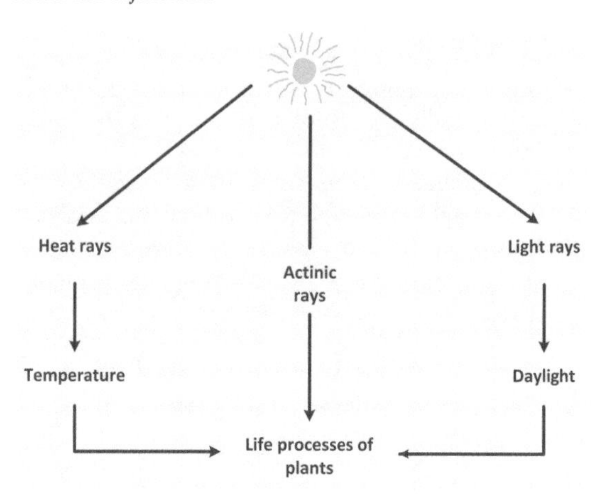

Figure 4.8 Sun's radiation

Radiation from the sun consists of rays of three differing wavelengths, heat rays, actinic rays and light rays. Heat rays and actinic rays are intercepted by solid bodies and produce peculiar effects in varying degrees according to the nature of the surface on which they fall. The light rays are responsible for daylight and both light rays and actinic rays are necessary for the life processes of plants. The heat ray's most important aspect is temperature and the amount of sunshine (and therefore the temperature) will depend on latitude and the length of day.

Radiant energy can be reflected from solid surfaces and intensified by that reflection. However, radiant energy can also cause damage to equipment as heat rays can warm the material and/or its environment to dangerous levels and photochemical degradation of materials can be caused by the ultraviolet content of solar radiation.

4.7.1 Photochemical degradation of material

One of the biggest problems caused by solar radiation is that it can bleach out colours in paints, textiles, paper, etc. (a major consideration when trying to read the colour coding of components!!) but by far the most important effect is the heating of materials.

Typical defects caused are:

- cracking and disintegration of cable sheathing;
- fading of pigments;
- rapid deterioration and breakdown of paints.

4.7.2 *Effects of irradiance*

As equipment (if fully exposed to solar radiation) can attain temperatures in excess of 60°C, one has to consider an equipment's outside surface. To guard against the effects of irradiance, the following guidelines should be considered when locating electrical equipment:

- the sun should be allowed to shine only on the smallest possible casing surfaces;
- windows should be avoided on the sunny side of rooms;
- heat sensitive parts must be protected by heat shields;
- air conditioning plant and cooling fans (should be efficient and reliable);
- convection flow should sweep across the largest possible surfaces of materials.

4.7.3 *Requirements from the regulations – Solar radiation*

The following points, shown in Table 4.7, are the most common environmental requirements concerning solar radiation:

Table 4.7 Requirements from the Regulations – Solar radiation

Air conditioning	Air conditioning plant and cooling fans (when used) in rooms housing electronic equipment should be efficient and reliable
Exposure	The sun should be allowed to shine only on the smallest possible casing surfaces and the convection flow should sweep across the largest possible surfaces of materials with good conduction properties
Heat shields	Heat sensitive parts shall be protected by heat shields made of (for instance) polished stainless steel or aluminium plate
Solar photovoltaic modules	PV modules must be installed so that adequate heat dissipation when the site is subject to condition of maximum solar radiation
Survivability	Equipment that is exposed to the effect of solar radiation should remain unaffected
Windows	Windows should be avoided on the sunny side of rooms housing electronic equipment
Wiring systems	Wiring system shall be selected, erected and shielded whenever significant solar radiation or ultraviolet radiation is experienced or anticipated

4.8 Pollutants and contaminants

Pollutants and contaminants come in many forms and can cause extensive damage when deposited on electronic and electromechanically equipment as well as equipment housing.

Sources of natural pollutants include:

- **sulphur** – emitted by volcanoes and from biological processes;
- **nitrogen** – from biological processes in soil and lightning and biomass burning;
- **hydrocarbons** – methane from fermentation of rice paddies, fermentation of the digestive tract of ruminants (e.g. cows and also released by insects), coal mining and gas extraction.

Sources of manmade pollutants include:

- **carbon dioxide and carbon** monoxide produced during the burning of fossil fuels;
- **soot formation accompanied by carbon monoxide** and generally due to inadequate or poor air supply;
- **hydrocarbons** – most boilers and central heating units burning fossil fuels result in very low emissions of gaseous hydrocarbons or oxygenated hydrocarbons such as aldehydes.

4.8.1 Pollutants

Although pollutant gases are normally only present in low concentrations, they can cause significant corrosion and a marked deterioration in the performance of contacts and connectors.

For example, hydrogen sulphide is caused by bacterial reduction of sulphates in vegetation, soil, stagnant water and animal waste on a worldwide basis and is likely to cause corrosion in electronic equipment. In one laboratory, it was estimated that about 20% of all electrical and electronic failures were caused by corrosion problems.

4.8.2 Contaminants

Contaminants are composed of dust, sand, smoke and other particles that are contained within the air and these can have an effect on electrical equipment in various ways, especially:

- abrasion of moving parts;
- adding mass to moving parts thereby causing unbalance;
- clogging of air filters;
- corrosion and mould growth;
- reduction in thermal conductivity;
- deterioration of dielectric properties;
- deterioration of electric insulation seizure of moving parts;
- interference with optical characteristics reduction of thermal conductivity;
- overheating and fire hazard surface abrasion by erosion and corrosion.

The presence of dust and sand in combination with other environmental factors such as water vapour can also cause corrosion and promote mould growth.

Effects of flora and fauna – Small animals and insects that feed from, gnaw at, eat into and chew at materials are particular problems as are termites cutting holes into material.

Larger animals can also cause damage by impact or thrust. These attacks can cause:

- physical breakdown of material, parts, units, devices;
- mechanical deformation or compression;
- surface deterioration;
- electrical failure caused by mechanical deterioration.

Deposits from fauna (especially insects, rodents, birds, etc.) can be caused by the presence of the animal itself, nest building, deposited feed stocks and metabolic products such as excrement and enzymes, etc.

Deposits from flora may consist of detached parts of plants (leaves, blossom, seeds, fruits, etc.) and growth layers of cultures of moulds or bacteria. These attacks can lead to:

- clouding of optical surfaces (including glass);
- deterioration of material;
- electrical failure;
- interruption of electrical circuits;
- malfunctioning of mechanical parts;
- mechanical failure of moving parts;
- metallic corrosion.

Mould – When equipment is exposed (in use, storage or transportation) to the atmosphere, without proper protective covering, mould growth will occur and mould can cause unforeseen damage to equipment, whether constructed from mould resistant materials or not!

Even where only a slightly harmful attack on a material occurs, the formation of an electrically conducting path across the surface due to a layer of wet mycelium (i.e. the vegetative part of fungus) can drastically lower the insulation resistance between electrical conductors supported by an insulation material. When the wet mycelium grows in a position where it is within the electromagnetic field of a critically adjusted electronic circuit, it can cause a serious variation in the frequency-impedance characteristics of that circuit.

4.8.3 *Requirements from the regulations – Pollutants and contaminants*

As shown in Table 4.8, the following are the most common environmental requirements concerning pollutants and contaminants:

Table 4.8 Requirements from the Regulations – Pollutants and contaminants

Pollutants	The effects of pollution must be • considered in the design of equipment and components • must be reduced by the effective use of protective devices • the requirements of ISO 14001 regarding environmental protection and the prevention of pollution have to be met
Contaminants	The following should be considered • chemical active substances • biologically active substances • flora and fauna • dust • sand
Mould	Insulating materials should be chosen to give as great a resistance to mould growth as possible

4.9 Mechanical

Mechanics is the branch of physics concerned with the motions of objects and their response to forces.

4.9.1 Mechanical stresses

Mechanical stresses are normally attributed to a moving mass and there is frequently a tendency to underestimate the effect of the mechanical environment on the reliability of static installations. Experience shows, however, that vibrations and shocks are a significant Reliability, Availability and Maintainability (RAM) factor, not only from the point of view of vehicle mounted equipment, but also with respect to permanent installations.

4.9.1.1 Acceleration

Equipment, components and electrotechnical products that are likely to be installed in moving bodies (e.g. rotating machinery) will be subjected to forces caused by steady accelerations.

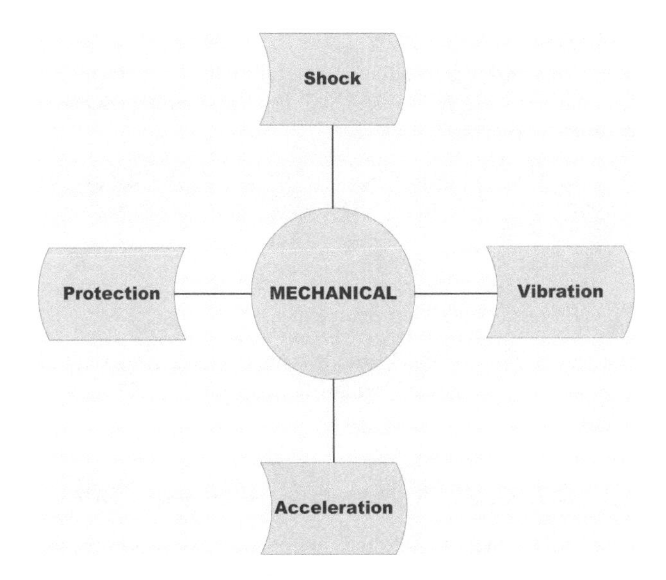

Figure 4.9 Effects of the mechanical environment

4.9.1.2 Shock

'Shock' is generally defined as *'an impact shock characterised by a simple acceleration and free impact on a firm base'* and is usually the result of a violent collision, or a heavy blow. Although it is difficult to design and install electrical and electromechanical equipment, components and systems so that they are completely immune to shock, precautions **should**, nevertheless, be taken to guard against potential problem areas.

For locations where wiring system could be exposed to impact and mechanical shock owing to vehicles and mobile agricultural machines, etc.:

- conduits shall provide a degree of protection against impact of 5 J according to BS EN 61386-2;
- cable trunking and ducting systems shall provide a degree of protection against impact of 5 J according to BS EN 50085-2-1.

4.9.1.3 Vibration

During transportation or whilst in use, electrical equipment and their associated parts may be subjected to vibration of a harmonic pattern caused by machinery and seismic incidents (such as earthquakes) and precautions **should** to be taken to prevent damage to electrical equipment and circuitry.

Note: Although rarely headline news, UK suffers 5–6 earthquakes a year – indeed in Sep 2020 at Leighton Buzzard, they experienced a 3.0 magnitude earthquake!:

- if stationary equipment is moved temporarily for the purposes of connecting, cleaning, etc., (e.g. a cooker) it shall be connected by flexible cables.

4.9.1.4 Protection against mechanical stresses

Vibration dampers and shock absorbers are often used as a form of protection against mechanical stresses but generally speaking, however, they offer little no protection against vibrations.

4.9.2 Mechanical and physical stresses on wiring systems

Wiring systems must be selected and erected so as to avoid (during installation, use or maintenance) having the sheath or insulation of its cables and their terminations damaged.

- A conduit system (or cable ducting system) that is going to be buried in a structure, must be completely installed between access points before any cable is drawn in.
- The radius of every bend in a wiring system must not cause conductors or cables to suffer damage or terminals subjected to stress.
- Cables and conductors shall be supported at appropriate intervals so that they:
    - do not suffer any damage because of their own weight;
    - (and their terminations) are not exposed to any undue mechanical strain.

Cables and conductors shall:

- not be damaged by the means of fixing;
- have adequate means of access to allow them to be drawn in and out of from a product's wiring system.

Buried cables, conduits and ducts shall:

- be at a sufficient depth to avoid being damaged by any reasonably foreseeable disturbance of the ground;
- be suitably identified by cable covers or marking tape;

Figure 4.10 Buried electric cable marking note

Other requirements include:

- the location of buried cables shall be marked by cable covers or a suitable marking tape;
- cables buried directly in the ground (i.e. not installed in a conduit or duct) shall include an earthed armour or metal sheath, suitable for use as a protective conductor;
- cable supports and enclosures shall not have any sharp edges that could damage the wiring system;
- cables, busbars and other electrical conductors which pass across expansion joints shall be installed so that any predictable movement does not cause damage to the electrical equipment.

See IEC 61386-24 for further details concerning underground conduits.

4.9.3 Electrical connections

Connections between conductors or between a conductor and other equipment shall ensure electrical continuity, mechanical strength and protection, by taking account of:

- the cross-sectional area of the conductor;
- the material of the conductor and its insulation;
- the number and shape of the wires forming the conductor;
- the number of conductors to be connected together;
- the provision of adequate locking arrangements in situations subject to vibration or thermal cycling;
- the temperature attained at the terminals in normal service.

4.9.3.1 Electrical connections in caravans and motor homes

In a caravan or motor caravan:

- as wiring will be subjected to vibration, all wiring needs to be protected against mechanical damage (particularly sharp edges and abrasive parts);
- wiring passing through metalwork must be protected by suitable bushes or grommets;

- all cables (unless enclosed in rigid conduit and all flexible conduit) shall be supported at intervals not exceeding 0.4 m for vertical runs and 0.25 m for horizontal runs.

See BS 7671:2018 Chapter 721 for more detailed requirements relating to electrical connections in caravans and motor homes.

4.9.3.2 Low voltage generating sets – Generators

- Electrical equipment associated with the generator should be mounted securely and, if necessary, on anti-vibration mountings.

See BS 7671:2018 Chapter 740 for more detailed requirements relating to generators.

4.9.3.3 Protection against fault current

- Electrical equipment, including conductors, shall be provided with mechanical protection against electromechanical stresses caused by fault currents to prevent injury or damage to persons, livestock or property.

4.9.3.4 Cross-sectional area of conductors

The cross-sectional area of conductors shall be determined for both normal operating conditions and, where appropriate, for fault conditions according to:

- the electromechanical stresses likely to occur due to short-circuit and Earth fault currents;
- other mechanical stresses to which the conductors are likely to be exposed.

4.9.4 Requirements from the regulations – Mechanical

As shown in Table 4.9, the following are the most common environmental requirements concerning solar radiation:

Table 4.9 Requirements from the Regulations – Mechanical

Closed rooms	Equipment located in closed room installations must be capable of withstanding self-induced vibrations
Encapsulated outdoor installations	Equipment contained in encapsulated outdoor installations must be capable of withstanding vibrations and shocks
In service	Equipment should be capable of withstanding, without deterioration or malfunction, all mechanical stresses that occur whilst in service
Long-term exposure	Equipment must be capable of withstanding long-term exposure to shocks
Mechanical shock	Equipment should be capable of withstanding shock pulses (e.g. a minimum of 20,000 shocks at a shock level of 20 g)
On or near the roadside	Equipment located on or near the roadside must be capable of withstanding vibrations and shocks
Random vibration	Equipment should be capable of withstanding random vibration
Vibrations and shocks	Any dampers or anti-vibration mountings must be integral with the equipment to prevent the unit being accidentally installed without them

4.10 Electromagnetic compatibility

Although most forms of interference are usually tolerated as being one of those things '*that you cannot do much about*', the design of modern sophisticated equipment has become so susceptible to EMI that some form of regulation has had to be agreed.

Within Europe, this regulation is contained in the Electromagnetic Compatibility Directive 2004/108/EC (which repealed the original Directive 89/336/EEC). Following Brexit, UK has now modified our Electromagnetic Compatibility Regulations 2016 to set out the requirements that must be met before products can be placed on the GB market with effect 1 January 2021.

The purpose of the legislation is to ensure **only** safe products may be placed on the UK market by requiring manufacturers to show how their products meet the following 'essential requirements', which are:

- equipment must be designed and manufactured to ensure that any electromagnetic disturbance generated do not exceed the level above which radio and telecommunications equipment (or other associated equipment) cannot operate as intended; and

- the apparatus has an adequate level of intrinsic immunity to electromagnetic disturbance.

Electromagnetic disturbances and EMI can seriously disrupt and even damage Information Technology (IT) systems and equipment, electronic components and circuits. Lightening, switching operations short-circuits and other electromagnetic phenomena can also cause overvoltages and electromagnetic interference.

These effects are potentially more severe:

- where large metal loops exist;
- where different electrical wiring systems are installed in common routes – such as power supply, signalling and/or data communication cables connecting IT equipment within a building.

This is of particular relevance in (or near) rooms that are used for medical purposes as electromagnetic disturbances can dramatically interfere with medical electrical equipment.

4.10.1 Electromagnetic interference

EMI is the disturbance caused by generated currents (typically due to lightning, switching operations and short-circuits, etc.) which cause overvoltages and electromagnetic interference, induction, coupling or conduction. This is a particular problem with sensitive equipment where signal transmission can become corrupted or distorted. Data transmission may also result in an increased error rate or total loss of data.

4.10.2 Sources of electromagnetic disturbances

Potential sources of electromagnetic disturbances within an installation typically include:

- electric motors;
- fluorescent lighting;
- frequency convertors/regulators including Variable Speed Drives (VSDs);
- lifts;
- power distribution busbars rectifiers;
- rectifiers;
- switchgear;
- switching devices for inductive loads;
- transformers;
- welding machines.

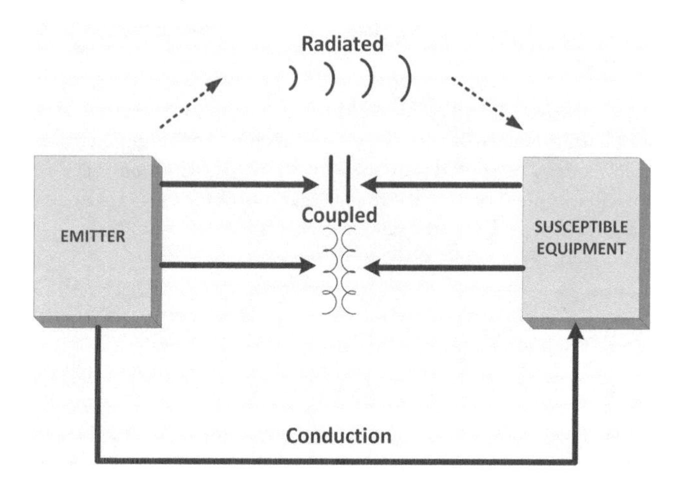

Figure 4.11 Electromagnetic sources

For further information regarding electromagnetic disturbances, please refer to the BS EN 50174 series of standards.

4.10.3 Measures to reduce EMI

To reduce the effects of EMI, the following measures should be considered:

- ensure that there is adequate separation between power and signal cables;
- ensure that the area of all wiring loops are as small as possible (in order to minimise voltages induced by lightning);
- include surge protective devices and/or filters to improve the electromagnetic compatibility of electrical equipment sensitive to electromagnetic disturbances;
- install an equipotential bonding network;
- install power cables close together in order to minimise cable loop areas;
- limit the amount of fault current from power systems flowing through the screens and cores of signal cables, or data cables, which are earthed;
- use surge protective devices and/or filters to improve electromagnetic compatibility.

4.10.3.1 TN system

To prevent electromagnetic fields caused by stray currents in the main supply system of an installation:

- the transfer from one supply to an alternative supply for any installation forming part of a TN system shall be via a multipole switching device that switches both the line conductors and the neutral conductor (if any).

To minimise electromagnetic disturbances, the following requirements should be met:

- Combined Protective and Neutral (PEN) conductors must not be used downstream of the origin of the installation;
- the installation must have separate neutral and protective conductors downstream of the origin of the installation.

BS 7671:2018 Requirement

In Great Britain, Regulation 8(4) of the Electricity Safety, Quality and Continuity Regulations 2002 prohibits the use of PEN conductors in consumers' installations.

4.10.3.2 TT system

If the live conductors of the supply into any of the buildings are less than 35 mm^2 in cross-sectional area the main protective bypass conductor shall have a minimum cross-sectional area of 10 mm^2.

See BS 7671:2018 Table 54.8 for other sized supply neutral conductors.

4.10.3.3 Separate buildings

- Where different buildings have separate equipotential bonding systems, metal-free optical fibre cables (or other non-conducting systems) should be used for signal and data transmission.
- Within a single building, all protective and functional earthing conductors of an installation shall normally be connected to the main earthing terminal.

- If a number of installations have separate earthing arrangements, then any protective conductor that is common to any of these installations shall either be:
    - capable of carrying the maximum fault current that is likely to flow through them; or
    - earthed within one of the installations, only.

4.10.3.4 Equipotential bonding networks

To avoid the possibility of electromagnetic disturbances affecting an installation:

- the impedance of equipotential bonding connections shall be as low as practicable;
- the minimum size and installation of an equipotential bonding ring network shall be:
    - flat cross-section: 25 mm × 3 mm;
    - round diameter: 8 mm.

Bare conductors shall be protected against corrosion at their supports and on their passage through walls.

The following parts shall be connected to the equipotential bonding network:

- armoured metal sheaths and screens;
- conductive screens, sheaths or armoured data transmission cables used for information technology equipment;
- the functional earthing conductors of antenna systems;
- the earthed pole's conductor of a d.c. supply used for IT equipment;
- functional earthing conductors;
- protective conductors.

For buildings with several floors, it is recommended that, on each floor:

- an equipotential bonding system be installed;
- the bonding systems of the different floors are interconnected, at least twice, by protective conductors.

See BS 7671:2018 Chapter 54 for more detailed requirements.

4.10.3.5 Earthing arrangements and equipotential bonding for it installations

IT installations should:

- be connected to the main earthing terminal by the shortest practicable route;
- use one or more earthing busbars;
- ensure that all bare conductors are protected to prevent corrosion.

4.10.3.6 Segregation of circuits

Cables that are used at voltage Band II (low voltage) and Band I (extra-low voltage) which share the same cable management system or the same route, shall be installed according to the following requirements:

- each part of a circuit shall be arranged such that the conductors are not distributed over different multicore cables, conduits, ducting systems, franking systems or tray or ladder systems;
- the line and neutral conductors of each final circuit are electrically separate from other final circuits;
- whenever multicore cables are installed in parallel, each cable only contains one conductor of each line;
- a voltage Band I circuit shall not be contained in the same wiring system as a Band II circuit unless:
    - each conductor of a multicore cable is insulated for the highest voltage;
    - the cores of the Band I circuit are separated from the cores of the Band II circuit by an earthed metal screen.
- if underground telecommunication or power cables are liable to cross each other than a minimum clearance of 100 mm shall be maintained; and
    - a fire-retardant partition shall be provided between the cables; and
    - mechanical protection between the cables shall be provided where the cables crossover.
- the minimum distance between IT cables and high-intensity discharge lamps shall be 130 mm;
- data wiring racks and electrical equipment shall always be separated.

4.10.3.7 *Protection against voltage disturbances and measures against electromagnetic disturbances*

Persons, livestock and property shall be protected against any harmful effects, caused by:

- a fault between live parts of circuits supplied at different voltages;
- overvoltages such as those originating from atmospheric events or from switching;
- undervoltage and subsequent voltage recovery.

The installation shall have an adequate level of immunity against electromagnetic disturbances so as to function correctly in the specified environment.

See also the BS EN 62305 series of standards for further information concerning protection against lightning strikes.

4.10.3.8 *Electrical installations*

- Electrical installations shall be arranged so that they do not interfere with other electrical (or non-electrical) installations in a building.
- With effect from 1 January 2021, all fixed installations shall be in accordance with the Electromagnetic Compatibility Regulations 2016 and immunity levels of equipment shall be chosen taking into account:
    - the electromagnetic influences that can occur when connected for normal use; and
    - the intended level of continuity of service necessary for the application. (See BS EN 50082).
- Equipment shall be chosen with sufficiently low emission levels so that it cannot cause unacceptable electromagnetic interference with other electrical equipment. (See BS EN 50081).
- The area of all wiring loops shall be as small as possible so as to minimise voltages induced by lightning.

4.10.3.9 *Wiring installations*

The choice of the type of wiring system and the method of installation shall include consideration of the following:

- accessibility of wiring (to persons and livestock);
- electromagnetic interference;

- electromechanical stresses;
- mechanical, thermal and other external influences to which the wiring is likely to be exposed;
- nature of the location;
- type of structure;
- voltage.

Every installation shall be divided into circuits, as necessary, to:

- minimise danger and inconvenience in the event of a fault;
- ensure safe inspection, testing and maintenance;
- prevent the indirect energising an isolated circuit;
- reduce unwanted tripping of RCD;
- reduce the effects of EMI.

4.10.3.10 Cables and conductors

Line conductors, neutral conductor and protective conductor should all be contained in the same enclosure.

In ferrous enclosures, conductors shall be collectively surrounded by ferrous material.

Single-core cables armoured with steel wire or steel tape shall **not** be used for an a.c. circuit.

4.10.3.11 Medical locations

Special considerations have to be made concerning EMI and EMC in medical locations.

It is recommended that radial wiring patterns are used to avoid 'Earth loops' that may cause electromagnetic interference.

4.10.4 *Requirements from the regulations – Electromagnetic compatibility*

Table 4.10 indicates the most common environmental requirements concerning Electromagnetic compatibility:

Table 4.10 Typical requirements – Electromagnetic compatibility

Equipment	The use of electronic equipment shall not interfere with the operation of other equipment All active electronic devices shall comply with the EMC Directive
Apparatus cases	Input/output connections from apparatus cases should always be of non-screened non-balanced signalling cable which are normally restricted in length
Atmospheric disturbances	To counteract the effects of storms, it is generally recommended that all equipment should be capable of withstanding (as a minimum) the following overvoltages Magnitude 2000 V Rise time 1.2 μS Middle voltage time 50 μS
Equipment immunity levels	Equipment should be immune to induced common-mode voltages Equipment should not experience a permanent loss of availability or suffer Component damage for any induced common-mode voltage within this range
Magnetic field	As low, frequency fields can influence cathode ray tubes equipment should be capable of withstanding the following intensities **Hz** **A/m** 5 0.8 50 3.0 250 1.5
Power supply lines	Equipment should be immune to the following high frequency bursts Initial peak to peak voltage 1 kV Burst repetition rate 5 kHz
Transients	All electronic equipment should be capable of withstanding: • transients (either directly induced or indirectly coupled) so that no damage or failure occurs during operation • without damage or abnormal operation, transient non-repetitive surges

4.11 Fire

A fire will normally start when sufficient thermal energy from – for example an electric short circuit or a burning cigarette – is supplied to a combustible material. Following ignition, the fire will then produce its own thermal energy. Some of which will be used as feedback to maintain combustion and some transferred via radiation and convection to other materials. These materials may also ignite and spread the fire.

The environmental conditions relating to the development and spread of fire within a building and its effect on electrotechnical products exposed to fire is primarily covered by Section 8 (Fire Exposure) of IEC 721.2. This section confirms that the development of the fire generally consists of four processes:

1 ignition;
2 growth;
3 fully development;
4 decay.

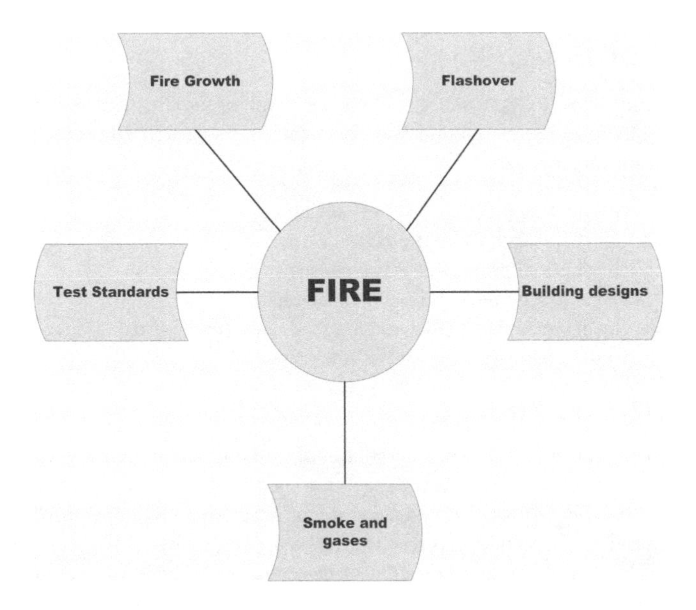

Figure 4.12 Fire exposure

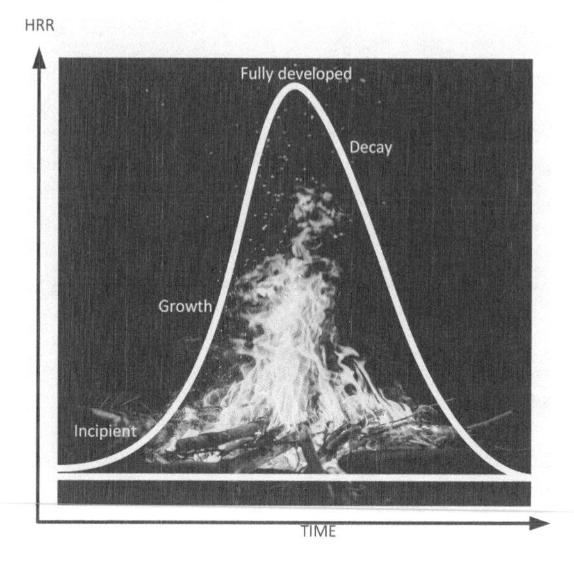

Figure 4.13 Heat release rate and fire development

4.11.1 *Ignition*

When heat, oxygen and a fuel source combine, a chemical reaction results in the start of a fire and once started, its growth and spread is determined by:

- the starting place;
- the shape and size of the place;
- the thermal properties of the place;
- the aerodynamic conditions of the place;
- the arrangement of fuel and fire load, its volume, distribution, continuity, porosity and combustion properties.

4.11.2 *Growth*

During the growth of a fire, a hot layer of gas builds up under the ceiling of the space and under certain conditions, this gas layer can give rise to a rapid-fire growth and *flashover* might occur when a temperature of around 500°C to 600°C is reached in the upper gas layer.

The availability and operation of detectors, alarm systems, associated cables and sprinklers, etc. are vital during the pre-flashover stage in maintaining the level of safety required for escape and rescue of people caught in a fire.

4.11.3 Fully developed

Following the initial growth stage when all combustible materials have ignited, a fire is considered to be fully developed and this is the hottest phase of a fire and the most dangerous for people and animals to be trapped within.

4.11.4 Decay

Smoke is a mixture of heated gases, small liquid drops and solid particles from the combustion. During a fire (pre and post flashover), smoke will be distributed within the building. This can have disastrous effects because hydrogen chloride is a substance of smoke which will not only damage property, it can also prevent the functioning of critical equipment, it can also destruct and damage electrotechnical products.

Metal surfaces, exposed to air under normal (non-fire) conditions, often have a chloride deposit up to 10 mg/m^2. Such an amount is generally not harmful. However, after exposure to smoke from a fire involving polyvinyl chloride (PVC), a surface contamination of up to thousands of milligrams per square metre can be found, often causing significant damage.

4.11.5 Electrical installations

In electrical installations, risk of injury may result from excessive temperatures to guard against this happening:

- equipment in areas that are susceptible to the risk of fire or explosion must be chosen, constructed and installed so as to prevent danger.

4.11.5.1 Choice of wiring system

The type of wiring system selected and its installation will depend on the following:

- the nature of the location;
- accessibility of the wiring to persons and livestock;
- the structure supporting the wiring;
- type of voltage;
- electromagnetic interference;
- electromechanical stresses likely to occur (e.g. due to short-circuit and Earth fault currents);

- external influences (such as mechanical, thermal and those associated with fire) which the wiring is likely to be exposed to during, erection, installation and whilst in service;
- precautions shall be taken such that a cable or wiring system cannot propagate flame.

4.11.5.2 Precautions where a particular risk of fire exist

- electrical equipment shall be so selected and erected so that its temperature in normal operation will not cause a fire;

A temperature cut-out device should always have a manual reset.

Where there is a risk of fire due to the manufacture, processing or storage of flammable materials (such as in barns) a fire risk will be present and the following precaution need be observed:

- cable shall, as a minimum, satisfy the test under fire conditions specified in BS EN 60332-1-2;
- cables not completely embedded in a non-combustible material (such as plaster or concrete) shall meet the flame propagation characteristics as specified in BS EN 60332-1-2;
- cable tray systems or a cable ladder shall meet the requirements specified in BS EN 61537;
- cable trunking or cable ducting systems shall fulfil the fire conditions specified in BS EN 50085;
- conduit and trunking systems shall meet the fire conditions specified in BS EN 61386-1 and BS EN 50085-1;
- powertrack systems shall satisfy the fire conditions specified in the BS EN 61534 series;
- where the risk of flame propagation is high the cable shall meet the flame propagation characteristics specified in the appropriate part of the BS EN 50266 series.

4.11.5.3 Protection against the risk of fire

The selection of protective and monitoring devices shall take into account the nature of the load and the likelihood of the device continuing to safely operate. Typical protective devices include:

- installing an RCD;
- installing an RCM (as an alternative to RCDs in IT systems);

- installing an IMD;
- Installing and AFDD (Arc Fault Detection Device).

4.11.5.4 Protection against thermal effects

Persons and livestock (particularly in agricultural and horticultural premises) **shall** be protected against harmful effects of heat or fire which may be generated or propagated in electrical installations. These effects include:

- external influences such as lightning causing electrical equipment such as protective devices, switchgear, thermostats, temperature limiters, seals of cable penetrations and wiring systems to fail;
- insulation faults or arcs, sparks and high temperature particles;
- harmonic currents;
- heat accumulation, heat radiation, hot components or equipment;
- overcurrent.

Electrical heating appliances used for the breeding and rearing of livestock shall:

- comply with BS EN 60335-2-71; and
- be fixed at an appropriate distance from livestock and combustible material, to minimise any risks of burns to livestock and of fire.

Note: See BS 7671:2018 Chapter 705 for more detailed requirements relating to agricultural and horticultural premises.

At exhibition, shows and stands which normally use temporary electrical installations:

- lighting equipment and appliances with high temperature surfaces shall be suitably guarded, installed and located;
- signs used for showcases and indicators shall be made from material with adequate heat-resistance, mechanical strength, electrical insulation;
- ventilation devices sufficient to overcome the combustibility of exhibits with high heat generation.

Note: See BS 7671:2018 Chapter 711 for more detailed requirements relating to exhibition, shows and stands.

4.11.5.5 Protection against fire caused by electrical equipment

According to Government statistics, electricity is a major cause of accidental fires in UK homes – in fact, over 20,000 each year!

4.11.5.5.1 FIREFIGHTER'S SWITCHES

A firefighter's switch shall be provided in the low voltage circuit supplying:

- exterior electrical installations operating at a voltage exceeding low voltage; and
- interior discharge lighting installations operating at a voltage exceeding low voltage.

For single premises:

- wherever practicable, outdoor installations and every internal installation shall be controlled by a single firefighter's switch.

For other buildings:

- the switch shall be outside the building and adjacent to the equipment for an exterior installation;
- the switch shall be in the main entrance to the building for an interior installation.

In all cases, every firefighter's switch should:

- be placed in a conspicuous position, that is reasonably accessible to firefighters and, at not more than 2.75 m from the ground;
- be easily accessible and clearly marked to indicate the installation or part of the installation which it controls.

A firefighter's switch shall:

- be coloured RED and have fixed on (or near it) a permanent nameplate marked with the words 'FIREFIGHTER'S SWITCH' or 'FIRE SWITCH' in lettering not less than 1-mm high;

- it can be seen clearly by a person standing on the ground at the intended site, without opening the enclosure;
- have its ON and OFF positions clearly indicated by lettering not less than 10 mm with the OFF position at the top.

4.11.6 *Requirements from the regulations – Fire*

The current version of the Wiring Regulations emphasises this point by stating that:

> **BS 7671:2018 Requirement**
>
> *Persons, livestock and property shall be protected against harmful effects of heat or fire which may be generated or propagated in electrical installations; and the risk of spread of fire shall be minimised by the selection of appropriate materials and erection.*

- Although Chapter 42 of BS 7671:2018 contains the most important aspects of fire precautions, further requirements are sprinkled throughout this British Standard.
- In most contracts, reference is made to the IEC 60695 series of standards which cover the assessment of electrotechnical products against a nominated fire hazard.
- CENELEC, on the other hand, show the requirement for equipment to operate in fire hazardous areas as three distinct clauses, as follows:
  - **Class FO** – no special fire hazard envisaged;
  - **Class F1** – equipment subject to fire hazard;
  - **Class F2** – equipment subject to external fire.
- Materials are normally expected to confirm to the requirements defined in EN 60721.3.3 and EN 60721.3.4.

BS 7671:2018 has been updated so as to maintain technical alignment with CENELEC harmonisation documents. One of the main changes concerned the requirements for safety services (e.g. emergency escape lighting, fire alarm systems, installations for fire pumps, fire rescue service lifts, smoke and heat extraction equipment) which now need to be observed.

Safety services have also been expanded in line with IEC standardisation.

4.11.6.1 Other related standards and specifications

IEC 60695 Series	Fire hazard testing – guidance, tests and specifications for assessing fire hazard of electrotechnical products
ISO 5657	Fire tests – Reaction to fire – Ignitability of building products
ISO 5658	Reaction to fire tests – Spread of flame on building products and vertical configuration
ISO 5660	Fire tests – Reaction to fire – Rate of heat release from building products
ISO 9705	Fire tests – Full scale room test for surface products
ISO TR 5924	Fire tests – Reaction to fire – Smoke generated by building products (dual-chamber test)
ISO TR9112.1	Toxicity testing of fire effluents – General

Author's End Note

Having looked at how important earthing and external influences are to electrical installations, the next thing to consider, before going any further, is the safety precautions that need to be taken into consideration.

With this in mind, Chapter 5 provides a complete resume of the mandatory requirements – not only from the Wiring Regulations but also the Building Regulations. It shows methods for protecting against electric shock, precautions and protections, safety service circuits and disconnecting devices

5 Safety protection

Author's Start Note

This chapter lists the mandatory and fundamental requirements for safety protection contained in both the Wiring Regulations and the Building Regulations 2010. It also includes information concerning protection against electric shock, fault protection, protection against direct and indirect contact, protective conductors, protective equipment and lists the test requirements for safety protection.

Over a thousand electrical accidents at work are reported to the UK's Health and Safety Executive (HSE) every year and about 30 people die of their injuries, indeed, electrocution is one of the top five causes of workplace deaths and everyone needs to be warned about these possibilities! (see Figure 5.1).

Many of these deaths and injuries arise from:

- use of poorly maintained electrical equipment;
- working near overhead power lines;
- contact with underground power cables during excavation work;
- work on or near 230 V domestic electricity supplies;
- use of unsuitable electrical equipment in explosive areas such as car paint spraying booths, etc.

For this reason, protection against electric shock and safety protection methods are an essential part of the Regulations.

DOI: 10.1201/9781003165170-5

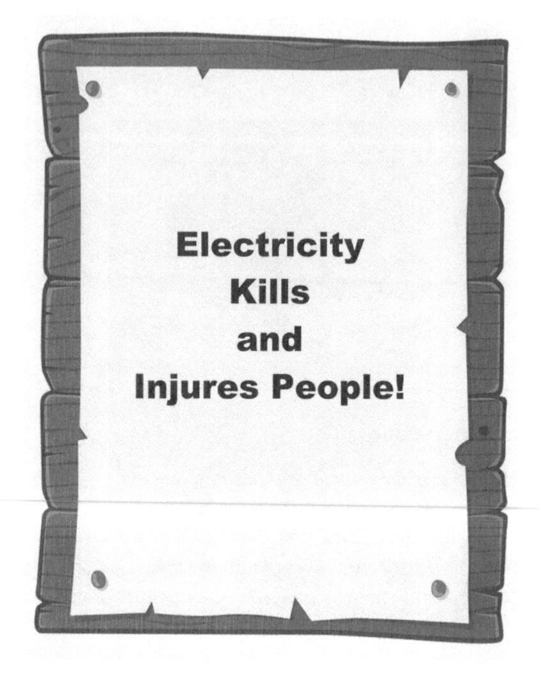

Figure 5.1 Safety protection

In electrical installations, risk of injury may result from:

- arcing or burning;
- excessive temperatures;
- ignition of a potentially explosive atmosphere;
- mechanical movement of electrically actuated equipment;
- power supply interruptions and/or interruption of safety services;
- shock currents;
- undervoltages, overvoltages and electromagnetic influences likely to cause or result in injury or damage.

5.1 Basic safety requirements

The fundamental safety requirements (as detailed in BS 7671:2018) are as follows:

5.1.1 Mandatory requirements

The following are amongst the most important mandatory notices in the Standard:

- protective safety measures shall be applied in every installation, part installation and/or equipment;
- installations shall comply with the requirements for safety protection in respect of:
  - electric shock;
  - fault current (i.e. overcurrent, thermal effects, undervoltage, isolation and switching).
- there shall be no detrimental influence between various protective measures used in the same installation, part installation or equipment.

5.1.2 Fundamental safety requirements

As shown in Figure 5.2, the following are précised details of the most important elements from BS 7671:2018 that meet these fundamental design requirements.

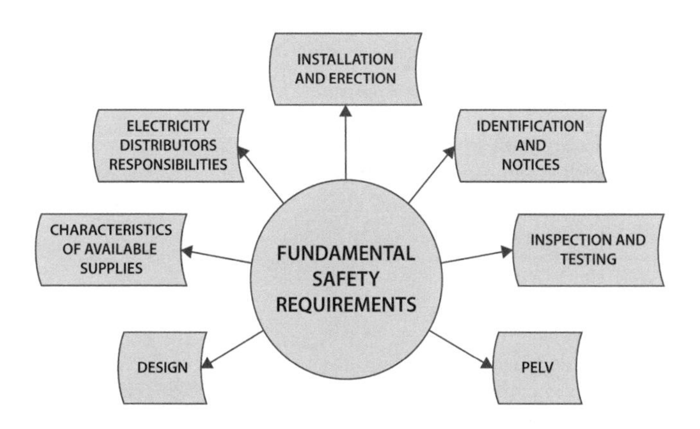

Figure 5.2 Basic safety requirements

5.1.2.1 Design

Electrical installations shall be designed for:

- the protection of persons, livestock and property;
- the proper functioning of the electrical installation;
- protection against mechanical and thermal damage; and
- protection of people from an electric shock or fire hazard.

5.1.2.2 Characteristics of available supply or supplies

Detailed design characteristics shall be available for all supplies and (as shown in Table 5.1) these shall include:

- nature of current (a.c. and/or d.c.);
- purpose and number of conductors;
- values and tolerances of:
    - Earth fault loop impedance;
    - nominal voltage and voltage tolerances;
    - nominal frequency and frequency tolerances;
    - maximum current allowable;
    - particular requirements of the distributor;
    - prospective short-circuit current;
    - protective measures included in the supply (e.g. Earth, neutral or mid wire).

5.1.2.3 Electricity distributors' responsibilities

The electricity distributor is responsible for:

- evaluating and agreeing proposals for new installations or significant alterations to existing ones;

Table 5.1 a.c. and d.c. supplies

For a.c.	*For d.c.*
• Phase conductor(s)	• Outer conductor
• Neutral conductor	• Middle conductor
• Protective conductor	• Earthed conductor
• PEN conductor	• Live conductor
	• Protective conductor
	• PEN conductor

- ensuring that their equipment on consumers' premises:
    - is suitable;
    - is safe;
    - clearly shows the polarity of the conductors;
- installing the cut-out and meter in a safe location;
- ensuring that the cut-out and meter is mechanically protected;
- providing an earthing facility;
- maintaining the supply;
- providing certain technical and safety information to the consumer.

5.1.2.4 Installation and erection

All electrical joints and connections shall meet stipulated requirements concerning conductance, insulation, mechanical strength and protection.

- Conductors shall be identified by colour, lettering and/or numbering.
- Connections and joints shall be accessible for inspection, testing and maintenance, unless they are in a compound filed or encapsulated joint.
- Design temperatures shall not be exceeded by the installation of electrical equipment.
- Electrical equipment shall be arranged so that they it is fully accessible.
- Safety services shall be arranged to allow easy access for periodic inspection, testing and maintenance.
- Exposed parts of electrical equipment shall be located (or guarded) so as to prevent accidental contact and/or injury to persons or livestock.
- Installed electrical equipment shall minimise the risk of igniting flammable materials.
- Installed equipment must be accessible for operational, inspection and maintenance purposes.
- Installations shall be divided into circuits in order to:
    - avoid danger;
    - minimise inconvenience in the event of a fault;
    - facilitate safe operation, inspection, testing and maintenance.

5.1.2.5 Identification and notices

Wiring shall be marked and/or arranged so that it can be quickly identified for inspection, testing, repair or alteration of the installation.

5.1.2.6 Inspection and testing

Precautions shall be taken to avoid danger to persons and livestock, and to avoid damage to property and installed equipment, during inspection and testing, appropriate safety equipment (see Figure 5.3) needs to be worn.

- Electrical installations shall be inspected and tested during, erection and on completion before being put into service.
- Details of the general design characteristics of the electrical installation must be made available.
- Information (e.g. diagrams, charts, tables and/or schedules) must be made available to the person carrying out the inspection and testing.
- Following inspection and tests, a signed Electrical Installation Certificate together with a Schedule of Inspections and a Schedule of Test Results (see Chapter 10 of this book) shall be provided to the person responsible for ordering the work.

Figure 5.3 Inspection and testing of electrical installations

5.1.2.7 Maintenance

An assessment shall be made of the frequency and type of maintenance (e.g. periodic inspection, testing and/or maintenance) that an installation can reasonably be expected to receive during its intended life.

5.2 Building regulations requirements

As shown in Figure 5.4, there are numerous standards, regulations and documents that have an effect on buildings and the following are précised details of the most important elements of the Building Regulations' Approved Documents and Standards concerning safety protection.

5.2.1 Design, installation, inspection and testing of electrical installations

All proposals to carry out electrical installation work must be notified to the local authority's Building Control Body before work begins, unless the proposed installation work:

- is undertaken by a person who is registered with an electrical self-certification scheme; and
- does not include the provision of a new circuit.

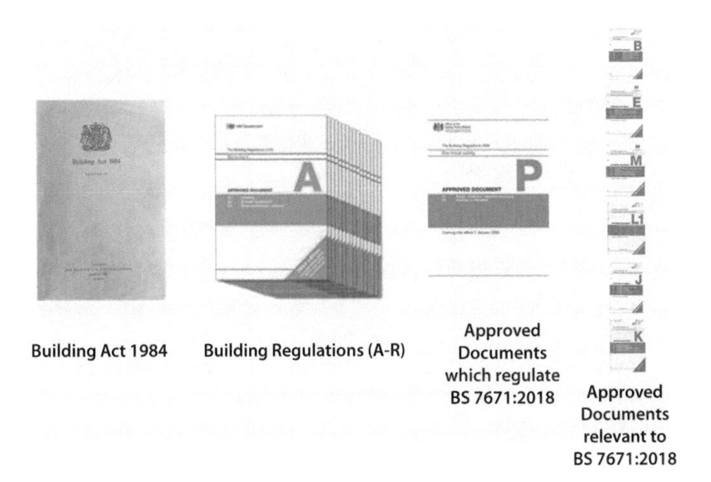

Figure 5.4 The interoperability of the Building Regulations with BS 7671:2018

Work involving any of the following will also have to be notified to the Building Control Body:

- electric floor or ceiling heating systems;
- extra-low voltage lighting installations (other than pre-assembled, lighting sets);
- garden lighting or power installations;
- hot air saunas;
- locations containing a bath tub or shower basin;
- small scale generators such as microchip units;
- solar PV power supply systems;
- swimming pools or paddling pools.

Note: Whilst Part P of the Building Regulations makes requirements for the safety of fixed electrical installations, this does not cover system functionality (such as electrically powered fire alarm systems, fans and pumps) which are covered in other Parts of the Building Regulations and Government legislation.

5.2.2 *Classifications of electrical equipment*

There are three basic classifications for electrical appliances.

- Class I and Class II appliances are all powered by mains voltages; whilst
- a Class III appliance is designed to be supplied from a separate SELV power source.

5.2.2.1 Class I appliances

With Class I appliances, the user has two levels of protection, basic insulation and the provision of an Earth connection.

The required PAT (Portable Appliance Testing) tests for Class I appliances are the Earth Continuity and Insulation Resistance tests which will check (as the name implies) the basic insulation and Earth connection.

Note: Typically; refrigerators, microwaves, kettles, irons, and toasters, are all Class I.

5.2.2.2 Class II appliances

The Double Insulated Class II appliance protects the user with at least two layers of protection – the plastic connector and the plastic casing. For this reason, Class II appliances are also known as double insulated appliances.

Note: Typical examples of Class II appliances are hair dryers, DVD players, televisions, computers, and photocopiers. Most plastic power tools would also be Class II.

5.2.2.3 Class III appliances

Class III appliances use an isolating transformer which has two separate coil windings called (not surprisingly!) the 'Primary Winding', which is connected to the power source, and the 'Secondary Winding', which is connected to the appliance itself.

Due to the lack of an Earth connection, if there is an electromagnetic problem, the current is cut off and cannot continue to flow and a person can safely come into contact with it without risk of electrical shock.

5.2.3 Conservation of fuel and power

Energy efficiency measures shall be provided which provide lighting systems that utilise energy-efficient lamps with manual switching controls (in the case of external lighting fixed to the building) or automatic switching.

The person responsible for achieving compliance should either provide a certificate, themselves, or obtain a certificate from the sub-contractor confirming that commissioning has been successfully carried out. The certificate should be made available to the client and the Building Control Body.

Responsibility for achieving compliance with these requirements rests with the person carrying out the work. That '*person*' may be, for example, a developer, a main (or sub) contractor, or a specialist firm directly engaged by a private client.

5.2.4 Extensions, material alterations and material changes of use

Where electrical installation work is classified as an extension, a material alteration or a material change of use, the work **must** consider and include:

- the amount of additions and alterations that will be required to the existing fixed electrical installation in the building;

- confirmation that the mains supply equipment is suitable and can carry the additional loads envisaged;
- the earthing and bonding systems are capable of meeting the requirements;
- the necessary additions and alterations to the circuits which feed them are acceptable;
- the protective measures required to meet the requirements are sufficient;
- the rating and the condition of existing equipment (belonging to both the consumer and the electricity distributor) is acceptable.

Appendix C to Part P of the Building Regulations offers guidance on some of the older types of installations that might be encountered during alteration work and Appendix D provides guidance on the application of the new harmonised European cable identification system.

5.2.5 *Access and facilities for disabled people*

In addition to the requirements of the '*Disability and the Equality Act 2010*' [which makes it unlawful to discriminate against employees (including workers) because of a mental or physical disability] precautions need to be taken to ensure that:

- new, non-domestic buildings and/or dwellings;
- extensions to existing non-domestic buildings; and
- non-domestic buildings that have been subject to a material change of use (e.g. so that they become a hotel, boarding house, institution, public building or shop);

are capable of allowing people, **regardless** of their disability, age or gender to be able to safely use the facilities of the buildings.

5.3 Protection from electric shock

Safety requirements for protection against electric shock need to be provided (see Figure 5.5):

- against both basic (i.e. direct contact) and fault (i.e. indirect contact);
- for persons and livestock against dangers that may arise from contact with:
  - exposed conductive-parts during a fault;
  - live parts of the installation.

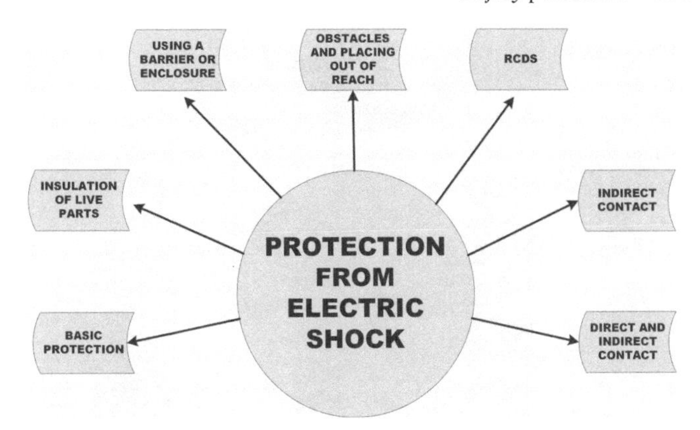

Figure 5.5 Protection from electric shock

Live parts should be completely covered with insulation which:

- can only be removed by destruction;
- are capable of strongly withstanding electrical, mechanical, thermal and chemical stresses normally encountered during service.

Live parts should also:

- be inside enclosures (or behind barriers);
- protected to at least IP2X or IPXXB).

Bare live parts (other than overhead lines) shall:

- not be within arm's reach;
- not be within 2.5 m of:
    - an exposed conductive part;
    - an extraneous conductive part;
    - a bare live part of any other circuit.
- be located (or guarded) so as to prevent accidental contact and/or injury to persons or livestock.

Failure to recognise these safety requirements can result in an accident similar to those shown in Figure 5.6.

1–2 mA	Barely perceptible, no harmful effects
5–10 mA	Throw off, painful sensation
10–15 mA	Muscular contraction, can not let go!
20–30 mA	Impaired breathing
50 mA and above	Ventricular fibrillation and death

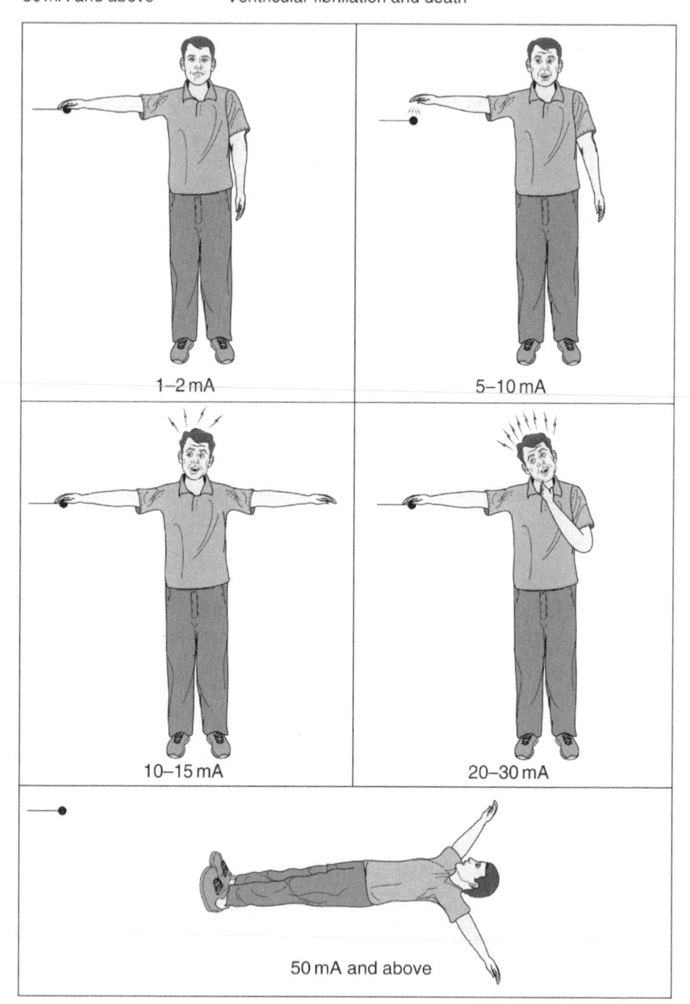

Figure 5.6 The effects of electric shock

Source: Courtesy Brian Scaddan

The following methods are used for protection against direct contact (i.e. basic protection) and for protection against indirect contact (fault protection).

5.3.1 *Basic protection against electric shock*

Basic protection against electric shock is taken as meaning that:

- the nominal voltage cannot exceed the upper limit of voltage Band I;
- all electrical equipment is protected by some form of basic insulation; barriers, enclosures; obstacles or placing out of reach;
- the SELV and/or PELV supply is from a recognised source such as a safety isolating transformer, battery, diesel-driven generator, insulation testing equipment, monitoring device; motor-generator; and
- exposed-conductive-parts of a SELV circuit have not been connected:
    - to Earth;
    - to protective conductors; or
    - to exposed-conductive-parts of another circuit.

A person may perform work involving direct contact with electrical parts **only** (see Figure 5.7) if the electrical part:

- is isolated from all sources of electricity;
- is tested to ensure its isolation from all sources of electricity; and
- is earthed if it is of high voltage.

In all cases, live parts must be completely covered with insulation which can **only** be removed by destruction.

To meet these requirements, the Regulations state that one of the following, basic, measures (see Figure 5.7) shall be used for protection against indirect contact:

- insulating live parts;
- using a barrier or an enclosure;
- using obstacles;
- placing equipment out of reach;
- using an RCD.

Figure 5.7 Basic protection against electric shock

5.3.2 *Insulation of live parts*

As the title suggests, this is a basic form of insulation protection and is intended to prevent contact with a live parts of an electrical installation from direct contact.

5.3.3 *Using a barrier or enclosure*

Live parts of any circuit or electromagnetic equipment should always be inside an enclosure or behind a barrier that provides a degree of protection at last that of IPXXB or IP2X and the barrier (or enclosure) must:

- be firmly secured in place;
- have sufficient stability and durability to maintain the required degree of protection;
- be separated from other live parts;
- restrict the removal of a barrier, opening an enclosure or removal of parts of enclosures.

This Regulation does not apply to:

- a ceiling rose complying with BS 67;
- a cord operated switch complying with BS 3676;
- a bayonet lampholder complying with BS EN 61184;
- an Edison screw lampholder complying with BS EN 60238.

5.3.4 *Protection by obstacles and placing out of reach*

Protection by obstacles and placing out of reach is primarily intended for installations that are controlled or supervised by skilled persons. It will only provide basic protection and the prime intention is to prevent unintentional contact with a live part, but **not** an intentional contact caused by deliberately circumnavigating the obstacle.

5.3.4.1 Obstacles

Obstacles are intended and shall be designed to:

- prevent unintentional bodily approach;
- prevent unintentional contact with to live parts during the operation of live equipment whilst in normal service;
- be secured to prevent unintentional removal;
- be incapable of being removed without using a key or tool.

5.3.4.2 Placing out of reach

Protection by placing out of reach is **only** intended to prevent unintentional contact with live parts. It also:

- ensures that simultaneously accessible parts at different potentials shall not be within arm's reach;
- ensures that a bare live part (other than an overhead line) is not within arm's reach or within 2.5 m of:
    - an exposed-conductive-part;
    - an extraneous-conductive-part;
    - a bare live part of any other circuit.

Bare (or insulated) overhead lines used for distribution between buildings and structures shall be installed in accordance with the Electricity Safety, Quality and Continuity Regulations 2002 (as amended).

It should also be noted that the protective measure of placing out of reach and installing obstacles is **not** permitted in:

- agricultural and horticultural premises;
- construction and demolition site installations;
- conducting locations with restricted movement;
- electrical installations in caravan/camping parks and similar locations;
- electrical installations **inside** caravans and motor caravans;
- exhibitions, shows and stands;
- floor and ceiling heating systems;
- locations containing a bath or shower;
- medical locations;
- mobile or transportable units;
- rooms and cabins containing sauna heaters;
- swimming pools and other basins.

5.3.5 *Protection by RCDs*

In electrical installations, an RCD or an RCCB (Residual Current Operated Circuit Breaker with integral overcurrent protection) is a circuit breaker that operates to disconnect a particular circuit whenever they detect that current leaking out of that circuit (such as current leaking to Earth through a ground fault) exceeds safety limits.

Figure 5.8 illustrates the construction of an RCD and works on the principal that in a normal (i.e. healthy) circuit, the magnetic effects of the phase and neutral currents will cancel out because the same current will pass through the phase coil, the load and then back through the neutral coil. In a faulty circuit where the phase or the neutral are to Earth, the currents will no longer be equal and the out of balance current will produce some residual magnetism in the core. As the magnetism will be alternating, it will link with the turns of the search coil and induce an Electromotive force (emf) in it which will drive a current through the trip coil and cause the tripping mechanism to operate.

Note: The construction and use of RCDs (see Figure 5.8) is not recognised as a sole means of protection and does not obviate the need to apply one of the other protective measures (such as automatic disconnection of supply, double or reinforced insulation, SELV or PELV).

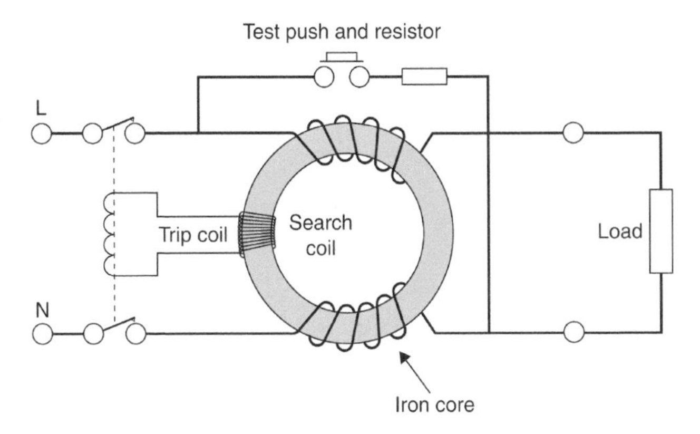

Figure 5.8 The construction of a basic RCD

Source: Courtesy Brian Scaddan

The use of RCDs with a rated residual operating current not exceeding 30 mA and an operating time not exceeding 40 ms is recognised in a.c. systems as providing additional protection in the event of failure of:

- one of the other methods of basic protection against electric shock;
- the provision for fault protection; or
- the carelessness by users.

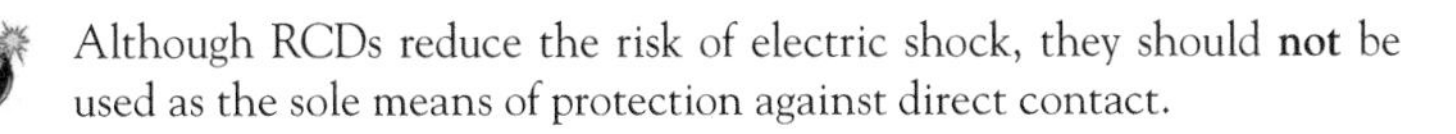

Although RCDs reduce the risk of electric shock, they should **not** be used as the sole means of protection against direct contact.

5.3.6 *Fault protection against indirect contact*

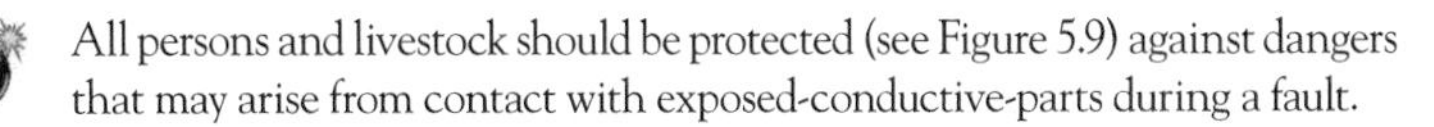

All persons and livestock should be protected (see Figure 5.9) against dangers that may arise from contact with exposed-conductive-parts during a fault.

To meet the requirements for protection against indirect contact, one of the following basic measures shall be used:

- Earthed Equipotential Bonding and Automatic Disconnection of Supply (EEBADS);
- protection by non-conducting location;
- protection by obstacles and placing out of reach;
- protection by Class II equipment or equivalent insulation.

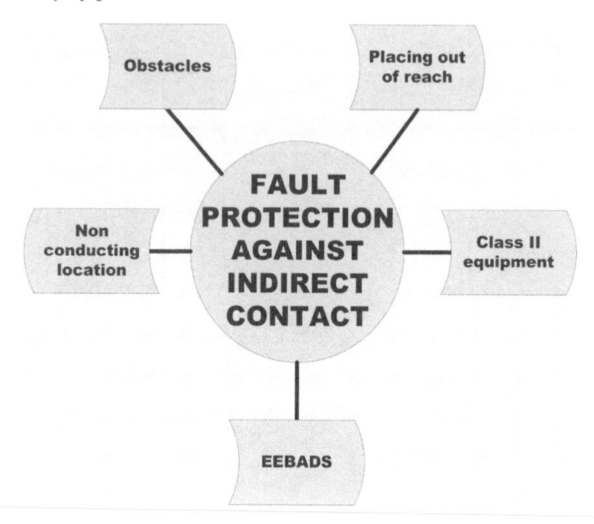

Figure 5.9 Protection against indirect contact

5.3.6.1 Protection by earthed equipotential bonding and automatic disconnection of supplies

EEBADS provides a very good form of protection against indirect contact by joining together all metallic parts together and connecting them to Earth.

Note: This ensures that all metalwork is at (or near) zero volts and so under fault conditions, all metalwork will rise to a similar potential and simultaneous contact with two metal parts will not result an electric shock as there is no significant potential difference between them.

The connection of a lightning protection system to the protective equipotential bonding shall be made in accordance with BS EN 62305.

5.3.6.2 Protection by non-conducting location

A 'non-conducting location' is a location where there is no earthing or protective system because:

- there is nothing which needs to be earthed;
- exposed conductive parts are arranged so that it is impossible to touch two of them at the same time.

This method of protection is **not** recognised for general application and may not be used in installations such as electrical installations in caravan and camping parks, outdoor or temporary electrical installations at fairgrounds, amusement parks, circuses, construction and demolition sites, marinas, swimming pools or rooms containing a bath or showers or medical locations that are subject to an increased risk of shock. (For complete details of these restrictions see to BS 7671:2018 Sections 701 to 753).

5.3.6.3 Protection by Class II equipment or equivalent insulation

Class II equipment is unique in that as well as providing the basic insulation for live parts, it also has a second layer of insulation, which can be used to either prevent contact with exposed conductive parts or to make sure that there can never be any contact between such exposed conductive parts and live parts.

Class II protection is provided by one or more of the following:

- electrical equipment having double or reinforced insulation;
- low voltage switchgear;
- low voltage control-gear assemblies;
- supplementary insulation;
- reinforced insulation applied to uninsulated live parts.

5.3.7 Fault protection against both direct and indirect contact

The Regulations state that one of the following, basic, measures shall be used for protection against both direct contact and indirect contact (also see Figure 5.10):

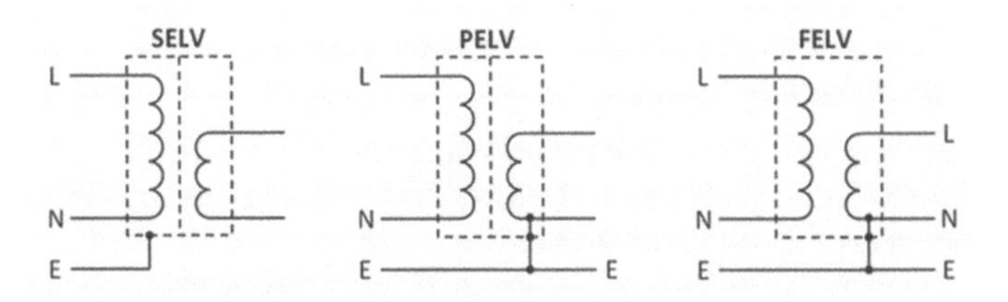

Figure 5.10 SELV PELV and FELV

5.3.7.1 SELV

SELV is an extra-low voltage system that is electrically separated from Earth and from other systems so that a single fault cannot give rise to the risk of electric shock. It is a term used to describe the highest voltage level that can be contacted by a person without causing injury. It is usually defined as 60 V d.c.

This type of protection may also be used for electric fences – provided that they are supplied from electric fence controllers complying with BS EN 61011 or BS EN 6101 1-1.

5.3.7.2 PELV

PELV on the other hand is an extra-low voltage system which is not electrically separated from Earth, but which otherwise satisfies all the requirements for SELV.

5.3.7.3 FELV

FELV (Functional Extra-Low Voltage) describes any other extra-low-voltage circuit that does not fulfil the requirements for an SELV or PELV circuit.

In medical locations, FELV is **not** permitted as a method of protection against electric shock.

The separation of the live parts from those of other circuits and from Earth needs to be confirmed by a measurement of the insulation resistance.

For details of more stringent requirements for wiring fire alarm systems in buildings, see BS 5839-1 and Approved Document P.

5.3.7.4 Medical locations

When using SELV and/or PELV circuits in Group 1 and/or Group 2 medical locations, protection by basic insulation of live parts, by barriers, or enclosures shall be provided.

In Group 2 medical locations (where PELV is used) operating theatre luminaires (and any other exposed-conductive-parts of equipment) will need to be connected to the circuit protective conductor.

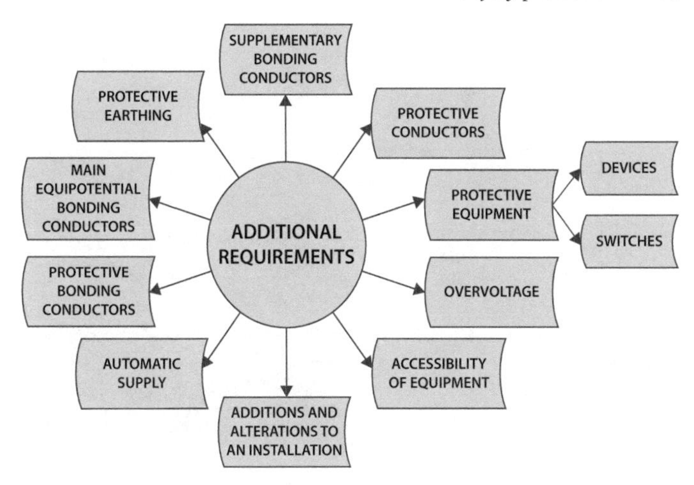

Figure 5.11 Additional requirements

5.4 Additional requirements

The following are additional requirements (see Figure 5.11) for installations and locations where the risk of electric shock is increased by a reduction in body resistance and/or by contact with Earth potential.

The type of protective equipment chosen will depend on the type of protection that is required (e.g. whether overcurrent, Earth fault current, overvoltage or undervoltage).

5.4.1 Protection against overvoltage

Overvoltage is a hazardous condition that occurs when the voltage in a circuit (or part of a circuit) is suddenly raised above its upper limit. This can be permanent or transient and it is often referred to as a 'voltage spike'.

Transient overvoltage might last microseconds and reach hundreds of volts – sometimes thousands – of volts, in amplitude.

A typical example of a naturally occurring transient overvoltage is by lightning as shown in Figure 5.12.

Other man-made sources are usually electromagnetic induction when switching on or off inductive loads (e.g. electric motors or electromagnets).

Figure 5.12 Lightning

Source: Photo by kind permission of Yann Allegre via Unsplash

In accordance with the requirements of BS 7671:2018, additional protection against overvoltages of an atmospheric origin is **not** necessary for:

- installations, supplied by low voltage systems (that do not contain overhead lines);
- installations, supplied by low voltage networks that might (or do) contain overhead lines – whose location is subject to less than 25 thunderstorm days per year;

provided that they meet the required minimum equipment impulse to withstand voltages shown in Table 5.2.

Suspended cables with insulated conductors that have earthed metallic coverings are considered to be an '*underground cables*'.

5.4.2 *Protective earthing*

Automatic disconnection of supply is a protective measure in which fault protection is provided by protective earthing and the sign shown in Figure 5.13 shall be included.

Table 5.2 Required minimum impulse withstand voltage kV

Nominal voltage of the installation	*Category IV* *Equipment with very high impulse voltage.* *(e.g. energy meter, or telecontrol systems)*	*Category III* *Equipment with high impulse voltage.* *(e.g. boards, switches and socket-outlets)*	*Category II* *Equipment with normal impulse voltage.* *(e.g. domestic appliances and tools)*	*Category I* *Equipment with reduced impulse voltage.* *(e.g. sensitive electronic equipment such as alarm panels, computers and home electronics)*
120/208 V	4	2.5		0.8
230/240 V	6	4	2.5	1.5
277/480 V	6	4	2.5	1.5
400/690 V	8	6	4	2.5
1000 V	12	8	6	4

Figure 5.13 Earthing and bonding notice

5.4.3 *Protective conductors*

A protective conductor is (as the name implies!) a conductor that provides a measure of protection against electric shock and is used to connect together:

- Earth electrodes;
- exposed-conductive-parts;
- extraneous-conductive-parts;
- the earthed point of the source;
- the main earthing terminal.

A circuit protective conductor on the other hand is an arrangement of conductors that join all of the exposed conductive parts together and connect them to the main earthing terminal. There are many types of circuit protective conductor such as:

- a separated conductor;
- a conductor included in a sheathed cable with other conductors;
- a conducting cable enclosure (such as conduit or trunking);
- exposed conductive parts (such as the conducting cases of equipment);
- the metal sheath and/or armouring of a cable.

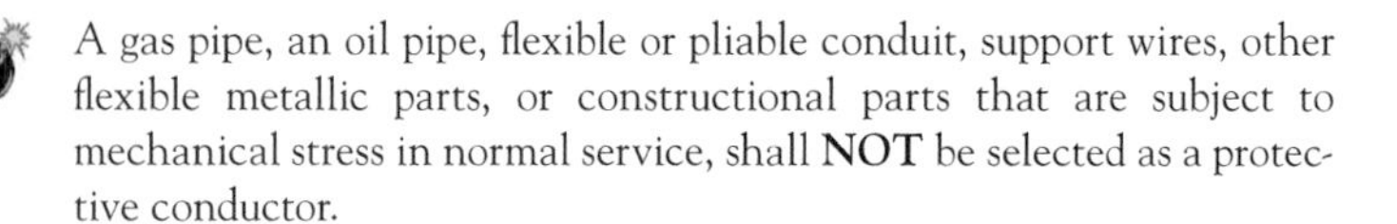

A gas pipe, an oil pipe, flexible or pliable conduit, support wires, other flexible metallic parts, or constructional parts that are subject to mechanical stress in normal service, shall **NOT** be selected as a protective conductor.

5.4.4 *Protective bonding conductors*

Equipotential bonding ensures that protective devices will remove all dangerous potential differences, before a hazardous shock can be delivered. This is done by making sure that all of the installation's earthed metalwork (i.e. exposed conductive parts) is connected to other metalwork (i.e. peripheral conductive parts) via the Earth conductor so as to provide an Earth fault current path that ensures dangerous potential differences cannot occur installations such as those shown in Figure 5.14 should be considered.

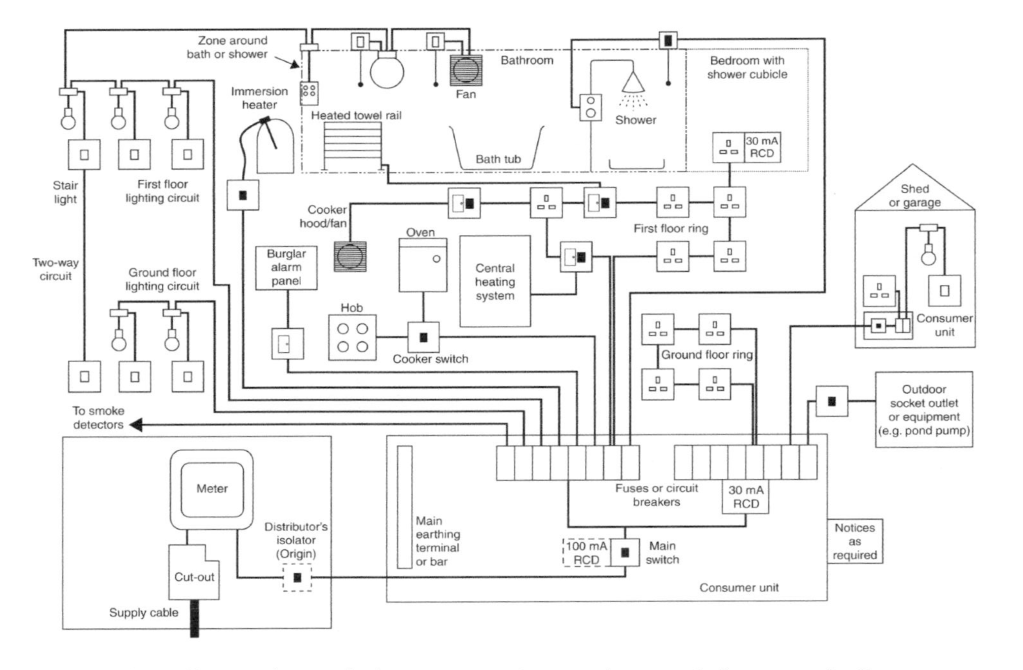

Figure 5.14 Typical fixed installations that might be encountered in new (or upgraded) existing dwellings

5.4.5 Main equipotential bonding conductors

Main equipotential bonding conductors (as shown in Figure 5.15) connect together the installation's earthing system and the metalwork of other services such as gas, electricity and water as close as possible to their point of entry to the building.

In accordance with the requirements of BS 7671:2018, main equipotential bonding conductors for every electrical installation **must** be connected to the main earthing terminal of that particular installation as shown in Figure 5.16.

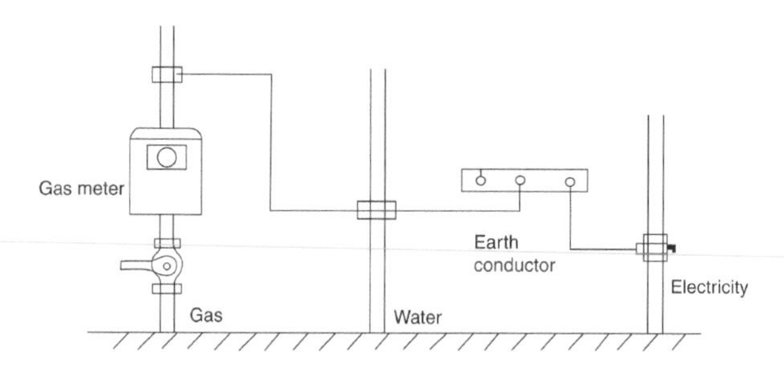

Figure 5.15 Main equipotential bonding

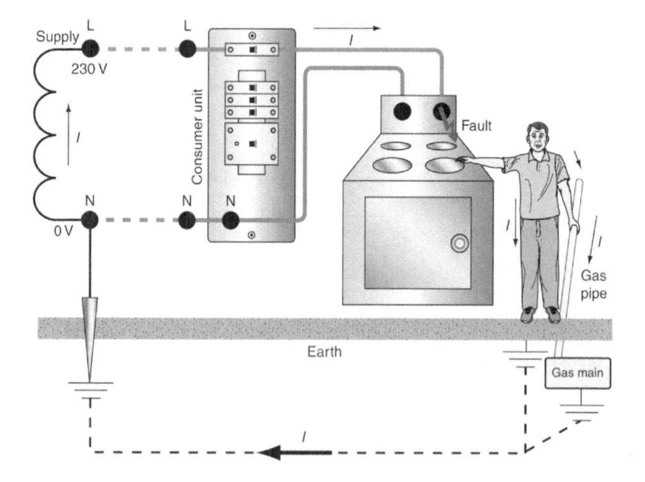

Figure 5.16 Earthed equipotential bonding

Source: Courtesy Brian Scaddan

5.4.6 Supplementary bonding conductors

For installations and locations where there is an increased risk of shock (such as agricultural and horticultural premises and building sites, etc.) additional measures may be required, such as reduction of maximum fault clearance time and the use of supplementary equipotential bonding.

Supplementary bonding conductors (see Figure 5.17) connect together conductive metalwork which is not associated with the electrical installation but which may provide a conducting path that could give rise to shock.

Locations which contain a bath or shower (and where body resistance is lowered as a result of water) are potentially very hazardous environments and it is important to ensure that no dangerous potentials exist between exposed and extraneous conductive parts.

For this reason, local supplementary equipotential bonding needs to be provided to connect together the terminals of the protective conductors of each circuit supplying Class I and Class II equipment with extraneous-conductive-parts in those zones, such as:

- metallic pipes supplying services and metallic waste pipes (e.g. water, gas);
- metallic central heating pipes;
- air conditioning systems;
- accessible metallic structural parts of the building;
- metallic baths and shower basins.

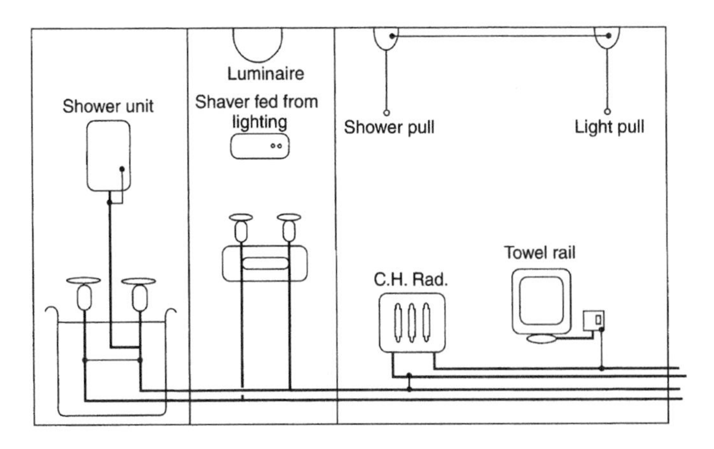

Figure 5.17 Supplementary equipotential bonding

Source: Courtesy Brian Scaddan

If mechanical protection is not provided, its cross-sectional area shall be not less than 4 mm^2.

In agricultural and horticultural premises, supplementary bonding conductors shall be protected against mechanical damage and corrosion, and shall be selected to avoid electrolytic effects.

5.4.7 *Automatic supply*

An automatic safety service must:

- be capable of maintaining a supply of acceptable duration;
- have equipment, either by construction or by erection, that is capable of fire-resistance of adequate duration.

Note: The safety source is generally additional to the normal source such as the public supply network.

5.4.8 *Accessibility of electrical equipment*

Electrical equipment should always be arranged so that:

- there is sufficient space for the initial installation and (if required) later replacement of an individual item of electrical equipment;
- the equipment is completely accessible for operation, inspection, testing, fault detection, repair and maintenance.

5.4.9 *Additions and alterations to an installation*

No addition or alteration, (temporary or permanent) shall be made to an existing installation, unless:

- it has been ascertained that the rating and the condition of any existing equipment (including that of the distributor) will be adequate for the altered circumstances;
- the earthing and bonding arrangements used as a protective measure for the safety of the addition or alteration, are adequate;
- it has been verified that alterations to an existing installation fully complies with the Regulations and does not weaken the safety of the existing installation.

5.5 Circuits for safety services

5.5.1 *Safety sources*

Protection against fault current and against electric shock in case of a fault shall be ensured whether the installation is supplied separately from one of two sources or by both in parallel.

Examples (see Figure 5.18) of safety services include:

- CO_2 detection and alarm systems;
- Emergency lighting;
- Essential medical systems;
- Fire detection and alarm systems;
- Fire evacuation systems;
- Fire pumps;
- Fire rescue service lifts;
- Fire services communication systems;
- Industrial safety systems;
- Smoke ventilation systems.

The following sources for safety services are recognised:

- storage batteries;
- primary cells;
- generator sets independent of the normal supply;
- a separate feeder of the supply network effectively independent of the normal feeder.

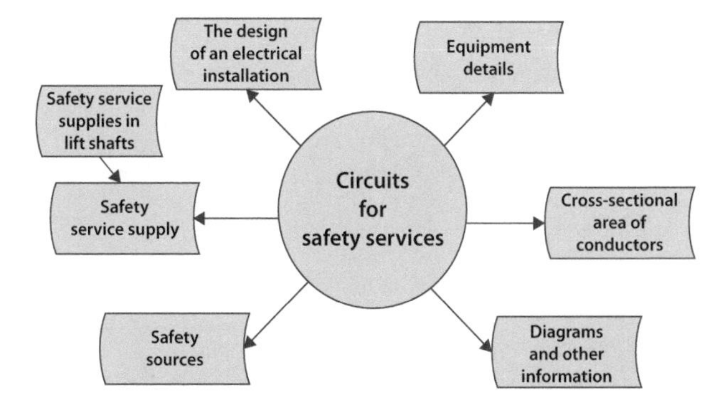

Figure 5.18 Circuits for safety services

Safety sources shall:

- be placed in a suitable location;
- be accessible only to skilled persons or instructed persons;
- be installed as fixed equipment;
- be properly and ventilated.

5.5.2 Electrical safety service supply

An electrical safety supply service is either:

- a non-automatic supply, initiated by an operator; or
- an automatic supply, that is opened independent of an operator.

It is, however, extremely important that all circuits to a safety service:

- are independent of other circuits;
- do not pass through zones exposed to an explosion risk;
- do not pass through locations exposed to fire risk – unless they are fire-resistant;
- enable easy access for periodic inspection, testing and maintenance;
- avoid an overcurrent in one circuit weakening the correct operation of other circuits;
- safety circuit cables (other than metallic screened, fire-resistant cables) must be separated from other safety circuit cables.

5.5.2.1 *Safety service supplies in lift shafts*

Safety service supplies should **not** be installed in lift shafts (or other flue-like openings) unless they have been specifically designed for lifts with special requirements (e.g. a Fire and Rescue Service lift).

5.5.3 The design of an electrical installation

The electrical installation shall be designed to provide:

- the protection of persons, livestock and property;
- the proper functioning of the electrical installation for the intended use.

5.5.4 *Equipment details*

A list (showing the nominal electrical power, rated nominal voltage, current and starting current, together with its duration) of all current-using equipment that is permanently connected to the safety power supply, shall be available.

5.5.4.1 *Cross-sectional area of conductors*

The cross-sectional area of conductors shall be determined for both normal operating conditions and, (where appropriate) for fault conditions according to:

- the admissible maximum temperature;
- the voltage drop limit;
- the electromechanical stresses likely to occur due to short-circuit and Earth fault currents;
- other mechanical stresses to which the conductors are likely to be exposed;
- the maximum impedance for operation of short-circuit and Earth fault protection;
- the method of installation;
- harmonics;
- thermal insulation.

5.5.5 *Diagrams and other information*

As well as a general schematic diagram, details of **all** electrical safety sources must be provided adjacent to the distribution board. In addition, drawing(s) of the electrical safety installations need to be available showing the exact location of all:

- electrical equipment and their distribution boards;
- safety equipment;
- special switching and monitoring equipment for the safety power supply.

Operating instructions for all safety equipment and electrical safety services must be made available.

5.6 Disconnecting devices

Disconnecting devices (see Figure 5.19) shall be provided to enable electrical installations, circuits and individual items of equipment to be switched off or isolated during operation, inspection, fault detection, testing, maintenance and repair.

5.6.1 *Earthing arrangements and protective conductors*

If protective bonding conductors are installed (especially in PV power supply systems) they must be parallel to and in close contact with d.c. cables, a.c. cables and accessories.

If overcurrent protective devices are used for fault protection, the protective conductor needs to be included in the same wiring system as the live conductors or in their immediate proximity.

5.6.2 *Earthing requirements for the installation of equipment having high protective conductor currents*

Equipment having a protective conductor current greater than 3.5 mA but less than 10 mA needs to be either permanently connected to an installation or via a plug and socket-outlet complying with BS EN 60309-2.

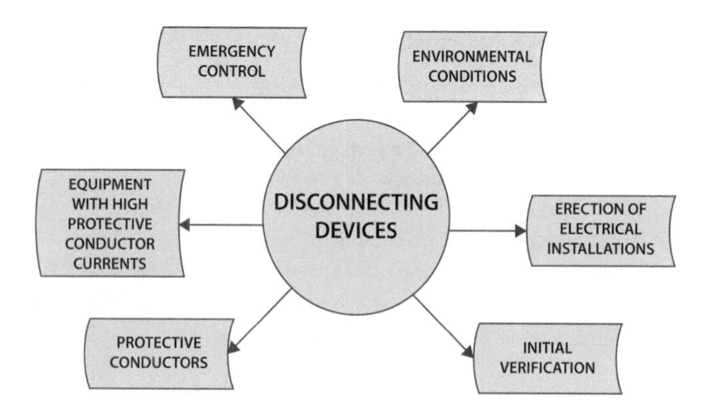

Figure 5.19 Disconnecting devices

Equipment which has a protective conductor current exceeding 10 mA shall be connected to the supply:

- permanently via the wiring of the installation;
- via a flexible cable with a plug and socket-outlet; or
- via a protective conductor with an Earth monitoring system.

The wiring for final distribution circuits supplying one or more items of equipment with a total protective conductor current exceeding 10 mA shall have one of the following:

- a single protective conductor (with a cross-sectional area greater than 10 mm^2);
- a single copper protective conductor (with a cross-sectional area greater than 4 mm^2);
- two individual protective conductors;
- an Earth monitoring system; or
- be connected to the supply by means of a double-wound transformer or equivalent unit.

Where two protective conductors are used, the ends of the protective conductors shall be terminated independently of each other at connection points, such as the distribution board, junction boxes and socket-outlets etc.

At the distribution board, circuits which have a high protective conductor current shall be listed.

In agriculture and horticultural premises, protective bonding conductors shall be protected against mechanical damage and corrosion to avoid electrolytic effects.

Electrical installations in caravan/camping parks (or similar locations) socket-outlet protective conductors shall **not** be connected to any PEN conductor of the electricity supply.

5.6.3 *Emergency control*

An interrupting device, that it can be easily recognised and effectively (and rapidly) operated in the case of danger, shall be installed.

5.6.4 *Environmental conditions*

Electrical installations shall take into account the environmental conditions to which it will be subjected – particularly where the equipment's surroundings are susceptible to risk of fire or explosion.

5.6.5 *Erection of electrical installations*

Electrical equipment shall be installed in accordance with the instructions provided by the manufacturer of the equipment to ensure that during the process of erection:

- design temperatures are not exceeded;
- electrical joints and connections are properly constructed (with regard to conductance, insulation, mechanical strength and protection);
- exposed parts of electrical equipment that could cause injury to persons or livestock are suitably protected;
- electrical equipment (and its characteristics) are not damaged;
- the positioning of electrical equipment (that is likely to cause high temperatures or electric arcs) minimises the risk of igniting flammable materials.

Note: Where necessary, suitable safety warning signs and/or notices shall be provided.

5.6.6 *Initial verification*

The person or persons responsible for the design, construction, inspection and initial testing of the installation shall provide the person ordering the work, with a Certificate of Work.

5.7 Precautions and protections

As can be seen from Figure 5.20, there are many safety hazards that need to be considered when working on or with electrical installations.

5.7.1 *Omission of devices for protection against overload for safety reasons*

Omission of devices for protection against overload is permitted for circuits supplying current-using equipment if unexpected disconnection of the circuit could cause danger or damage.

Figure 5.20 Precautions and protections

Consideration should be given to installing an overload alarm.

5.7.2 *Precautions within a fire-segregated compartment*

The spread of fire can be restricted by sub-dividing buildings into a number of discrete compartments. These fire compartments are separated from one another by fire resistant compartment walls and floors.

5.7.3 *Preservation of the electrical continuity of protective conductors*

Protective conductors should:

- be suitably protected against mechanical and chemical deterioration and electrodynamic effects;
- ensure that every joint in metallic conduit is mechanically and electrically continuous;
- ensure that every connection and joint is accessible for inspection, testing and maintenance.

See Regulation 526.3 of BS 7671:2018 for exceptions to this rule.

- be protected by insulating sleeving complying with BS EN 60684;
- ensure that any exposed, or conductive, part of an equipment is not used to form a protective conductor for another equipment.

See Regulation 543.2 of BS 7671:2018 for exceptions to this rule.

5.7.4 Prevention of harmful effects

Electrical equipment shall not cause any harmful effects to other equipment or interfere with the supply during normal service – particularly during switching operations.

5.7.5 Protection against fault current

All electrical equipment, including conductors, need some kind of mechanical protection against electromechanical stresses caused by fault currents in order to prevent injury or damage to persons, livestock or property.

5.7.6 Protection against overcurrent

All persons and livestock shall be protected against injury due to excessive temperatures or electromechanical stresses caused by any overcurrents that might develop in live conductors.

5.7.7 Protection against power supply interruption

Installations and equipment shall be protected against damage, or causing any danger, due to an interruption of supply.

5.7.8 Protection against thermal effects

Electrical installations shall be installed so that:

- the risk of ignition of flammable materials due to high temperature or electric arc is minimised;
- there is minimal risk of burns to persons or livestock during normal use;
- persons, livestock, fixed equipment and fixed materials adjacent to electrical equipment are protected against the effects of any heat or thermal radiation emitted by the equipment.

5.7.9 *Protection against voltage disturbances and measures against electromagnetic influences*

Persons and livestock shall be protected against injury and property protected against any harmful effects, caused by:

- a fault between live parts of circuits supplied at different voltages;
- overvoltages (e.g. from atmospheric events or switching;
- undervoltage and any subsequent voltage recovery.

The design of installations' shall ensure that:

- they are reasonably immune to electromagnetic disturbances;
- any electromagnetic emissions, generated by the installation or the installed equipment has been taken into effect.

5.7.10 *Seismic effects*

Wiring systems shall be selected and erected with due regard to the seismic hazards.

Where the seismic hazards of the physical location of the installation are low severity (i.e. AP2) or higher, particular attention shall be paid to:

- the fixing of wiring systems to the building structure;
- the connections between the fixed wiring and all items of essential equipment.

5.8 Testing and inspection

Regular testing and inspection of electric equipment and installations needs to be considered and some of the most important are shown in Figure 5.21.

5.8.1 *Initial inspection*

Initial inspection shall precede testing and is aimed at verifying that the installed electrical equipment is:

- in compliance;
- correctly selected and erected; and
- not visibly damaged or defective so as to impair safety.

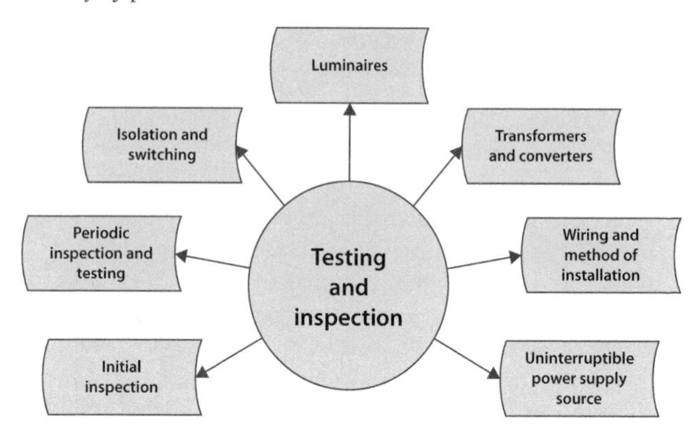

Figure 5.21 Testing and inspection

5.8.2 *Periodic inspection and testing*

Periodic inspections should be carried to confirm that the requirements for protective devices are sufficient to ensure:

- safety of persons and livestock against the effects of electric shock and burns;
- protection against damage to property by fire and heat arising from an installation defect;
- that there are no installation defects;
- confirmation that the installation remains safe and has not been damaged or weakened.

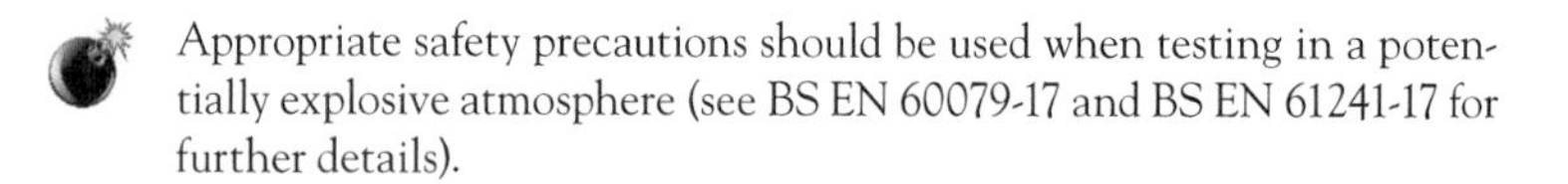

Appropriate safety precautions should be used when testing in a potentially explosive atmosphere (see BS EN 60079-17 and BS EN 61241-17 for further details).

5.8.2.1 *Isolation and switching*

It is essential that:

- all voltage can be cut off (when required) from every installation, circuit and equipment, so as to prevent or remove danger;
- fixed electric motors should be provided with an instantly accessible, easily operated, easily located and efficient means of switching off as to prevent danger.

5.8.2.2 *Luminaires*

A luminaire is the construction around the light source such as the mounting, lampholder, reflector, shade or glass cover.

Any light source which (on failure) could eject flammable materials should be equipped with a safety protective shield and the cable between the luminaire and the fixed source shall be installed so that any expected stresses in the conductors, terminals and terminations do not interfere with the safety of the installation.

5.8.2.3 *Transformers and converters*

Safety isolating transformers for an extra-low voltage lighting installations shall comply with BS EN 61558-2-6 and:

- either the transformer shall be protected on the primary side by a protective device; or
- the transformer shall be short-circuit proof (both inherently and non-inherently).

5.8.2.4 *Type of wiring and method of installation*

The type of wiring system used and installation method should be based on:

- the accessibility of the wiring system to persons and livestock;
- electromagnetic interference;
- the electromechanical stresses likely to occur due to short-circuit and Earth fault currents;
- the nature of the location;
- the nature of the structure supporting the wiring;
- the working voltage;
- other external influences such as mechanical, thermal and fire.

5.8.2.5 *Uninterruptible power supply sources*

A static type Uninterruptible Power Supply (UPS) source shall be capable of:

- operating distribution circuit protective devices; and
- starting safety devices when operating in the emergency condition.

Author's End Note

Now that we are aware of the mandatory and fundamental requirements for safety protection contained in both the Wiring Regulations, Building Regulations and their associated British, European and International Standards, in the next chapter we will look at the different types of equipment, components, accessories and supplies for electrical installations that are currently available to satisfy these requirements.

6 Electrical equipment, components, accessories and supplies

Author's Start Note

The amount of different types of equipment, components, accessories and supplies for electrical installations currently available is enormous and any attempt to cover every type, model and/or manufacture would prove an impossible task for a book such as this.

The intention of this chapter, therefore, is to provide a catalogue of all the different types identified and referred to in the Wiring Regulations (e.g. luminaires, RCDs, plugs and sockets, etc.) and then make a list of the specific requirements that are sprinkled throughout the Regulations.

6.1 Supplies

6.1.1 *Generators*

There are two basic types of Generators:

- **large generator sets** – mainly used by commercial and industrial customers as a stand-by supply source for safety services such as emergency escape lighting, fire detection and alarm systems, etc.;
- **low voltage portable standby generating sets** (such as an electric motors or electrochemical accumulators, etc.) which are designed to automatically (or manually) keep power running during an emergency. They can be used as a cost-effective standby power supply for domestic buildings or other purposes such as camping, caravanning, catering, computers and laptops.

DOI: 10.1201/9781003165170-6

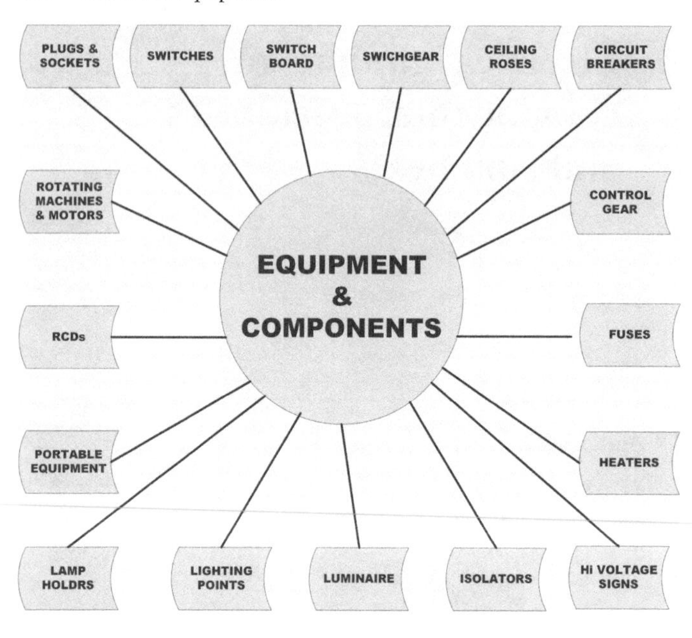

Figure 6.1 Electrical equipment and components

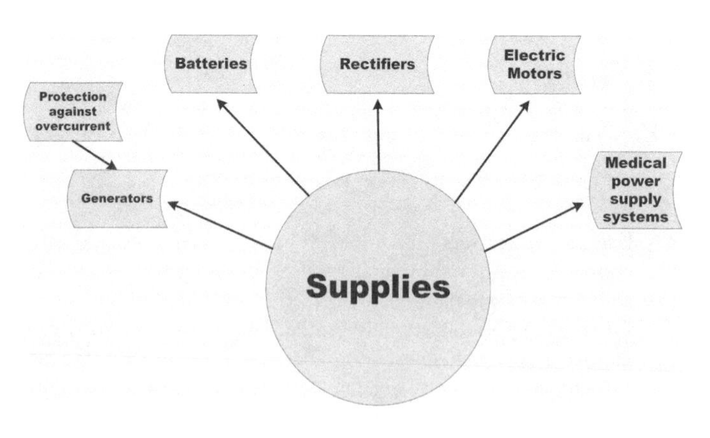

Figure 6.2 Types of supplies

6.1.1.1 BS 7671:2018 requirements and recommendations

As noted in BS EN 50438, in addition to The Electricity Safety, Quality and Continuity (Amendment) Regulations (ESQCR) requirements, if a generating set with an output exceeding 16 A is to be connected in parallel with a public distribution system:

- it shall be possible to automatically disconnect such parts of the installation if the capacity of the generating set is exceeded;
- the neutral (star) point of the secondary windings of a three-phase generator must always be connected to Earth;
- the prospective short-circuit current and Earth fault current shall be assessed for each source of supply (or combination of sources);
- the requirements of the electricity distributor should be ascertained before the generating set is connected;
- the safety and proper functioning of other supply sources shall remain unaffected by the generating set;
- the short-circuit rating of protective devices within the installation shall not be exceeded;
- it shall not be possible for the generator to operate in parallel with a public distribution system.

6.1.1.2 Basic fault protection

Basic fault protection shall ensure that:

- the nominal voltage does not exceed the upper limit of voltage Band I;
- the supply is from a recognised source (such as a diesel-driven generator or motor-generator).

The following sources may be used for SELV and PELV systems:

- a motor-generator with windings that provide a similar amount of isolation as a safety isolating transformer;
- a source independent of a higher voltage circuit (such as a diesel-driven generator).

A conductor connecting a generator to its control panel does not require a device for protection against fault current:

- provided that the protective device is placed in the panel;
- provided that the wiring is installed so that:
  - the risk of a fault is reduced to a minimum;
  - the risk of fire or danger to persons is minimised.

Note: The connection of a generating set shall not affect the effectiveness of any RCD used in the installation regardless of its supply source.

6.1.1.3 Protection against overcurrent

The generating set will require overcurrent protection and the effects of circulating harmonic currents may be limited by one or more of the following:

- confirmation that any switches which interrupt the circulatory circuit are all interlocked;
- the generating set has compensated windings;
- there is a capable impedance at the generator star connection points;
- the provision of filtering equipment.

6.2 Batteries

In 1957 the first known 'battery' was discovered in Bagdad!

It consisted of a ceramic pot, a tube of copper, and a rod of iron (see Fig 6.3). It was made by the Parthians, who ruled Bagdad from 247BC to 224 AD and was assumed to function as a galvanic cell that was used to electroplate silver!

But back to today's world …

BS 7671:2018 requirements and recommendations

- A battery may be used as a source for SELV and PELV systems.
- All batteries should have some form of basic protection against misuse (e.g. via insulation or an enclosure).
- Large industrial batteries should be installed in a secure location.

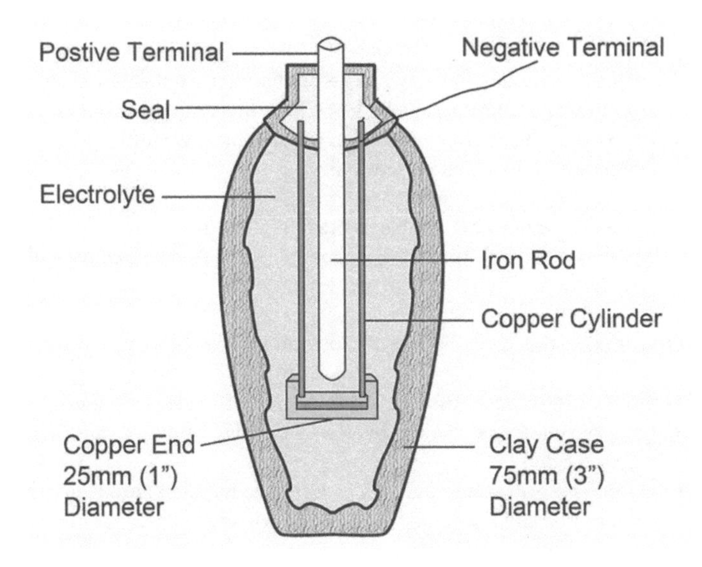

Figure 6.3 The construction of a typical 2000-year-old Bagdad battery

Batteries which are used as a central power source shall be a maintenance-free, heavy duty, industrial design that complies with BS EN 60623.

- a battery-operated emergency lighting circuit:
  - shall be connected to the same RCD that protects the lighting circuit;
  - should have a minimum life span of 10 years and be either vented or valve-regulated;
  - if the battery is a low power supply source.
- the connection from the battery to the control panel does not need to be protected against fault current.

A static, uninterruptible power supply source shall be able to operate all distribution circuits and safety devices provided that it meets the requirements of BS EN.

6.2.1 Rectifiers

A rectifier is a device that converts an oscillating two-directional alternating current (a.c.) into a single-directional direct current (d.c.) and is widely used in the electronic and electromechanical world.

Protection against fault current only:

- the device must be capable of breaking (and for a circuit-breaker, making) the fault current up to and including the prospective fault current.

Protection against fault current is not required provided:

- the wiring reduces the risk of a fault to an absolute minimum;
- the wiring is installed to reduce to the risk of fire or danger to persons; and
- booths (e.g. in fairgrounds, etc.) containing rectifiers are adequately ventilated and the vents are not be obstructed.

6.2.2 Electric motors

An electric motor is a machine that converts electrical energy into mechanical energy and is capable of being automatically or remotely controlled and unsupervised provided that:

- a motor with star-delta starting is protected against excessive temperature in both the star and delta configurations;
- all circuits carrying the motors are capable of accepting a current at least equal to the full-load current rating of the motor;
- automatic restart following a stoppage due to a drop in voltage or a failure of the supply is prevented;
- electric motors with a rating in excess of 0.37 kW are protected against overloading the motor;
- electric motors are safeguarded against excessive temperature by a protective device with manual reset;
- if the electric motor is subject to frequent starting and stopping, the probability of this causing the equipment to rise in temperature has been taken into account;
- fixed electric motors have been provided with a readily accessible, easily operated means of switching off;
- protection against reverse rotation at the end of braking, or reverse operation is provided.

6.2.3 *Medical power supply systems*

In medical locations a safety power supply system is required for some installations, which will energise that installation for continuous operation in case of a failure in the general power system.

This safety power supply system shall automatically take over if the voltage of one or more incoming live conductors of the main distribution board of the building of the main power supply has dropped for more than 0.5 s and by more than 10% with regard to the nominal voltage.

The normal power supply for life-supporting medical electrical equipment **shall** be restored within a changeover period *not exceeding* 0.5 s.

Other services which may require a safety service supply with a changeover period not exceeding 15 sec include, for example, the following:

- electrical equipment used in Group 2 medical locations for surgical procedures of vital importance;
- electrical equipment for medical gas supplies – including compressed air, vacuum supply and narcosis (anaesthetics) exhaustion, in addition to their monitoring devices;
- fire detection and fire alarms to BS 5839;
- fire extinguishing systems;
- selected lifts for firefighters;
- ventilation systems for smoke extraction;
- paging systems.

A few other things to remember are that:

- a circuit which connects the power supply source for safety services to the main distribution board shall be considered a safety circuit;
- primary cells are not allowed as safety power sources;
- the availability of safety power sources shall be monitored and indicated at a suitable location.

A list of examples with suggested reinstatement times is given in Table A710, of Annex A710 of BS 7671:2018.

6.2.3.1 *Automatic changeover devices*

Automatic changeover devices shall be arranged so that safe separation between supply lines is maintained. (See BS EN 60947-6-1 for further information).

In Group 1 and Group 2 medical locations, at least two different sources of supply shall be provided – one of which shall be connected to the electrical supply system for safety services.

Equipment, which is required for the maintenance of healthcare installations [such as sterilisation equipment, technical building installations (e.g. air conditioning, heating and ventilation) and storage battery chargers, etc.], shall be connected to a safety power supply source capable of maintaining it for a minimum period of 24 h.

Where socket-outlets are supplied from the safety power supply source they shall be readily identifiable according to their safety services classification.

6.3 Transformers

A transformer is a passive electrical device that transfers electrical energy from one electrical circuit to another. Transformers are most commonly used for increasing low a.c. voltages at high current (a step-up transformer) or decreasing high a.c. voltages at low current (a step-down transformer) in electric power applications, and for coupling the stages of signal-processing circuits. Transformers can also be used for isolation, where the voltage in equals the voltage out, with separate coils not electrically bonded to one another.

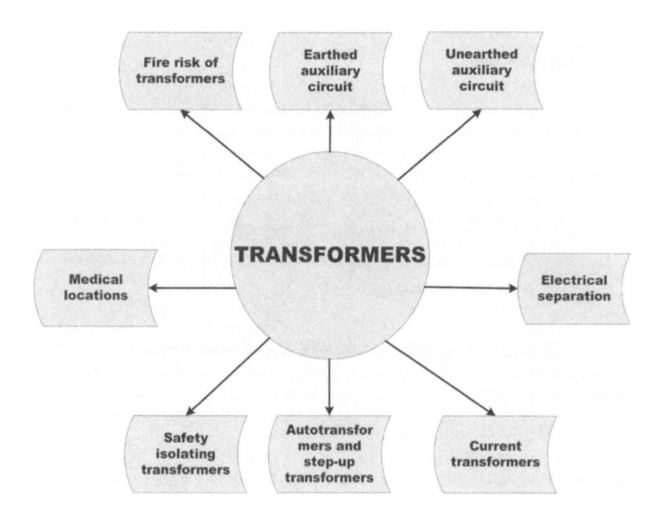

Figure 6.4 Transformers

6.3.1 *Things to remember*

- the neutral (star) point of the secondary windings of three-phase transformers (or the midpoint of the secondary windings of single-phase transformers) **must** be connected to Earth.
- for SELV and PELV systems, a safety isolating transformer in accordance with BS EN 61558-2-6 or BS EN 61558-2-8 may be used.
- where an auxiliary circuit is supplied by more than one transformer, they shall be connected in parallel both on the primary and secondary sides.

6.3.1.1 *Fire risk of transformers*

Transformers must either be protected on the primary side (by a suitable protective device) but the following are permitted to be mounted on a flammable surface:

- a 'class P' thermally protected ballast/transformer;
- a temperature declared thermally protected ballast/transformer.

6.3.1.2 *Earthed auxiliary circuit*

An earthed auxiliary circuit supplied via a transformer shall only be connected to Earth at one point on the secondary side of the transformer and the connection to Earth shall be situated as close to the transformer as possible and be easily accessible.

6.3.1.3 *Unearthed auxiliary circuit*

So long as the circuit is not a SELV or PELV, an auxiliary circuit can operate unearthed via a transformer, **provided** that an Insulation Monitoring Device (IMD) according to BS EN 61557-8 is installed on the secondary side.

6.3.1.4 *Electrical separation*

An IT system at mobile and/or transportable units can be provided by:

- an isolating transformer (or a low voltage generating set) with an insulation monitoring device installed; or

- a transformer offering simple separation (e.g. in accordance with BS EN 61558-1 and providing:
    - automatic disconnection of the supply in case of a first fault between live parts and the frame of the unit; or
    - the use of an RCD.

6.3.2 *Current transformers*

Where a measurement device is connected to the main circuit via a current transformer, the following requirements need to be taken into account:

- conductors on the secondary side of the transformer shall either be insulated or protected by a short-circuit protective device, so that they cannot come into contact with other live parts such as busbars;
- the secondary side of the transformer in a low voltage installation is not earthed;
- protective devices interrupting the circuit are not be used on the secondary side of the transformer;
- terminals for temporary measurements shall be provided.

Note: The use of self-resetting overcurrent protective devices is permitted only for transformers up to 50 VA.

6.3.3 *Autotransformers and step-up transformers*

Where an autotransformer is connected to a circuit having a neutral conductor, the common terminal of the winding shall be connected to the neutral conductor.

Where a step-up transformer is used, a linked switch shall be provided for disconnecting the transformer from all live conductors of the supply.

A step-up autotransformer shall **not** be connected to an IT system.

6.3.4 *Safety isolating transformers*

A safety isolating transformer used for an extra-low voltage lighting installation (such as at a booth in a Fairground); or a fixed onshore isolating transformer that is used to prevent galvanic currents circulating between the hull of the vessel and metallic parts on the shore side, etc.:

- shall comply with BS EN 61558-2-6; and
- either the transformer is protected on the primary side by a protective device; or
- the transformer is short-circuit proof (both inherently and non-inherently).

In addition:

- a manual reset protective device shall safeguard the secondary circuit of each transformer or electronic convertor;
- transformers are mounted out of arm's reach; and
- the protective conductor of the supply to the isolating transformer shall not be connected to the Earth terminal of the socket-outlet supplying the inland navigation vessel.

6.3.5 *Medical locations*

In medical locations, transformers shall be in accordance with BS EN 61558-2-15 and:

- should be installed in close proximity to the medical location;
- ensure that the leakage current does not exceed 0.5 mA;
- use at least one single-phase transformer per room (or functional group of rooms) to form the IT systems for mobile and fixed equipment;
- if more than one transformer is required to supply equipment in one room, they shall not be connected in parallel;
- if the supply of three-phase loads via an IT system is also required, a separate three-phase transformer shall be provided for this purpose;
- overload current protection shall not be used in either the primary or secondary circuit of the transformer of a medical IT system; however;
- overcurrent protection against short-circuit and overload current is required for each final circuit.

Capacitors shall **not** be used in transformers for medical IT systems.

6.4 Fuses

6.4.1 Things worth remembering

Single-pole fuses shall:

- only be inserted in the line or earthed neutral conductor;
- provide an indication of its intended rated current.

Fuses that provide protection against fault current shall **only** be installed where overload protection is achieved by other means.

6.4.2 Devices for isolation and switching

For TN systems the means of isolation can be provided by a suitably rated fuse carrier As an example, Table 6.1 below provides an indication

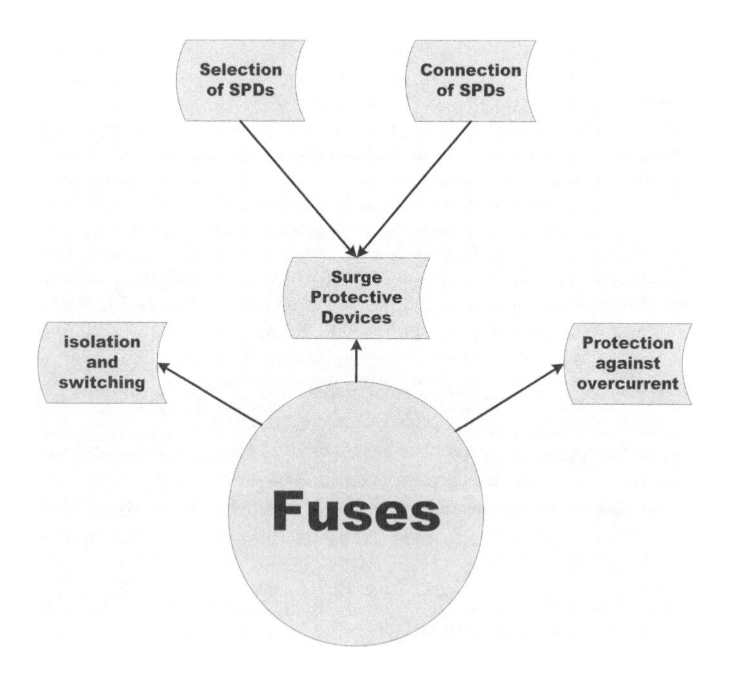

Figure 6.5 Fuses

Table 6.1 Maximum Earth fault loop impedance (Z_s) for circuit breakers

	General purpose fuses							
Rating (amperes)	6	10	16	20	25	32	40	50
Z_s(ohms)	12.8	7.19	4.18	2.95	2.30	1.84	1.35	1/04

of the maximum values of Earth fault loop impedance corresponding to a disconnection time of 5 s for a nominal voltage to Earth (U_o) of 230 V.

If the device is a general-purpose type fuse to BS 88-2.1, BS 88-6 or BS 1361 (or a semi-enclosed fuse to BS 3036), the conditions shown in Table 6.2 shall apply.

All low voltage fused plug and socket-outlets shall conform to the applicable British Standard listed in Table 6.3.

Table 6.2 Coordination between conductor and protective device

Design current (I_B) of the circuit	*Type of fuse*	*Lowest current-carrying capacity (I_z) of any conductor in a circuit*	*Operating current of any protective device (I_2)*
Shall be greater than the nominal current or current setting (I_n) of a protective device		Shall be greater than the nominal current or current setting (I_n) of a protective device	Shall not exceed 1.45 times the lowest of the current-carrying capacities (I_z) of any of the conductors of the circuit
	BS 88-2.1 fuses	Yes	Yes
	BS 88-6 fuses	Yes	Yes
	BS 1361 fuses	Yes	Yes
	BS 3036 (semi enclosed) fuse		Yes provided (I_n) does not exceed 0.725 (I_z)

Table 6.3 Plugs and socket-outlets for low voltage circuits

Type of plug and socket-outlet	*Rating (amperes)*	*Applicable British Standard*
Fused plugs and shuttered socket-outlets, 2-pole and Earth, for a.c.	13	BS 1363 (fuses to BS 1362)
Plugs, fused or non-fused, and socket-outlets, 2-pole and Earth	2, 5, 15, 30	BS 546 (fuses, if any, to BS 646)
Plugs, fused or non-fused, and socket-outlets, protected-type, 2-pole and Earth	5, 15, 30	BS 196
Plugs and socket-outlets (industrial type)	16,32,63,125	BS EN 60309-2

6.4.3 *Devices for protection against overcurrent*

Fuses shall:

- preferably be of the cartridge type;
- shall provide an indication of its intended rated current applicable to the circuit it protects;
- restrict the possibility of a fuse carrier making contact with the conductive parts of adjacent fuse bases;
- preferably use a screw-in fuse base;
- fuses that are likely to be removed or replaced shall be clearly marked with the type of replacement fuse link that should be used.

fuses should **only** be capable of being removed or replaced provided this can be achieved without contact with live parts!

In electrode water heaters and boilers:

- a line conductor fuse may be used as a substitute for the circuit-breaker, provided the boiler or water heater is not in physical contact with any earthed metal;
- a fuse need not be fitted in the neutral conductor of single-phase water heaters and boilers that have an uninsulated heating element immersed in the water.

Note: If a semi-enclosed fuse is used, it **must** either be fitted with a single element of tinned copper wire whose diameter is shown in Table 6.4.

Table 6.4 Sizes of tinned copper wire for use in semi-enclosed fuses

Rated current of fuse (A)	*Diameter of tinned copper wire* (mm)
3	0.15
5	0.2
10	0.35
15	0.5
20	0.6
30	0.85
60	1.53
80	1.8
100	2.0

6.4.4 *Surge protective devices*

SPDs are intended:

- for protection against overvoltages;
- to limit short-lived overvoltages of atmospheric origin that are transmitted via the supply distribution system;
- to protect against transient overvoltages caused by direct lightning strikes;
- protect sensitive and critical equipment (for example hospital equipment and fire/security alarm systems, etc.);
- to fail safely.

SPDs shall be selected:

- in accordance with the impulse withstand voltage of the equipment to be protected;
- to provide a voltage protection level lower than the impulse withstand capability of the equipment or lower than the impulse immunity of the equipment;
- SPDs shall be selected and erected so as to ensure coordination in operation;
- which will protect equipment from failure, remain operational during surge activity and withstand most temporary overvoltage conditions.

Where required SPDs shall be installed:

- as close as possible to the origin of the installation; or
- in the main distribution assembly close to the origin of an installation;
- If more than one SPD is connected on the same conductor, they should be done so in a coordinated manner.

6.4.4.1 Connection of SPDs

SPDs at or near the origin of the installation shall be connected between specific conductors according to Table 6.5.

6.4.4.2 Selection of SPDs

The maximum continuous operating voltage U_c of SPDs shall be equal to (or be higher than) that required by Table 6.6.

Table 6.5 Types of protection for various LV systems

SPDs connected between	*TN-C-S, TN-S or TT* — *Installation in accordance with*		*IT without distributed neutral*
	CT 1	*CT 2*	
Each line conductor and neutral conductor	Optional	SPD required	Not applicable
Each line conductor and PE conductor	SPD required	Not applicable	SPD required
Neutral conductor and PE conductor	SPD required	SPD required	Not applicable
Each line conductor and PEN conductor	Not applicable	Not applicable	Not applicable
Line conductors	Optional	Optional	Optional

Table 6.6 Minimum required continuous operating voltage of the SPD dependent on supply system configuration

SPDs connected between	*TN-C-S, TN-S or TT*	*IT without distributed neutral*
Line conductor and neutral conductor	1.1 U_{aspd}	Not applicable
Each line conductor and PE conductor	1.1 U_{aspd}	Line to line voltage
Neutral conductor and PE conductor	U_{aspd}	Not applicable
Each line conductor and PEN conductor	Not applicable	Not applicable

6.4.4.3 *Selection with regard to nominal discharge current (I_{nspd}) and impulse current (I_{imp})*

The following are recognised as amongst the most important aspects to consider when selecting an SPD:

- their withstand capability, as classified in BS EN 61643-11 (for power systems) and BS EN 61643-21 for telecommunication systems;
- that the nominal discharge current I_{nspd} of the SPD shall be not less than 5 kA with a waveform characteristic 8/20 for each mode of protection;
- type 1 SPDs need to be installed where a lightning protection system is fitted;
- where Type 1 SPDs are required, the value of the lightning impulse current (I_{imp}) shall be not less than 12.5 kA for each mode of protection;
- where Connection Type 2 (CT 2) installations are required, I_{imp} shall be calculated in accordance with BS EN 62305-4.

If the current value cannot be established, the value of I_{imp} must be not less than 50 kA for three-phase systems and 25 kA for single-phase systems.

6.4.4.4 *Selection with regard to the prospective fault current and the follow current interrupt rating*

When a 'follow current interrupt rating' (i.e. potential short-circuit current that an SPD is able to interrupt without operation of a disconnector) is declared by the manufacturer, it shall be equal to, or higher than the prospective line to neutral fault current at the point of installation. In addition:

- the short-circuit resistance of a combined SPD and OCPD (Overcurrent Protective Device) shall be equal to, or higher than, the maximum prospective fault current expected at the point of installation;
- SPDs connected between the neutral conductor and the protective conductor in TT or TN systems, shall have a follow current interrupt rating greater than (or equal to) 100 A;
- In IT systems, the current follow interrupt rating for SPDs connected between the neutral connector and the protective conductor shall be the same as for SPDs connected between line and neutral.

The OCPD (which may be either internal or external to the SPD) provides protection against overcurrent and consequences of SPD's end of life.

6.4.4.5 *Fault protection integrity*

Fault protection shall remain effective in the protected installation even in case of failures of SPDs.

- In TN systems, automatic disconnection of supply shall be obtained by correct operation of the overcurrent protective device on the supply side of the SPD.
- In TT systems this shall be obtained by either the installation of SPDs on the load side or the supply side of an RCD.

Notes:

- SPDs installed on the load side of an RCD, should be immune to surge currents of at least 3 kA.
- SPDs shall be provided with a status indicator (e.g. electrical, visual or audible alarm system) to indicate when the SPD no longer provides overvoltage protection.

6.4.4.6 *Critical length of connecting conductors*

To gain maximum protection:

- current loops shall be avoided;
- supply conductors shall be kept as short as possible in order to minimise additional inductive voltage drops across the conductors;

- the total lead length of supply conductors should not exceed 1.0 m;
- the protective conductor of an in-line SPD shall preferably be no longer than 0.5 m.

To ensure that SPD connections are as short as possible and their inductance as low as feasible. SPDs may be connected to the main earthing terminal (or to the protective conductor) via the metallic enclosures of the assembly being connected to the protective conductor.

6.4.4.7 *Cross-sectional area of connecting conductors*

The connecting conductors of SPDs shall have a cross-sectional area:

- not less than 4 mm^2 copper; or
- not less than that of the line conductors if the line conductor's cross-sectional area is less than 4 mm^2.

Note: The minimum cross-sectional area for Type 1 SPDs shall be 16 mm^2 copper, or equivalent, where there is a structural lightning protection system.

6.5 Wiring systems

The following are probably the most important requirements of wiring systems:

- equipment should always be fixed, so that connections between wiring and equipment are **not** be subject to undue stress or strain during normal;
- unenclosed equipment shall be mounted in a suitable mounting box or enclosure;
- socket-outlets, connection units, plate switches and similar accessories should be fitted to a mounting box;
- wherever equipment is fixed on or in cable trunking, skirting trunking or in mouldings it shall not be fixed on covers which can accidentally be removed.

Group 2 medical location wiring systems shall be for the exclusive use of equipment and accessories within those locations.

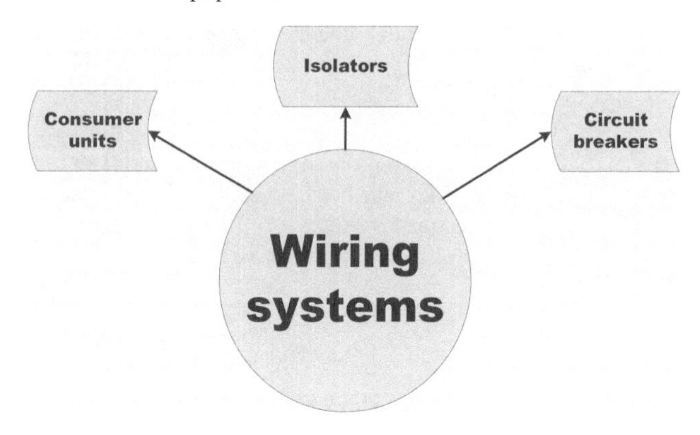

Figure 6.6 Wiring systems

6.5.1 Consumer units

A consumer unit is a particular type of distribution board for the control and distribution of electrical energy, primarily in domestic premises.

The two main requirements of the Regulations are that:

- all consumer units shall comply with the requirements of BS EN 61439-3; and specifically;
- **all** circuits and final circuits **shall** be provided with a means of switching for interrupting the supply on load.

6.5.2 Isolators

The location of each isolator (frequently described as the 'disconnector') shall include a suitable notice to avoid any possibility of confusion.

6.5.3 Circuit breakers

Every circuit shall be provided with a means of isolation from all live supply conductors by a linked switch or a linked circuit-breaker which shall:

- provide protection against both overload and fault current;
- be capable of 'making' any overcurrent up to and including the maximum prospective fault current at the point where the device is installed;

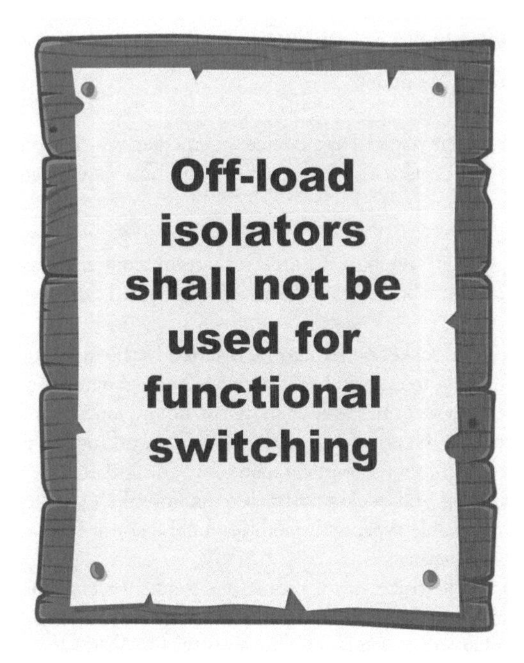

Figure 6.7 Location of off-load isolators

- be inserted in the line conductor only;
- be inserted in an earthed neutral conductor (except where linked);
- control the supply to an electrode water heater or electrode boiler;
- be designed and installed so that it is not possible to modify the setting or the calibration of its overcurrent release without the use of either a key or a tool which results in a visible indication of its modified setting or calibration.

A main linked switch or linked circuit-breaker shall be provided as near as practicable to the origin of every installation as a means of switching the supply on load and as a means of isolation.

Linked circuit-breakers inserted in an earthed neutral conductor must be capable of breaking all of the related line conductors.

6.6 Plug and socket outlets

A plug and socket-outlet:

- shall **not** be selected as a device for emergency switching;
- shall **not** be used as a device for connecting a water heater or boiler to the supply;

The following are amongst the most important requirements as stated in BS 7671:2018 and other BS, EN and ISO associated Standards:

- a plug and socket-outlet may be inserted in the main supply circuit to enable it to be switched off for mechanical maintenance;
- a socket-outlet on a wall or similar structure must be mounted sufficiently high enough above the floor or a working surface to minimise the risk of mechanical damage to the socket;
- every plug and socket-outlet (except for SELV) shall be of the non-reversible type, with facilities for the connection of a protective conductor;
- equipment which has a protective conductor current exceeding 3.5 mA (but not exceeding 10 mA) must be connected by means of a plug and socket complying with BS EN 60309-2;
- if the rating of the plug and socket-outlet does not exceed 16 A, it shall be capable of cutting off the full load current of the relevant part of the installation;

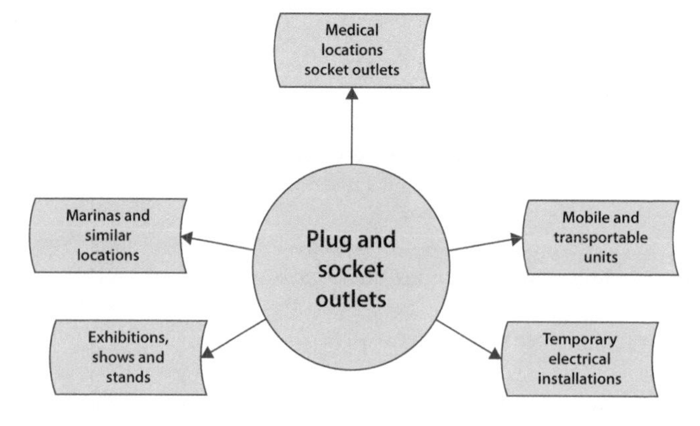

Figure 6.8 Plug and socket outlets

- socket outlets for household and similar use shall be of the shuttered type and, for an a.c. installation, shall preferably be of a type complying with BS 1363;
- a plug and socket-outlet **not** complying with BS 1363, BS 546, BS 196 or BS EN 60309-2, may be used in single-phase a.c. or two-wire d.c. circuits operating at a nominal voltage not exceeding 250 volts for:
    - the connection of an electric clock;
    - the connection of an electric shaver;
    - a circuit having special characteristics.

Except for the plug of a plug and socket-outlet identified in Table 55.1 of BS 7671:2018, equipment of overvoltage categories I and II should **not** be used for isolation. Table 6.7 provides an indication of other applicable British Standards.

For final circuits containing a number of socket-outlets the circuit shall be provided with a high integrity protective conductor connection.

Equipment which has a protective conductor current exceeding 10 mA shall be connected to the supply:

- permanently via the wiring of the installation; or
- via a flexible cable with a plug and socket-outlet complying with BSEN 60309-2; or
- via a protective conductor with an Earth monitoring system.

For more information about the requirements for plug and socket-outlets in special installations and locations, please see Table 6.8 below and the relevant sections contained in BS 7671:2018 Chapters 701–740.

Table 6.7 Plugs and socket-outlets for low voltage circuits

Type of plug and socket outlet	*Rating (amperes)*	*Applicable British Standard*
Fused plugs and shuttered sockets-outlets	13	BS 1363 and BS 1362 (for fuses)
Plugs, fused and non-fused, and socket outlets 2 pole sand Earth	2,5,15,30	BS 546 and BS 646 (for the fuses)
Plugs and sockets of an industrial type	16,32,63,125	BS EN 60309-2

Table 6.8 Plug and socket outlets at different locations

Category	*Main requirements*	*BS 7671:2018 detailed requirements*
Caravan and camping parks	It is recommended that at least one socket-outlet is provided for each caravan pitch provided that it: • complies with BS EN 60309-2 • is provided with individual overcurrent protection • is protected individually by an RCD; • has a current rating not less than 16 A. • meets the degree of protection of at least IP44 in accordance with BS EN 60529 The socket-outlets shall be • placed at a height of 0.5 m to 1-.5 m from the ground shall not be connected to any PME earthing service	See BS 7671:2018 Chapters 708, 721 and 722 for more detailed requirements
Exhibitions, shows and stands	At exhibitions, shows and stands • sufficient socket-outlets shall be installed to allow user requirements to be met safely • floor mounted socket-outlets shall be protected from accidental ingress of water and have sufficient strength to be able to withstand the expected traffic load.	See BS 7671:2018 Chapter 711 for more detailed requirements

(*Continued*)

Table 6.8 (Continued)

Category	*Main requirements*	*BS 7671:2018 detailed requirements*
Marinas and similar locations	At Marinas and similar locations • 200 V–250 V, 16 single-phase socket-outlets shall be provided • a maximum of four socket-outlets shall be grouped together in one enclosure • one socket-outlet shall supply just one pleasure craft or houseboat • socket-outlet protective conductors shall not be connected to a PME earthing facility Every socket-outlet shall • meet the degree of protection of at least IP44 or be protected by an enclosure • be located as close as practicable to the berth to be supplied • be installed in the distribution board or in a separate enclosure • comply with BS EN 60309-1 above 63 A and BS EN 60309-2 up to 63 A • be placed at a height of not less than 1 m above the highest water level	See BS 7671:2018 Chapter 709 for more detailed requirements **Note:** Figure 709.3 indicates the recommended instruction notice that should be placed in marinas adjacent to each group of socket-outlets.

(Continued)

Table 6.8 (Continued)

Category	*Main requirements*	*BS 7671:2018 detailed requirements*
Medical locations Socket outlets	In medical locations, **all** socket-outlets intended to supply medical electrical equipment shall be unswitched • in Group 1 & 2 medical locations, the resistance between the Earth terminals of a protective conductor's socket-outlets and any extraneous conducive part (such as the equipotential bonding busbar) shall not exceed 0.2 Ω • socket-outlets and switches installed below any medical-gas outlets shall be at least 0.2 m from the outlet (centre to centre), in order to minimise fire risk • socket outlets at a patient's medical location (e.g. their bedhead) shall get their supply from a protected circuit	See BS 7671:2018 Chapter 710 for more detailed requirements
	Socket-outlets used on medical IT systems, **must** be coloured blue and clearly and permanently marked '*Medical equipment only*'	
Mobile and transportable units	Socket outlets, plugs and connecting devices used to connect mobile equipment to the supply shall comply with BS EN 60309-2 and meet the following requirements • plugs shall be within an enclosure of insulating material • connecting devices located outside the unit shall provide a degree of protection of not less than IP44 when in use • enclosures containing the connector shall provide a degree of protection of at least IP55 when not connected • additional protection by means of an RCD shall be provided for mobile equipment with a current rating not exceeding 32 A when used outdoors	See BS 7671:2018 Chapter 717 for more detailed requirements

(Continued)

Table 6.8 (Continued)

Category	*Main requirements*	*BS 7671:2018 detailed requirements*
Temporary electrical installations	A plug and socket-outlet shall **not** be used as a device for connecting a water heater and/or boiler to the supply The design and installation of temporary electrical installations should always take into account that • socket-outlets dedicated to lighting circuits that are placed out of arm's reach shall be marked according to their purpose • when used outdoor, plugs, socket-outlets and couplers must comply with BS EN 60309 • all plug and socket-outlets (except for SELV) must be of the non-reversible type • all socket-outlets deigned for household and similar use must be of the shuttered type. If a final circuit has a number of socket-outlets (or connection units intended to supply two or more items of equipment and where it is known that the total protective conductor current in normal service will exceed 10 mA) the circuit must be provided with • a high integrity protective conductor connection • a radial final circuit with a ring protective conductor • a radial final circuit with a single protective conductor	See BS 7671:2018 Chapter 740 and BS 7671:2018 Table 537.4 for more detailed requirements

6.7 Switches

There are a number of rules and regulations in BS 7671:2018 regarding switches and switching operations, including:

- some form of disconnecting device (i.e. 'switch') must be provided that allows electrical installations (circuits, and individual items of equipment) to be switched off, or isolated, to enable their operation, inspection, fault detection, testing, maintenance and repair;
- switches shall only be inserted in the line conductor;
- switches should be fitted in a suitable mounting box or enclosure;
- non-linked switches need to be inserted in an earthed neutral conductor;
- the purpose and identification of each item of switchgear and controlgear shall be identified by a label or some other form of indicator.

6.7.1 *Isolation and switching*

From a safety point of view, isolation and switching procedures for the prevention or removal of dangers associated with electrical installations

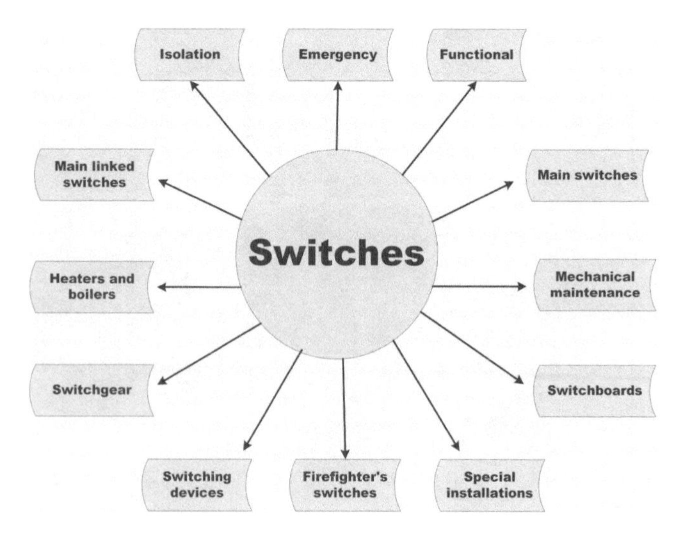

Figure 6.9 Switches

are an absolute necessity and BS 7671:2018 covers these aspects in great detail throughout the standard by providing advice on:

- emergency stopping;
- emergency switching;
- live conductors (including the neutral conductor);
- live supply conductors;
- protection against overcurrent;
- supply distribution cabinets.

BS 7671:2018 Table 537.4 summarises the functions provided by the devices for isolation and switching, together with indication of the relevant product standards.

6.7.2 *Emergency switching*

Emergency switching:

- is basically an operation intended to remove, as quickly as possible, danger, which may have occurred unexpectedly;
- shall be capable of breaking the full load current of the relevant part(s) of the installation.

An emergency switching device may consist of:

- a hand-operated switching devices for direct interruption of the main circuit;
- a switch in the main circuit or pushbuttons in the control (auxiliary) circuit.

The means of operating (handle, push-button, etc.) an emergency switching device shall:

- be easily identifiable (e.g. by red marking) and convenient for their intended use;
- be readily accessible at places where a danger might occur or at any additional remote position from which that danger can be removed;
- be capable of being kept in the 'OFF' or 'STOP' position;
- have priority over any other utility relative to safety;
- shall **not** re-energise the relevant part of the installation;
- shall not be erected where they are accessible to livestock or in any position where access may be impeded by livestock.

Fuses and links shall **not** be used for functional switching.

6.7.3 Functional switching devices

Functional switching is an operation intended to switch 'ON' or 'OFF' or vary the supply of electrical energy to all, or part of, an installation.

Functional switching devices shall:

- be provided for each part of a circuit that might need to be controlled independently from other parts of the installation;
- control the current without necessarily opening the corresponding poles;
- ensure the safe changeover of supply from alternative sources.

6.7.4 Firefighter's switches

All firefighter switches shall comply with the requirements of BS EN 60669-2-6 or BS EN 60947-3 and shall be provided in the low voltage circuit supplying:

- exterior electrical installations operating at a voltage exceeding low voltage;
- interior discharge lighting installations operating at a voltage exceeding low voltage;
- for an interior installation, the switch shall be independent of the switch for any exterior installation and in the main entrance to the building;
- the switch shall be placed in a conspicuous position, which is reasonably accessible to firefighters and, at not more than 2.75 m from the ground or from a person standing beneath the switch.

Author's Note:

Initially I couldn't understand why the switch should be 'not more than 2.75 m (9 ft 0 in) from the ground' as average firefighters are not that tall as far as I am aware! I, therefore, sought advice from the Local Fire Service and was told that the reason why the switch was positioned so high was to stop an unauthorised person from switching it off. Firefighters use a 'Ceiling Hook' to operate the switch. I am now no longer confused!!

- where more than one switch is installed on any one building, each switch shall be clearly marked to indicate the installation that it controls.

A firefighter's switch shall:

- be coloured red;
- have a permanent nameplate marked with the words 'FIREFIGHTER'S SWITCH' or 'FIRE SWITCH'; and
- have its ON and OFF positions clearly indicated (with the OFF position at the top);
- be prevented from the switch being inadvertently returned to the ON position.

6.7.5 *Main switches*

Where an installation is supplied from more than one source:

- a main switch shall be provided for each source of supply;
- a notice shall be fixed in a prominent position warning persons of the need to operate **all** such switches to achieve isolation of the installation; or
- a suitable interlock system shall be provided.

6.7.6 *Main linked switches*

An electrical installation must be capable of isolating itself from each supply via a main linked switch.

If this main switch is intended to be operated by ordinary persons (e.g. a householder) it shall interrupt both live conductors of a single-phase supply.

Note: In a TN-S or TN-C-S system the neutral conductor need not be isolated or switched where it can be regarded as being reliably connected to Earth by suitably low impedance.

6.7.7 *Mechanical maintenance*

Where a switch is provided for switching off for mechanical maintenance, it shall:

- be inserted in the main supply circuit;
- require manual operation;

- be capable of cutting OFF the full load current of the relevant part of the installation;
- be designed and installed so as to prevent inadvertent or unintentional switching ON;
- be in accordance with the requirements of BS EN 60204.

The open position of the contacts of the device should always be visible or be clearly and reliably indicated by the use of the symbols '**0**' and '**I**' to indicate the open and closed positions, respectively.

6.7.8 *Heaters and/or boilers*

A heater or boiler must be permanently connected to the electricity supply through a double-pole linked switch, which is either from or within easy reach of the heater or boiler or a part of the whole equipment.

If the heater or boiler is installed in a room containing a fixed bath, the switch must comply with Section 701 of BS 7671:2018.

Where a step-up transformer is used, a linked switch shall be provided for disconnecting the transformer from all live conductors of the supply.

6.7.9 *Switchboards*

Passageways and working platforms which have access to an open type switchboard or an equipment that has dangerous exposed live parts, need to allow persons to operate and maintain the equipment with ease.

In 2006, to avoid any confusion, the UK changed their cable colours so that they were the same as the European cable colours.

Note: Before commencing any work on an old UK property it is always wise to check the installation to see if the wiring colours are out of date, have deteriorated over time and, above all, are in compliance with the BS 7671:2018 Wiring Regulations.

BS 7671:2018 Table 51 provides more detailed information regarding the identification of conductors.

Figure 6.10 Certainly NOT the right sort of installation!

Source: Courtesy Stingray

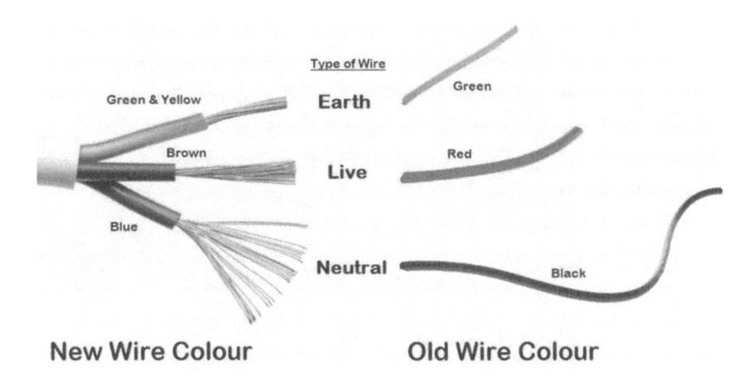

Figure 6.11 The new versus the old wiring colours

6.7.10 *Switchgear*

All switchgear:

- shall be accessible **only** to authorised persons;
- shall be subjected to regular functional tests to show that they are properly mounted, adjusted and installed in accordance the Regulations;

- when placed in an escape route, shall be enclosed in a cabinet or an enclosure constructed of non-combustible material;
- shall be installed outside the location (unless it is installed in an enclosure providing a degree of protection of at least IP4X or in a cable trunking system in compliance with the BS EN 50085 series).

6.7.11 *Switching devices*

A switching device:

- shall be protected against overcurrent;
- shall **not** be inserted in a protective conductor unless the switch:
    - has been inserted in the connection between the neutral point and the means of earthing;
    - is a linked switch that disconnects and connects the earthing conductor at substantially the same time as the related live conductors;
    - is a multipole linked switch or plug-in device in which the protective conductor circuit has not been interrupted before the live conductors and re-established not later than when the live conductors are reconnected.

Single-pole switching devices:

- shall not be inserted in the neutral conductor of a multiphase circuit; or
- isolated in the neutral conductor single-phase circuits.

Where single-pole switching devices are not permitted in the neutral conductor, a test should be made to verify that all such devices are connected in the line conductor(s) only.

6.7.12 *Special installations and locations*

The main requirements for special installations and locations is shown in Table 6.9.

Table 6.9 Special installations and locations

Category	*Main requirements*	*BS 7671:2018 detailed requirements*
Agricultural and horticultural premises	• The electrical installation of a building shall be isolated by a single, clearly marked, isolation device Emergency stopping devices or emergency switching shall not be erected where they are accessible to livestock	See BS 7671:2018 Chapter 705 for more detailed requirements
Construction and demolition site installations	• Each ACS (Assembly for Construction Sites) shall include a suitable device for switching and isolating the incoming supply • Ideally these devices should be capable of being secured in the OFF position by padlock or be located inside a lockable enclosure	See BS 7671:2018 Chapter 704 for more detailed requirements
Electrical installations in caravan, camping parks and similar locations	• All switchgear and controlgear assemblies used in caravan (or tent) area supplies shall comply with the requirements of BS EN 61439-7 • If an installation consists of just one final circuit, the isolating switch may be the overcurrent protective device A permanent notice **must** be fixed near the main isolating switch inside the caravan	See BS 7671:2018 Chapter 721 for more detailed requirements

(Continued)

Table 6.9 (Continued)

Category	*Main requirements*	*BS 7671:2018 detailed requirements*
Exhibitions, shows and stands	• Switch and controlgear need to be placed in closed cabinets which can only be opened by the use of a key or a tool • An emergency switch (easily visible, accessible and clearly marked) shall be installed to control a separate circuit supplying signs, lamps or exhibits	See BS 7671:2018 Chapter 711 for more detailed requirements
Extra-low voltage lighting installations	• The primary circuits of transformers that are operated in parallel, should be permanently connected to a common isolating device	See BS 7671:2018 Chapter 715 for more detailed requirements

Locations containing a bath or shower	Unless switches and controls are part of a fixed current-using equipment
	• zone 0 – switchgear or accessories shall not be installed
	• zone 1 – only switches of SELV circuits supplied at a nominal voltage not exceeding 12 VAC rms or 30 V ripple-free d.c. shall be installed
	• zone 2 – switchgear, accessories incorporating switches or socket-outlets shall not be installed with the exception of
	  • switches and socket-outlets of SELV circuits; and
	  • shaver supply units complying with BS EN 61558-2-5
		See BS 7671:2018 Chapter 701 for more detailed requirements

(Continued)

Table 6.9 (Continued)

Category	*Main requirements*	*BS 7671:2018 detailed requirements*
Marinas and similar locations	There shall be one isolating switching device for a maximum of four socket-outlets and these • shall be installed in each distribution cabinet • shall disconnect all live conductors including neutral	See BS 7671:2018 Chapter 709 for more detailed requirements
Medical locations	• Socket-outlets intended to supply ME equipment shall be unswitched ***Note:*** The requirements for medical electrical equipment subject to flammable gases and vapours are contained in BS EN 60601	See BS 7671:2018 Chapter 710 for more detailed requirement
Rooms and cabins containing sauna heaters	• Switchgear which forms part of the sauna heater equipment may be installed within the sauna room or cabin • Lighting and other switchgear will have to be placed outside the cabin	See BS 7671:2018 Chapter 703 for more detailed requirements
Solar photovoltaic power supply systems	To allow maintenance of the PV convertor • A switch-disconnector shall be provided on the d.c. side of the PV convertor • Switchgear assemblies shall be in compliance with BS EN 60439-1 and appropriate parts of BS EN 61439-1 Junction boxes containing a PV generator or PV array, must carry a warning label indicating that parts inside the boxes may still be live after isolation from the PV convertor	See BS 7671:2018 Chapter 712 for more detailed requirements

(Continued)

Table 6.9 (Continued)

Category	*Main requirements*	*BS 7671:2018 detailed requirements*

Swimming pools and other basins	• switchgear, controlgear or a socket-outlet shall **not** be installed in zones 0 and 1
	• a socket-outlet or a switch in zone 2 is permitted only where the supply circuit is protected by
	  • SELV (if its supply circuit is protected by an RCD
	  • automatic disconnection of supply using an RCD
		See BS 7671:2018 Chapter 702 for more detailed requirements

Temporary electrical installations	• Every electrical installation shall have its own means of isolation and switching • Switchgear shall be placed in cabinets which can be opened only by the use of a key or a tool • Temporary electrical installations for amusement devices shall be provided with its own means of isolation • A separate circuit shall be used to supply luminous tubes, signs or lamps, which will be controlled by an emergency switch	See BS 7671:2018 Chapter 740 for more detailed requirements

6.8 Luminaires

A 'Luminaire' (or light fixtures) distributes filters, or transforms the light transmitted from one or more lamps. Luminaires fall into six categories: recessed, ceiling-mounted, suspended, architectural, wall-mounted, and plug-in. They distribute, filter, or transform the light transmitted from one or more lamps.

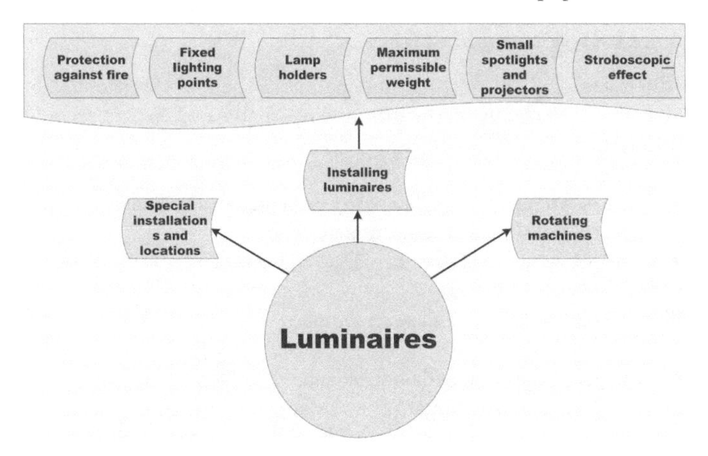

Figure 6.12 Luminaires

6.8.1 *Installing luminaires*

Every luminaire shall:

- be provided with an enclosure that has a degree of protection of at least IP5Xd;
- be used at a reasonable (i.e. sufficient) distance from combustible materials;
- have a limited surface temperature in accordance with BS EN 60598-2-24.

Luminaires shall:

- be appropriately controlled;
- be protected from foreseeable mechanical stresses;
- be suitable for the temperatures likely to be encountered;
- have adequate means of fixing.

Electronic switching devices should, where possible, include a neutral conductor.

6.8.1.1 *Protection against fire*

When selecting and erecting a luminaire, the thermal effects of radiant and converted energy on the surroundings shall be taken into account, including:

- the maximum permissible power dissipated by the lamps;
- the fire-resistance of adjacent material;
- the minimum distance to combustible materials.

Electrical installations shall be so arranged that:

- the risk of ignition of flammable materials due to high temperature or electric arc is minimised;
- during normal operation of the electrical equipment, there shall be minimal risk of burns to persons or livestock.

A luminaire with a lamp that could eject flammable materials in case of failure should be equipped with a safety protective shield.

6.8.1.2 *Fixed lighting points*

At each fixed lighting point one of the following Standards shall be used:

- a ceiling rose to BS 67;
- a luminaire supporting coupler to BS 6972 or BS 7001;
- a batten lampholder or a pendant set to BS EN 60598;
- a luminaire to BS EN 60598;
- a suitable socket-outlet to BS 1363-2, BS 546 or BS EN 60309-2;
- a plug-in lighting distribution unit to BS 5733;
- a connection unit to BS 1363-4;
- appropriate terminals enclosed in a box complying with the relevant part of BS EN 60670 series or BS 4662;
- a device for connecting a luminaire outlet according to BS IEC 61995-1.

Note: In suspended ceilings one plug-in lighting distribution unit may be used for a number of luminaires.

6.8.1.3 *Lamp holders*

A lampholder for a filament lamp shall **not** be installed in any circuit operating at a voltage exceeding 250 volts.

Bayonet lampholders B15 and B22 shall comply with BS EN 61184 and shall have the temperature rating T2 described in that standard.

In circuits of a TN or TT system (except for E14 and E27 lampholders that comply with BS EN 60238) the outer contact of every Edison screw or single centre bayonet cap type lampholder shall be connected to the neutral conductor. This regulation also applies to track mounted systems.

Insulation piercing lampholders shall **not** be used unless the cables and lampholders are compatible, and providing the lampholders are non-removable once fitted to the cable.

In exhibitions, shows and stands, lighting circuits incorporating B15, B22, E14, E27 or E40 lampholders shall be protected by an overcurrent protective device of maximum rating 16 A.

Lampholders with ignitability characteristic 'P' as specified in BS 476 Part 5 shall not be connected to any circuit where the rated current of the overcurrent protective device exceeds the appropriate value stated in Table 6.10, unless the wiring is enclosed in earthed metal or insulating material.

6.8.1.4 *Maximum permissible weight*

The weight of luminaires and their eventual accessories shall not exceed the mechanical capability of the ceiling, suspended ceiling or supporting structure where installed.

Note: In places where the fixing means is intended to support a pendant luminaire, the fixing shall be capable of carrying a mass of more than 5 kg.

Table 6.10 Overcurrent protection of lampholders

Type of lampholder			*Maximum rating of the overcurrent device protecting the circuit*
Bayonet lampholder	B15	SBC	6 A
(BS EN 61184)	B22	BC	16 A
	E14	SES	6 A
Edison screw lampholder	E27	ES	16 A
(BS EN 60238)	E40	GES	16 A

6.8.1.5 *Small spotlights and projectors*

Small spotlights or projectors should be installed at the following minimum distance from combustible materials:

- rating up to 100 W – 0.5 m;
- over 100 and up to 300 W – 0.8 m;
- over 300 and up to 500 W – 1.0 m.

6.8.1.6 *Stroboscopic effect*

Lighting for premises where machines with moving parts are in operation should consider the stroboscopic effects which can give a misleading impression of moving parts being stationary. Such effects may be avoided by selecting luminaires with a suitable lamp controlgear (such as high frequency controlgear) or by distributing lighting loads across all phases of a three-phase supply.

6.8.2 *Lumiere requirements at special installations and locations*

The main requirements for luminaires at special installations and locations are shown in Table 6.11.

6.8.3 *Rotating machines*

With a rotating machine:

- the increased amounts of starting or braking currents will cause a temperature rise in the electric motor which needs to be taken into account;
- every electric motor having a rating in excess of 0.37 kW shall include a system for protection against overload of the motor;
- every motor shall be provided with means to prevent automatic restarting after a stoppage due to a drop in voltage (or failure) of supply;
- lighting used in premises which have machines with moving parts in operation, should consider the stroboscopic effects that can give a misleading impression of the moving parts being stationary.

Side effects like this can be avoided by using a luminaires which includes a lamp with a high frequency controlgear.

Table 6.11 Luminaire requirements

Category	*Main requirements*	*BS 7671:2018 detailed requirements*
Underwater luminaires for swimming pools	• A luminaire in contact with water shall be fixed and shall comply with BS EN 60598-2-18 • Underwater lighting located behind watertight portholes, shall comply with the appropriate part of BS EN 60598	See BS 7671:2018 Chapter 702 for more detailed requirements
Luminaires in caravans and motor caravans	• Each luminaire in a caravan shall preferably be fixed directly to the structure or lining of the caravan • If a pendant luminaire is installed in a caravan, it shall be made secure so as to prevent damage when the caravan is in motion	See BS 7671:2018 Chapter 721 for more detailed requirements
Luminaires and lighting installations in exhibitions shows and stands	• Stands with a concentration of electrical equipment, luminaires and lamps that is liable to generate excessive heat, shall not be installed unless adequate ventilation is available • ELV lighting systems for filament lamps shall comply with BS EN 60598-2-23 • Insulation piercing lampholders shall **not** be used • Luminaires mounted below 2.5 m (arm's reach) from floor level shall be fixed	See BS 7671:2018 Chapter 711 for more detailed requirements
Luminaires in fountains	• A luminaire installed in zones 0 or 1 of a fountain shall be fixed and shall comply with BS EN 60598-2-18 • Electrical equipment in zones 0 or 1 shall be provided with mechanical protection to medium severity (AG2)	See BS 7671:2018 Chapter 702 for more detailed requirements

(continued)

Table 6.11 (Continued)

Category	*Main requirements*	*BS 7671:2018 detailed requirements*
Electric discharge lamp installations	If a luminous tube, sign or lamp is used to illuminate a stand, or an exhibit, which has a nominal power supply voltage higher than 230/400 V a.c., it shall meet the following regulations • *location* – the sign or lamp shall be installed out of arm's reach or be protected • *installation* – the facia or stand fitting material behind luminous tubes, signs or lamps shall be non-ignitable • *emergency switching devices* – the separate circuit used to supply signs, lamps or exhibits shall be controlled by an emergency switch This switch must be easily visible, accessible and clearly marked	See BS 7671:2018 Chapter 711.559 for more detailed requirements
Luminaires in temporary installations	For temporary installations every luminaire and decorative lighting chain shall • be securely attached to the structure or support intended to carry it • its weight shall not be carried by the supply cable (unless it has been selected and erected for this purpose) • be installed with supply cables that are flexible and protected against mechanical damage • be mounted greater than 2.5 m (arm's reach) above floor level	See BS 7671:2018 Chapter 704 or 740 for more detailed requirements

(continued)

Table 6.11 (Continued)

Category	*Main requirements*	*BS 7671:2018 detailed requirements*
Medical locations	In Group 2 medical locations, where PELV is used • exposed-conductive-parts of equipment (e.g. operating theatre luminaires) shall be connected to the circuit protective conductor • the changeover period to the safety services source (in the event of mains power failure) shall not exceed 15 s Escape route luminaires shall be arranged on alternate circuits	See BS 7671:2018 Chapter 710 for more detailed requirements
Outdoor lighting installation	An outdoor lighting installation comprises one or more luminaires, a wiring system and accessories and includes lighting installations for • places open to the public • sporting areas • illumination of monuments and floodlighting • telephone kiosks, bus shelters, advertising panels and town plans • road signs • temporary festoon lighting For an outdoor lighting installation, the protective measure for the whole installation is by double or reinforced insulation Suspension devices for extra-low voltage luminaires, shall be capable of carrying five times the mass of the luminaires and the amount of support provided shall **not** be less than 5 kg	For more information about the requirements for outdoor lighting, please see BS 7671:2018 Chapter 714.

(continued)

Table 6.11 (Continued)

Category	*Main requirements*	*BS 7671:2018 detailed requirements*
Highway power supplies and street furniture	Access to luminaire light sources that are mounted less than 2.80 m above ground level, shall only be possible after removing a barrier or an enclosure requiring the use of a tool The protective measures of placing out of reach and obstacles shall not be used The earthing conductor of a street electrical fixture shall have a minimum copper equivalent cross-sectional area not less than that of the supply neutral conductor at that point or not less than 6 mm^2, whichever is the smaller	See BS 7671:2018 Chapter 714 for more detailed requirements

6.9 Heaters

Electrical heaters need to be prevented from exceeding the following temperatures:

- 90°C under normal conditions; and
- 115°C under fault conditions.

6.9.1 *Electrode water heaters and boilers*

Electrode water heaters and boilers:

- shall **only** be connected to an a.c. system that is controlled by a linked circuit-breaker;
- must include an RCD:
  - if its supply voltage exceeds low voltage;
  - If the circuit is supplying Class II heating equipment;

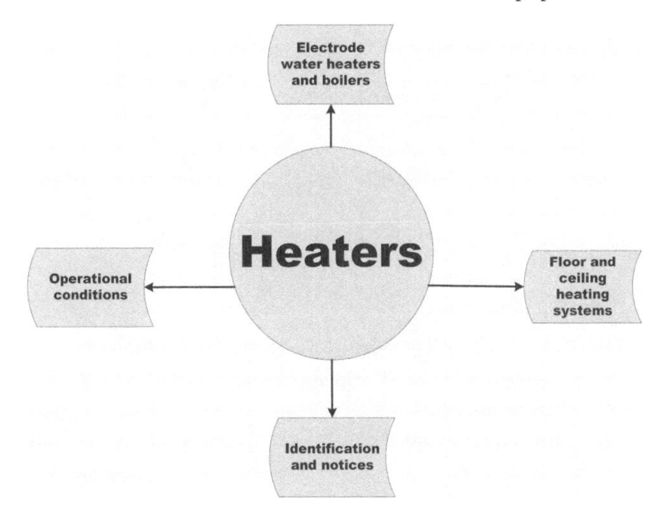

Figure 6.13 Heaters

RCDs that are time delayed are not permitted.

- the shell of the equipment needs to be:
    - bonded to the metallic sheath, or armour, of the incoming supply cable;
    - connected to the neutral of the supply as well as to the earthing conductor in a three-phase low voltage supply.

Liquid heaters shall have an automatic device to prevent a dangerous rise in temperature.

If the heater or boiler is installed in a room containing a fixed bath, it is essential that the switch complies with Section 701 of BS 7671:2018.

6.9.2 *Floor and ceiling heating systems*

- The protective measure 'protection by electrical separation' is not permitted for underfloor installations.
- Unless, the SELV protective measure is provided for the floor heating system, bathrooms need heating cables that either have a metal sheath, metal enclosure or a fine mesh metallic grid.

- Flexible sheet heating elements shall comply with the requirements of BS EN 60335-2-96.
- Heating cables shall comply with IEC 60800.
- Heating units used for installation:
    - in ceilings shall have a degree of protection of not less than IPX1;
    - in a floor of concrete (or similar material) shall have a degree of ingress protection not less than IPX7.
- Heating units shall be:
    - designed to avoid overheating the floor or ceiling;
    - connected to the electrical installation via cold tails or suitable terminals;
    - inseparably connected to cold tails (e.g. by a crimped connection).

6.9.3 *Identification and notices*

The designer of the installation/heating system or installer shall provide a plan (fixed to, or adjacent to, the distribution board of the heating system) for each heating system, containing the following details:

- heated area;
- layout of the heating units in the form of a sketch, a drawing, or a picture;
- length/area of heating units;
- manufacturer and type of heating units;
- number of heating units installed;
- position and depth of heating units;
- position of cables, earthed conductive, shields, etc.;
- position of junction boxes;
- product information and instructions for their installation;
- rated current of overcurrent protective device;
- rated residual operating current of RCD rated power;
- rated resistance (cold) of heating units;
- rated voltage;
- surface power density;
- the insulation resistance of the heating installation and the test voltage used;
- the leakage capacitance.

6.9.4 *Operational conditions*

- Precautions should be taken to avoid mechanically stressing the heating unit.
- Heating units shall not cross expansion joints of the building or structure.
- In floor areas where contact with skin or footwear is possible, the surface temperature of the floor shall be limited, for example, to 35°C).
- The following protective measures are not permitted:
    - placing out of reach;
    - non-conducting location and Earth-free local equipotential bonding;
    - electrical separation.
- Water heaters and boilers shall be permanently connected to the electricity supply via a double-pole linked switch which is either:
    - separate from and within easy reach of the heater/boiler; or
    - part of the boiler/heater.

Author's End Note

One of the difficulties in working with the 18th Edition of the Wiring Regulations is trying to find the full requirements for a particular type of cable or conductor that you might want to use. The main reason for this, of course, is that as there are so many different types of cables, conductors (not to mention, conduits, cable ducting, and cable trunking) for you to choose from and these are spread throughout the Standard's Chapters and Appendices. You could, therefore, quite easily miss an essential requirement in your research.

Granted, BS 7671:2018's enormous Index covers virtually everything, but unfortunately, as previously mentioned, there are so many cross references between different Sections of the Standard, that relevant points could easily be overlooked.

The aim of the next chapter (i.e. Seven), therefore, is an attempt to simplify the problem by listing the essential requirements into three main headings – namely 'cables', 'conductors and 'conduits'.

7 Cables, conductors and conduits

Author's Start Note

Within BS 7671:2018 there is frequent reference to different types of cables (e.g. single core, multicore, fixed, flexible, etc.) conductors (such as live supply, protective, bonding, etc.) and conduits, cable ducting, cable trunking and so on. Unfortunately, similar to equipment and components, the requirements for these items are liberally sprinkled throughout the Standard. The aim of this chapter, therefore, is to provide a catalogue of all the different types identified and referred to in the Wiring Regulations into three main headings (e.g. 'cables', 'conductors' and 'conduits') and then make a list of their essential requirements.

7.1 Cables

All cables should comply with the requirements of BS EN 50265-2-1 or 2-2 and a suitable notice board should erected as shown in Figure 7.1.

7.1.1 *Types of cable*

There are five main types of cables found in electrical installations, namely:

1 single core cables;
2 multicore cables;
3 flexible cables;
4 heating and warm floor cables;
5 fixed wiring.

DOI: 10.1201/9781003165170-7

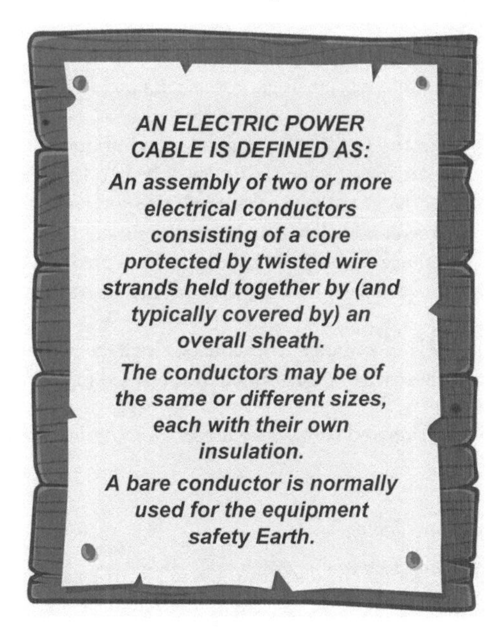

Figure 7.1 Definition of an electric power cable

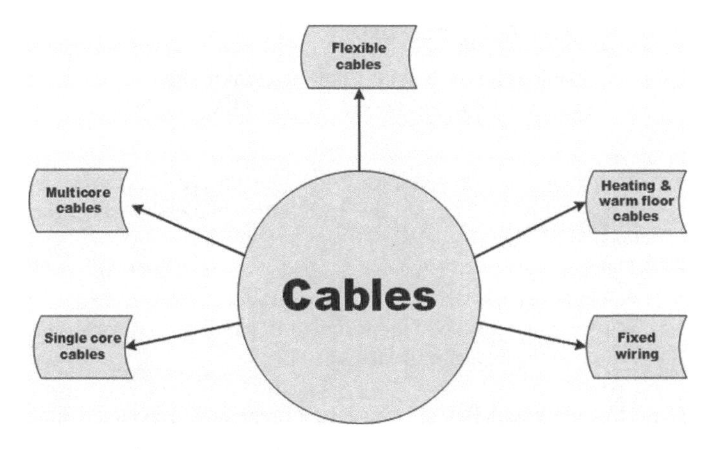

Figure 7.2 Cables

7.1.1.1 Single core cables

The general requirements for single core cables include:

- metallic sheaths and/or the non-magnetic armour of single-core cables that are in the same circuit, must be bonded together;
- single-core cables may be used as protective conductors but must be coloured **green**-and-**yellow** throughout their length;
- the conductance of the outer conductor of a concentric single core cable shall not be less than the internal conductor.

Owing to possible electromagnetic effects, single-core cables that are armoured with steel wire or tape should **not** be used for a.c. circuits.

For single-core armoured cables, the use of aluminium armour may be considered.

7.1.1.2 Multicore cables

For telecommunication and data transfer circuits, the possibility of electromagnetic and electrostatic, electrical interference (see BS EN 50081 and BS EN 50082 for further details) should be taken in to consideration. In addition:

- A Band I circuit must **not** be contained in the same wiring system as a Band II voltage circuit unless it is part of a multicore cable and the cores of the Band I circuit are:
  - insulated for the highest voltage present in the Band II circuit;
  - separated from the cores of the Band II circuit by an earthed metal screen.
- Separated Extra Low Voltage (SELV) circuit conductors that are contained in a multicore cable with other circuits should be insulated for the highest voltage present in that cable.
- The conductance of the outer conductor of a multicore cable should not be less than that of the internal conductors.
- Where multicore cables are installed in parallel, each cable shall contain one conductor of each line.

Note: In caravans and motor caravans, all protective conductors must be included in a multicore cable (or in a conduit) together with the live conductors. (For more details see BS 7671:2018 Chapter 721).

7.1.1.3 Flexible cables

- These cables may be used for equipment that is intended to be temporally moved whilst in use or for the purposes of cleaning, etc.
- They shall be of a heavy-duty type.
- They shall be suitably protected against mechanical damage.
- They shall include a protective bonding conductor.
- Flexible cables that are liable to mechanical damage shall be visible throughout their length.
- Flexible cables that are used as an overhead, low voltage, line shall comply with the relevant British or Harmonised Standard.

Non-flexible cables which are sheathed with lead, PVC or an elastomeric material, may include an overhead wire for aerial use or when the cable is suspended.

Note: In caravans and motor caravans, flexible cables shall **not** be laid in areas accessible to the public unless they are protected against mechanical damage (for more detailed requirements see BS 7671:2018 Chapter 711).

7.1.1.4 Floor warming and heating cables

Underfloor heating is a very cost-effective way to warm a room because it provides and distributes heat evenly and gently. The maximum conductor operating temperature for floor-warming cable is between 70 and 85°C depending on the type of cable used and there are no cold spots and very little heat is wasted.

The general requirements for heating cables contained in BS 7671:2018 include the following:

- Heating cables passing through (or in close proximity to) a fire hazard:
  - shall be enclosed in material with an ignitability characteristic 'P' as specified in BS 476 Part 5;
  - shall be protected from mechanical damage.
- A heating cable laid directly in soil, a road or the structure of a building, shall be installed so that it:
  - is completely embedded in the substance it is intended to heat;
  - does not suffer damage in the event of movement by the substance in which it is embedded;
  - is capable of withstanding external mechanical damage;
  - is resistant to damp and/or corrosion.

Note: In locations containing a bath or a shower, **only** heating cables (or thin sheet flexible heating elements of electric floor heating systems) which have either a metal sheath, metal enclosure or a fine mesh metallic grid, shall be provided.

7.1.1.5 Fixed wiring cables

- These are designed to be installed in a static position, fastened to a support or laid in a specific location.
- These are mostly used as power supply cables for sockets, switches and light fittings across residential, commercial and industrial environments.
- They are not suitable for mobile/static equipment.
- They shall be enclosed in conduit, ducting or trunking if non-sheathed cables are used for fixed wiring.

A full description of light, ordinary and heavy-duty types of fixed cables is contained in BS EN 50565-1.

Although BS 7671:2018 does not actually define 'fixed wiring', it does, nevertheless, stipulate a number of specific requirements as detailed below:

- equipment with a protective conductor current exceeding 3.5 mA (but not exceeding 10 mA) shall either be permanently connected to the fixed wiring of an installation or via a plug and socket-outlet complying with BS EN 60309-2;
- the termination of the wiring system at a fixed lighting point shall use one of the following:
  - a batten lampholder or a pendant set complying with BS EN 60598;
  - a box (enclosing the terminals of the fixed wiring) complying with the BS EN 60670 series or BS 4662;
  - a ceiling rose complying with BS 67;
  - a connection unit complying with BS 1363-4;
  - a device for connecting a luminaire (DCL) outlet complying with BS IEC 61995-1;
  - an installation coupler complying with BS EN 61535;
  - a luminaire complying with BS EN 60598;
  - a luminaire supporting coupler (LSC) complying with BS 6972 or BS 7001;

- a plug-in lighting distribution unit complying with BS 5733;
- a suitable socket-outlet complying with BS 1363-2, BS 546 or BS EN 60309-2;
- the cross-sectional areas of the fixed wiring should be such that the permissible voltage drop is not exceeded.

Note: In a caravan the connection between its internal fixed wiring should be via a terminal block, with a protective cover. (Complete wiring details for caravans are contained in BS 7671:2018 Chapter 721).

7.1.2 Cable Construction and manufacture

Cables and conduits are integral parts of an electrical system. Whilst the electrical cables are used for transmitting power, electrical conduits keep a bunch of cables safely together. Cable trunking on the other hand is an enclosure used to protect cables and provide space for other electrical equipment.

BS 7671:2018 contains the following requirements for electrical conduits, trunking and ducts:

- cable conduits shall comply with the appropriate part of BS EN 61386 series;
- cable tray and ladder systems shall comply with BS EN 61537;
- cable trunking or ducting shall comply with the appropriate part of the BS EN 50085 series;
- the cross-sectional area of each conductor shall be not less than the 1.5 mm;
- the cores of cables shall be by:
  - colour (see BS 7671:2008 Regulation 514.4); and/or
  - lettering and/or numbering (see BS 7671:2008 Regulation 514.5).

The two-colour combination of green-and-yellow shall **only** be used for identifying a protective conductor.

7.1.2.1 Mechanical stresses

- Underground distribution cables should be buried at a minimum depth of 0.5 m and shall include a protective conductor.
- A cable buried in the ground (but not installed in a conduit or duct) shall incorporate an earthed armoured or metal sheath suitable for use as a protective conductor.

- All buried cables need to avoid being damaged by any disturbance of the ground and should be clearly marked with cable covers or marking tape.
- Cables shall not be installed across a site road or a walkway unless adequate protection against mechanical damage is provided.
- Cables shall be supported so that it is not exposed to undue mechanical strain.

See IEC 61386-24 for further details concerning underground conduits.

A wiring system:

- shall be selected and erected to avoid damage to the sheath or insulation of cables and their terminations;
- shall ensure that the radius of every bend in the wiring system does not suffer damage and terminals are not stressed;
- intended to be drawn in or out of conductors shall have adequate means of access to allow this operation.

Note: Cables, busbars and other electrical conductors which pass across expansion joints should not cause damage to the electrical equipment.

7.1.3 Cables for special installations and locations

Where an electrical service is located in close proximity to one or more non-electrical services, it shall meet the following conditions:

- wiring systems shall be suitably protected against any hazards that are likely to arise from the presence of the other services during their normal use;
- cables and cable connections of a wiring system supported by (or fixed to) a structure or equipment that is subject to vibration, shall be suitable for such conditions;
- fault protection shall be provided by automatic disconnection of supply.

Cables that meet the requirements of BS EN 60332-1-2 can be installed without any special precautions.

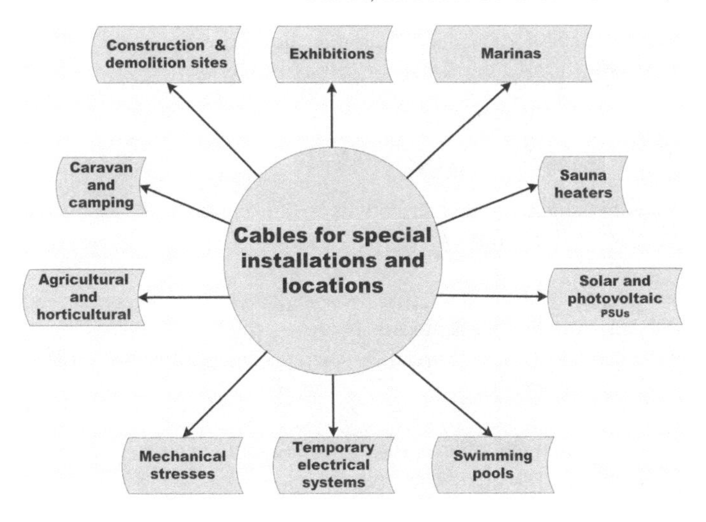

Figure 7.3 Cables for special installations and locations

7.1.3.1 Agricultural and horticultural premises

- In agricultural and horticultural premises, the external influences shall be classified AG3 and:
  - conduits shall provide a degree of protection against impact of 5 J according to BS EN 61386-2;
  - cable trunking and ducting systems shall provide a degree of protection against impact of 5 J according to BS EN 50085-2-1.
- Wiring systems shall be inaccessible to livestock and be suitably protected against mechanical damage and the presence of different kinds of fauna (e.g. rodents).
- Where vehicles and mobile agricultural machines are operating:
  - cables shall be buried in the ground at a depth of at least 0.6 m with added mechanical protection;
  - cables in arable or cultivated ground shall be buried at a depth of at least 1 m;
  - self-supporting suspension cables shall be installed at a height of at least 6 m.

See BS 7671:2018 Chapter 705 for more detailed requirements relating to cabling in agricultural and horticultural premises.

7.1.3.2 Caravan and camping parks

Note: In order not to mix regulations on different subjects, two sections are now available in BS 7671:2018:

- Section 708, for the electrical installations in caravan parks, camping parks and similar locations; and
- Section 721, for electrical installations designed for the caravans and motor caravans.

The following is a resume of the most important requirements form BS 7671:2018 for caravans and camping parks.

In a caravan and camping site, underground cables:

- shall be buried at a depth of at least 0.6 m;
- be placed outside any caravan pitch;
- be away from any surface where tent pegs or ground anchors are expected to be present;
- be protected against mechanical damage.

No more than four socket-outlets should be grouped in one location.

In caravans and motor caravans, (and to meet the requirements of BS EN 61386) wiring systems shall be installed using one or more of the following options:

- cables and flexible conduits shall be supported at 0.4 m intervals for vertical runs and 0.25 m for horizontal runs;
- cables shall, meet the requirements of BS EN 60332-1-2;
- non-metallic conduits shall comply with BS EN 61386-21;
- low voltage cable systems shall be run separately from the cables of extra-low voltage systems;
- cables that have to run through a compartment shall be installed within a conduit or ducting system so as to protect them against mechanical damage.

See BS 7671:2018 Chapters 708 and 721 for more detailed requirements.

7.1.3.3 Construction and demolition site installations

- Reduced low voltage systems low temperature, thermoplastic cable shall be used.
- Cables shall **not** be installed across a road or a walkway unless sufficiently protected against mechanical damage.
- Surface-run and overhead cables shall also be protected against mechanical damage.
- For applications exceeding reduced low voltage, a heavy-duty flexible cable shall be used.

See BS 7671:2018 Chapter 704 for more detailed requirements relating to construction and demolition site installations.

7.1.3.4 Exhibitions, shows and stands

To protect users, temporary electrical installations in exhibitions, shows and stands, shall meet the following requirements:

- cables shall be protected at their origin by an RCD;
- cables that are protected against mechanical damage shall be used;
- flexible cable shall **not** be laid in areas accessible to the public;
- insulation piercing lampholders shall **not** be used;
- joints shall not be made in cables except where necessary. Where joints are made, these shall be in enclosures with a degree of protection of at least IP4X or IPXXD;
- if no fire alarm system has been installed in a building which is going to be used for exhibitions, etc. The cable systems shall be either:
    - flame retardant to BS EN 60332-1-2 and low smoke to BS EN 61034-2; or
    - single-core or multicore unarmoured cables enclosed in metallic or non-metallic conduit or trunking, providing a degree of fire protection of at least IP4X.
- where strain can be transmitted to terminals the connection shall incorporate suitable cable anchorage(s);
- wiring cables shall be copper and have a minimum cross-sectional area of 1.5 mm^2.

See BS 7671:2018 Chapter 711 for more detailed requirements.

7.1.3.5 Marinas and similar locations

The following requirements do **not** apply to houseboats if these are directly supplied from the public network.

The following wiring systems are considered most suitable for marina distribution circuits:

- cables with copper conductors with thermoplastic or elastomeric insulation and sheaths;
- mineral-insulated cables with PVC protective covering;
- overhead cables or overhead insulated conductors;
- underground cables.

The following wiring systems shall **not** be used on or above a jetty, wharf, pier or pontoon:

- cables in free air suspended from a support wire;
- non-sheathed cables;
- cables with aluminium conductors;
- mineral insulated cables;
- Underground distribution cables (unless provided with additional mechanical protection, and buried at a depth of 0.5 m.

In addition:

- all overhead conductors shall be insulated;
- overhead conductors must be at least 6 m above ground level in all areas subjected to vehicle movement and 3.5 m in all other locations;
- poles and other supports for overhead wiring shall be located or protected so that they are unlikely to be damaged by unforeseen vehicle movement.

See BS 7671:2018 Chapter 709 for more detailed requirements Mobile or transportable units.

- flexible cables used to connect the unit to the supply shall have a minimum cross-sectional area of 2.5 mm^2 copper;
- flexible cables shall enter the unit by an insulated inlet;

- the wiring system shall be installed using either:
  - sheathed flexible cable thermoplastic or thermosetting insulation; or
  - unsheathed flexible cable installed in a conduit, trunking or ducting.
- Where cables have to run through a compartment, they shall be installed within a conduit system.

See BS 7671:2018 Chapter 717 for more details.

7.1.3.6 Rooms and cabins containing sauna heaters

In zone 3 of a room (or cabin containing a sauna heater) the insulation and sheaths of cables shall be capable of withstanding a minimum temperature of 170°C.

See BS 7671:2018 Chapter 703 for more detailed requirements

7.1.3.7 Solar and photovoltaic power supply systems

- If an electrical installation includes a PV power supply system without a simple separation between the a.c. and the d.c. side, an RCD shall be installed.
- The PV supply cable on the a.c. side shall be protected against fault current by an overcurrent protective device.
- Where protective bonding conductors are installed, they shall be parallel to, and in as close contact with, d.c. cables, a.c. cables and their accessories.
- Wiring systems shall be capable of withstanding expected external influences such as wind, ice formation, temperature and solar radiation.

7.1.3.8 Swimming pools

- In zones 0, 1 and 2 of a swimming pool, wiring systems with a metallic sheath or metallic covering shall be connected to the supplementary equipotential bonding.
- Cables should preferably be installed in conduits made of insulating material.

For a fountain, the following additional requirements shall be met:

- a cable for electrical equipment in zone 0 shall be installed as far outside the basin rim as is reasonably practicable;
- in zone 1, a cable shall be selected, installed and provided with mechanical protection to medium severity and the expected submersion in water depth.

See BS 7671:2018 Chapter 702 for more detailed requirements Swimming pools and other basins.

7.1.3.9 Temporary electrical systems

All parts of these installations shall conform to the relevant standards:

- all cables shall meet the requirements of BS EN 60332-1-2;
- cable trunking systems and cable ducting systems shall comply with part 2 of BS EN 50085;
- conduit systems shall comply with the BS EN 61386 series;
- tray and ladder systems shall comply with BS EN 61537.

Other wiring system requirements include:

- buried cables shall be protected against mechanical damage;
- joints shall **not** be made in cables except where necessary;
- the routes of cables buried in the ground shall be marked at suitable intervals;
- insulation-piercing lampholders shall **not** be used;
- the routing of cables in safe zones should be regularly checked for protection against mechanical damage.

The following applies to all temporary electrical systems in amusement parks, circuses and fairgrounds:

- cable trunking systems and cable ducting systems shall comply with the relevant part 2 of BS EN 50085;
- conduit systems shall comply with BS EN 61386 series.

See BS 7671:2018 Chapter 740 for more detailed requirements Inspection of cables.

7.1.3.10 Mechanical stresses

A wiring system:

- shall be selected and erected to avoid damage to the sheath, insulation of cables and their terminations during installation, use or maintenance.

A cable:

- buried in the ground shall include a protective conductor;
- buried cables should be clearly marked with cable covers or marking tape;
- shall be supported so that it is not exposed to undue mechanical strain.

See IEC 61386-24 for further details concerning underground conduits.

7.2 Conductors and conduits

Conductors:

- intended to operate at temperatures above 70 C shall **not** be connected to switchgear, protective devices, accessories or other types of equipment;
- shall **not** be subjected to excessive mechanical stress;
- **shall** be capable of withstanding all foreseen electromechanical forces during service.

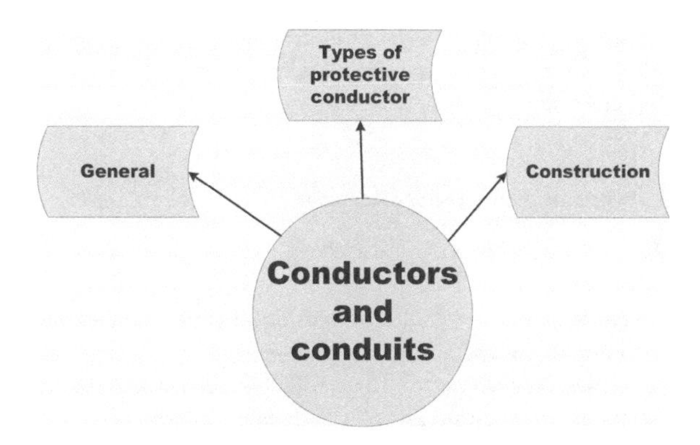

Figure 7.4 Conductors and conduits

7.2.1 General

The metallic covering of a cable, trunking, ducting and conduit, may be used as a protective conductor for the associated circuit.

A protective conductor may consist of one or more of the following:

- a conductor in a cable;
- a fixed bare or insulated conductor;
- a metal conduit, metallic cable management system or other enclosure;
- a single-core cable;
- an insulated or bare conductor in a common enclosure with insulated live conductors;
- an extraneous-conductive-part (provided that electrical continuity can be assured);
- a cable that has been suitably adapted, if necessary, to suit its use;
- the sheath, screen or armouring of a cable.

If the cross-sectional area of the protective conductors listed above is 10 mm^2 or less, it shall be of copper.

- If a metal enclosure or frame of a low voltage switchgear or controlgear assembly or busbar trunking system is used as a protective conductor its cross-sectional area shall be at in accordance with BS EN 60439-1.
- A protective conductor shall be identified by a bi-colour combination of green-and-yellow with one colour covering 30% of the surface and the other 70%.
- If a bare conductor or busbar is used as a protective conductor, they will be identified by equal green and yellow stripes between 15 and 100 mm wide.

7.2.2 Types of protective conductor

The Wiring Regulations emphasise that:

- gas pipes, oil pipes, flexible conduits or other flexible metallic parts shall **not** be used protective conductors;
- protective conductors shall be protected against mechanical and chemical deterioration and electrodynamic effects;
- exposed-conductive parts of equipment shall **not** be used as a protective conductor for other equipment.

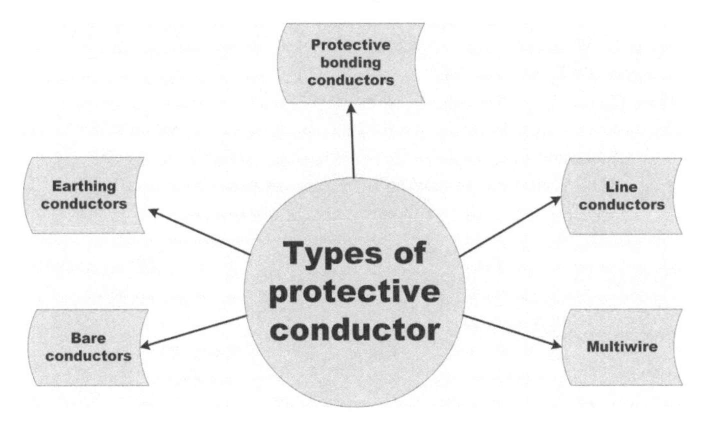

Figure 7.5 Types of protective conductor

If the protective conductor is not:

- a vital part of a cable; or
- ducting or trunking; or
- part of an enclosure formed by a wiring system;

then the cross-sectional area should not be less than:

- 2.5 mm^2 copper equivalent (if protection against mechanical damage is provided); and
- 4 mm^2 copper equivalent (if mechanical protection is not provided).

Where a protective conductor is common to two or more circuits, its cross-sectional area shall be:

- selected so as to correspond with the cross-sectional area of the largest line conductor of those circuits.

A protective conductor may consist of one or more of the following:

- a single-core cable;
- a conductor in a cable;

- an insulated or bare conductor in a common enclosure with insulated live conductors;
- a fixed bare or insulated conductor;
- a metal covering (for example, the sheath, screen or armouring) of a cable;
- a metal conduit connected by a separate protective conductor to an earthing terminal in the same box or enclosure.

Other flexible metallic parts, or constructional parts that are subject to mechanical stress whilst in normal service, shall **NOT** be used as a protective conductor.

7.2.2.1 Bare conductors

Bare conductors shall be protected against corrosion at their supports and whilst going through walls.

The minimum cross-sectional area of phase conductors in a.c. circuits and of live conductors in d.c. circuits shall be as shown in Table 7.1.

Bare conductors shall be painted or identified by a coloured tape, sleeve or disc as per Table 51 of BS 7671:2018.

If a bare conductor or busbar is used as a protective conductor, then it shall be identified by equal green and yellow stripes throughout the length of the conductor if adhesive tape is used, it shall also be bi-coloured in a similar manner.

Table 7.1 Minimum cross-sectional area of bare conductors

		Conductor	
Type of wiring system	*Use of circuit*	*Material*	*Minimum cross-sectional area*
Bare conductors	Power circuits	Copper	10 mm^2
		Aluminium	16 mm^2
	Signalling and control circuits	Copper	4 mm^2

7.2.2.2 Earthing conductors

In every installation, a main earthing terminal shall be provided to connect the following to the earthing conductor:

- circuit protective conductors;
- functional earthing conductors;
- lightning protection system bonding conductor;
- protective bonding conductors.

If a protective conductor forms part of a cable, then this shall **only** be earthed in the installation (See BS 7671:2018 Table 511 for further advice).

BS 7430 also includes the requirement that the connection of an earthing conductor to an Earth electrode shall be:

- soundly made;
- electrically and mechanically satisfactory;
- labelled; and
- suitably protected against corrosion.

7.2.2.3 Main protective bonding conductors

- A main protective bonding conductor shall have a cross-sectional area not less than half the cross-sectional area required for the earthing conductor of the installation not less than 6 mm^2.

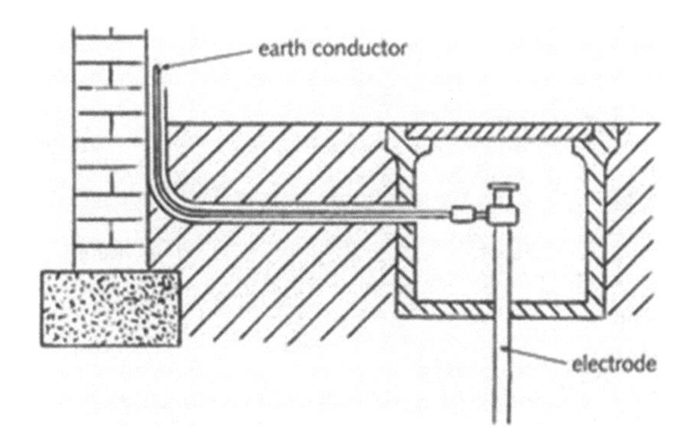

Figure 7.6 Typical earthing conductor

Source: Courtesy Stingray

Except for highway power supplies and street furniture the main protective bonding conductor shall be selected in accordance with the neutral conductor of the supply and BS 7671:2018 Table 54.8.

7.2.2.4 Line conductors

All line conductors will include an overcurrent detection facility live conductors

From the point of view of safety:

- bare live conductors shall be installed on insulators;
- conductors shall be able to carry fault current without overheating;
- live supply conductors **shall** be capable of being isolated from circuits;
- the supply to all live conductors **shall** be automatically interrupted in the event of an overload or fault current;
- persons and livestock **shall** be protected against injury; and
- property shall be protected against damage, due to excessive caused by any overcurrents likely to arise in live conductors.

7.2.2.5 Multiwire, fine wire and very fine wire conductors

- Terminals shall be used at the end of a conductor to avoid the possible separation or spreading of individual wires of multiwire, fine wire or very fine wire conductors.

7.2.3 Construction

7.2.3.1 Automatic disconnection in case of a fault

- Automatic Disconnection of Supply (ADS) is the most widely used method for achieving protection against electric shock in an installation.

Disconnection is not just required for protection against electric shock but may be required for other reasons, such as protection against thermal effects, (see BS 7671:2018 Table 41.1) but where automatic disconnection cannot be achieved in the time required, supplementary equipotential bonding shall be provided.

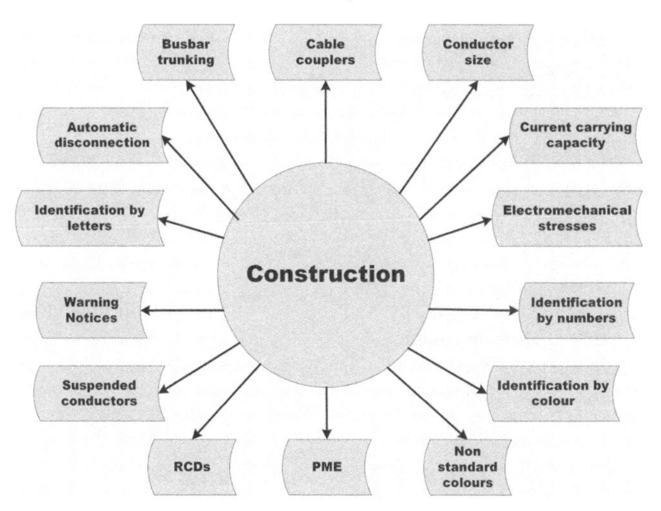

Figure 7.7 Construction

7.2.3.2 Busbar trunking

Where a busbar trunking system is used as a protective conductor:

- it shall be protected against insulation faults;
- its electrical continuity shall be assured;
- its cross-sectional area shall be at in accordance with BS EN 60439-1.

7.2.3.3 Cable couplers

- A cable coupler shall be arranged so that the connector of the coupler is fitted at the end of the cable which is remote from the supply.

7.2.3.4 Cross-sectional area of conductors

The cross-sectional area of conductors shall be according to:

- the admissible maximum temperature;
- the admissible voltage drop limit;
- the electromechanical stresses likely to occur due to short-circuit and Earth fault currents;

- other mechanical stresses to which the conductors are likely to be exposed;
- the method of installation;
- thermal insulation.

The cross-sectional area of a phase conductor in an a.c. circuit or of a live conductor in a d.c. circuit shall be as shown in Tables 52.3 and 55.2 of BS 7671:2018.

- The neutral conductor shall have a cross-sectional greater than that of the line conductor.
- In caravans and motor caravans, the cross-sectional area of every conductor shall be not less than 1.5 mm^2.

7.2.3.5 Current-carrying capacities of conductors

- The current (including any harmonic current) to be carried by a conductor during normal operation varies according to the type of insulation used.

For example, the thermoplastic insulation temperature limit would be 70°C at the conductor, whilst mineral based insulation would be 105°C dependent on whether it was exposed or not exposed to touch.

7.2.3.6 Electromechanical stresses

- Every conductor or cable shall have adequate strength and be installed so as to withstand the electromechanical forces that may be caused by any current (including fault current) it may have to carry whilst in service.

7.2.3.7 Identification of conductors by letters and/or numbers

The lettering or numbering system identifying conductors:

- shall be clear, legible and durable;
- shall be identified by numbers (0 is reserved for the neutral or mid-point conductor);
- shall be given in letters or Arabic numerals (in order to avoid confusion, unattached numerals 6 and 9 shall be underlined).

7.2.3.8 Identification of conductors – By colour

Whilst Table 51 of BS 7671:2018 provides full details of conductor colours the main ones to remember are conductors with **green**-and-**yellow** colour identification shall **not** be numbered other than for the purpose of circuit identification.

The single colour **green** shall **not** be used for live conductors, protective conductors, functional earthing or bonding conductors.

- There are other certain variations in conductor colours, and these are shown in Table 7.2.

All other conductors shall be identified by colour in accordance with BS 7671:2018 Table 51.

7.2.3.9 Installations with non-standard colours

If wiring additions or alterations are made to an installation so that some of the wiring complies with the current Regulations but there is also wiring to previous versions of these Regulations:

- a warning notice shall be affixed at or near the appropriate distribution board with the wording shown in Fig 7.8:

Table 7.2 Variance in conductor colours

Type of conductor	*Identification*
Bare conductors	A bare conductor is identified by the use of tape, sleeve or disc of the appropriate colour prescribed in BS 7671:2018 Table 51 or by painting it with the same colour
Neutral or midpoint conductor	Where a circuit includes a neutral or midpoint conductor, the colour used shall be **blue**
PEN conductor	PEN conductors shall be marked by one of the following methods: • green-and-yellow throughout its length and with blue markings at the terminations • blue throughout its length with green-and-yellow markings at the terminations

Figure 7.8 Warning notice – non-standard colours

7.2.3.10 *Protective multiple earthing*

- Where protective multiple earthing (PME) exists, the minimum cross-sectional area (see Table 7.3) of the main equipotential bonding conductor in relation to the neutral shall be in accordance with BS 7671:2018 Table 54.8.

For example,

In some cases, the local distributor's network conditions may require a larger conductor.

Table 7.3 Cross-sectional area of the main equipotential bonding conductor

Copper equivalent cross-sectional area of the supply neutral conductor	*Minimum copper equivalent cross-sectional area of the main equipotential bonding conductor*
35 mm² or less	10 mm²
over 35 mm² up to 50 mm²	16 mm²

7.2.3.11 RCDs

- An RCD shall be selected and erected so as to limit the risk of unwanted tripping and should be capable of disconnecting all the line conductors of the circuit at substantially the same time.

7.2.3.12 Suspended conductors

- Extra-low voltage luminaires suspension devices and supporting conductors shall be capable of carrying five times the mass of the luminaires including their lamps intended to be supported, but not less than 5 kg.
- The suspended system shall be fixed to walls or ceilings by insulated distance cleats.

7.2.3.13 Conductor warning notices

7.2.3.13.1 SAFETY EARTH

- A warning notice (as shown in Figure 7.9) shall be permanently fixed at the point of connection of every earthing conductor to an Earth electrode and every bonding conductor to an extraneous-conductive-part, as well as the main Earth terminal (when separated from the main switchgear).

Figure 7.9 Warning notice – earthing and bonding

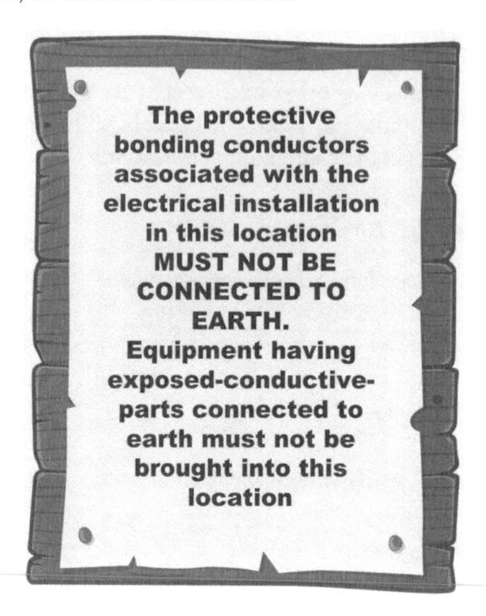

Figure 7.10 Warning notice – protective bonding conductors

7.2.3.13.2 ELECTRICAL SEPARATION

Where electrical separation from the supply to more than one current using equipment is used, the warning notice shall read as follows as shown in Fig 7.10:

7.2.3.13.3 ALTERNATIVE SUPPLIES

Where an installation includes alternative or additional sources of supply, a warning notice shall be displayed at the following locations:

- at all points that isolate the supply source;
- the origin of the installation;
- the meter position, if remote from the origin;
- the consumer unit or distribution board to which the alternative or additional sources are connected.

7.3 Installation

7.3.1 Installation of cables

- Cables should be strong enough to withstand the electromechanical forces that, it may have to carry whilst in service.

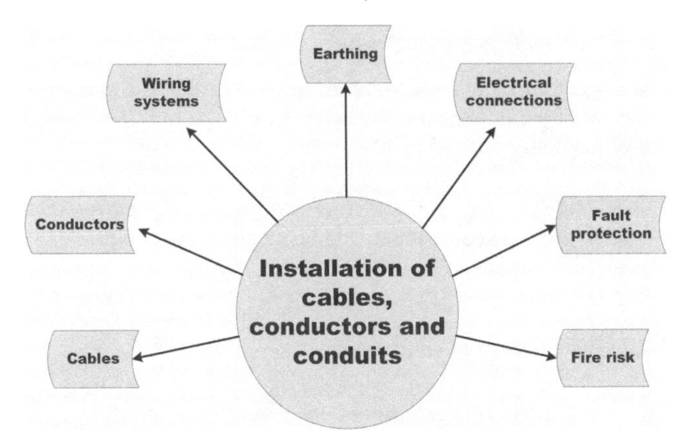

Figure 7.11 Installation of cables, conductors and conduits

All bare, live, cables **must** be installed on insulators.

- Non-sheathed cables: for fixed wiring, shall be enclosed in conduit, ducting or trunking which provides a minimum of IP4X or IPXXD protection.

7.3.1.1 Cable couplers

If a cable coupler has been fitted at the end of the cable that is remote from the supply, then every cable coupler of a reduced low voltage system shall:

- have a protective conductor contact; and
- be dimensionally incompatible with all other couplers used for other systems in that particular premises.

On a FELV system the coupler:

- shall have a protective conductor contact; and
- shall not be dimensionally compatible with those used for any other system used in the same premises.

Note: For a SELV system, or a Class II circuit, a cable coupler shall be non-reversible and shall comply (where appropriate) with BS 6991, BS EN 61535, BS EN 60309-2 or BS EN 60320-1.

7.3.1.2 Current-carrying capacities of cables

The current carrying capacity of an insulated cable is the maximum current that it can continuously carry without exceeding its temperature rating, known as '*ampacity*':

- in compliance with the Wiring Regulations, the cross-sectional areas of the fixed wiring should be such that the permissible voltage drop is not exceeded.

7.3.2 Installation of conductors

7.3.2.1 Electrical connections to bare connectors and/or busbars

Where a cable is to be connected to a bare conductor or busbar, the type of insulation and/or sheath shall be chosen with respect to the maximum operating temperature of the bare conductor or busbar.

A busbar trunking or a powertrack system:

- shall take account of external influences. (See Appendix 8 of BS 7671:2018 for further details);
- shall comply with BS EN 60439-6; and
- a powertrack system shall comply with BS EN 61534 series.

7.3.2.2 Bare conductors

If the nominal voltage does not exceed 25 V a.c. or 60 V d.c., bare conductors may be used for extra-low voltage lighting installations, provided that:

- that the risk of a short-circuit is reduced to a minimum;
- the conductors have a cross-sectional area of at least 4 mm^2;
- the conductors are not placed directly on combustible material.

7.3.2.3 Bonding conductors

In agricultural and horticultural premises, protective bonding conductors shall be protected against mechanical damage and corrosion and selected to avoid electrolytic effects.

- A warning notice – see Figure 7.12 – shall be permanently fixed at or near the bonding conductor's connection point to an extraneous part.

Figure 7.12 Warning notice – where protective bonding conductors are installed

7.3.2.4 Connecting conductors

Connecting conductors shall meet the following requirements:

- bare live conductors shall be installed on insulators;
- circuit protective conductors shall be run to (and terminated at) each point in a wiring circuit supplying one or more items of Class II equipment;
- an IMD shall be connected between Earth and a live conductor of the monitored equipment;
- the 'line' terminal(s) of an IMD shall be connected as close as practicable to the origin of the system.

For d.c. installations, the 'line' terminal(s) of the IMD must either be connected to the midpoint or one or the supply conductors.

In some d.c. two-conductor IT installations, a passive IMD may be used, provided that:

- the insulation of all live distributed conductors is monitored;

- all exposed-conductive-parts of the installation are interconnected;
- circuit conductors are selected and installed so as to reduce the risk of an Earth fault to a minimum.

7.3.2.5 Conductors in parallel

Where two or more live conductors or PEN conductors are connected in parallel in a system;

- they shall be of the same material, correctional area and achieve equal load current sharing between them.

Typical PEN conductors are:

- multicore cables;
- twisted single-core cables;
- non-sheathed cables;
- non-twisted single-core cables;
- non-sheathed cables in flat formation and where the cross-sectional area is greater than 50 mm² in copper or 70 mm² in aluminium.

7.3.2.6 Connection of multi-wire, fine wire and very fine wire conductors

To avoid undesirable separation or spreading of individual wires, terminals with suitably treated conductor ends should be used.

BS 7671:2018 stipulates that:

- non-sheathed cables at the termination of a conduit, ducting or trunking, shall be enclosed;
- soldering (tinning) the end of conductors is not permitted if screw terminals are used.

7.3.3 *Installation of a wiring system*

The installation of wiring systems will depend on the type of conductor or cable used (see Table 4A1 of Appendix 4 of BS 7671:2018 for compete details) and:

- The minimum cross-sectional area of the extra-low voltage conductors shall normally be 1.5 mm² copper.

A voltage Band I circuit shall not be contained in the same wiring system as a Band II circuit unless:

- every cable (and each conductor of a multicore cable) is insulated against the highest voltage present;
- for a multicore cable, the cores of the Band I circuit are separated from the cores of the Band II circuit by an earthed metal screen of equivalent current-carrying capacity of the largest core of a Band II circuit.

7.3.4 *Earthing*

Earth electrodes are specifically designed and installed to improve a systems earthing and as previously covered in Part 3 of this book ('Earthing'), the following types of Earth electrode are recognised as being suitable for the purposes of these Regulations:

- Earth rods or pipes;
- Earth tapes or wires;
- Earth plates;
- underground structural metalwork embedded in foundations;
- welded metal reinforcement of concrete (except pre-stressed concrete) embedded in the Earth;
- lead sheaths and other metal coverings of cables;
- other suitable underground metalwork.

If the lead sheath or other metal covering of a cable is used as an Earth electrode, it will be subject to all of the following conditions:

- precautions have been taken to prevent excessive deterioration by corrosion;
- the sheath or covering shall be in effective contact with Earth.

Further information on Earth electrodes can be found in BS 7430.

7.3.4.1 Earthing arrangements for protective conductors

The earthing of protective conductors shall ensure that:

- they meet the protective and functional requirements of the installation;
- are continuously effective;

- Earth fault currents and protective conductor currents that occur are carried without danger;
- the protective conductors are sufficiently robust against external influences.

Where the supply to an installation is at high voltage, protection against faults between the high voltage supply and Earth **shall** be provided.

If a number of installations have separate earthing arrangements, protective conductors common to another installation shall either:

- be able to carry the maximum fault current;
- have only one installation earthed and be insulated from the earthing arrangements of any other installations.

For a TN-S system, the main earthing terminal of the installation needs to be connected to the earthed point of the source of energy.

For a TN-C-S system, where protective multiple earthing is provided; the main earthing terminal of the installation needs to be connected by the distributor to the neutral of the source of energy.

For a TT or IT system, the main earthing terminal shall be connected via an earthing conductor to an Earth electrode.

7.3.4.2 Earthing terminal

- The earthing terminal of each accessory shall be connected by a separate protective conductor.
- The earthing conductor needs to be easily disconnected during maintenance and/or repair.

7.3.4.3 Earthing requirements for the installation of equipment having high protective conductor currents

Equipment with a protective conductor current:

- above 3.5 mA but less than 10 mA, may either be permanently connected to the fixed wiring of the installation or connected by means of a plug and socket-outlet complying with BS EN 60309-2;
- exceeding 10 mA shall be permanently connected to the supply or via a protective conductor with an Earth monitoring system.

Note: Information must be provided indicating which circuits have a high protective conductor current at the distribution board.

7.3.5 Electrical connections

Electrical connections and joints must be accessible for inspection, except for:

- a joint buried in the ground;
- a joint made by welding, soldering or brazing;
- an encapsulated joint;
- floor heating systems;
- equipment complying with the requirements of BS 5733 for a maintenance free accessory and marked with an MF symbol.

Terminations and joints in live conductor or a PEN conductor shall be made within an enclosure that provides adequate mechanical protection and protection against relevant external influences.

There must be no appreciable mechanical strain on the connections of conductors.

7.3.5.1 Temperature

To reduce the risk of injury resulting from excessive temperatures emanating from electrical installations:

- cables or flexible cords must be selected which are suitable for the temperatures likely to be encountered;
- cables should preferably **not**, be installed in a location where they are liable to be covered by thermal insulation – such as in a loft.

7.3.5.2 Electric floor heating systems

- Heating cables in locations such as a bathroom shall have either a metal sheath, metal enclosure or a fine mesh metallic grid connected to the protective conductor of the supply circuit.

The protective measure 'protection by electrical separation' is **not** permitted in electric floor heating systems in locations containing a bath or shower.

7.3.5.3 Electrode water heaters and boilers

- the protective conductor shall be connected to the shell of the electrode water heater or electrode boiler; and
- the shell shall be bonded to the metallic sheath or armour of the incoming supply cable; and

if an electrode water heater or electrode boiler

- has an uninsulated heating element immersed in the water, it shall not have a single-pole switch, non-linked circuit-breaker or fuse fitted in the neutral conductor, in any part of the circuit between the heater or boiler, or in the origin of the installation.
- is connected to a three-phase low voltage supply, the equipment's shell shall be connected to the supplies neutral as well as to the earthing conductor.
- is not in physical contact with any earthed metal, a fuse in the line conductor may be substituted for the circuit-breaker and the equipment's shell need not be connected to the neutral of the supply.

7.3.5.4 Enclosures

- All conductive parts within an insulating enclosure shall be connected to a protective conductor.
- Enclosures containing rectifiers and transformers shall be adequately ventilated.

7.3.6 Fault protection

- Exposed-conductive-parts of the separated circuit shall be connected to the protective conductor, the exposed-conductive-parts of other circuits, or to the Earth.
- Fault protection by automatic disconnection of supply shall be provided either by an overcurrent protective device in each line conductor or by an RCD.
- Live parts of a separated circuit shall **not** be connected to any other circuit, Earth or to a protective conductor.

7.3.6.1 *Protection against fault current*

Any fault occurring in a circuit must be stopped as soon as possible so that the fault current does not exceed the agreed conductor or cable temperature limits.

- so long as the protective device operates immediately the fault occurs, a single protective device can be used to may protect parallel conductors.

Provided, that the wiring minimises the risk of a fault occurring, a device for protection against fault current does **not** need to be provided for:

- a conductor connecting a generator, transformer, rectifier or an accumulator battery to a control panel which already has a protective device installed;
- a circuit (if disconnection would cause danger for the operation of the installation);
- certain measuring circuits;
- the origin of an installation.

7.3.6.2 *Ferromagnetic enclosures: Electromagnetic effects*

- a.c. circuit conductors that are installed in ferromagnetic enclosures must be organised so that the line conductors, the neutral conductor (if any) and the protective conductor are all kept in the same enclosure.

7.3.6.3 *Final circuits*

- Final circuit are deemed to be acceptable if:
    - a ring final circuit has a ring protective conductor;
    - a radial final circuit has a single protective conductor.
- Final circuits containing a number of socket-outlets or connection units that are intended to supply two or more items of equipment shall be provided with a high integrity protective conductor connection.

7.3.7 *Fire risk*

A fire risk assessment (see Figure 7.13) is conducted in five key steps:

1. identify the fire hazards;
2. identify equipment and/or systems at risk;
3. evaluate, remove or reduce the risks;
4. record your findings, prepare an emergency plan and provide training;
5. review and update the fire risk assessment regularly.

It is recommended that, wherever possible, every termination of a live conductor or connection or joint between live conductors is contained within an enclosure. (BS 7671:2018 para 422.3.13).

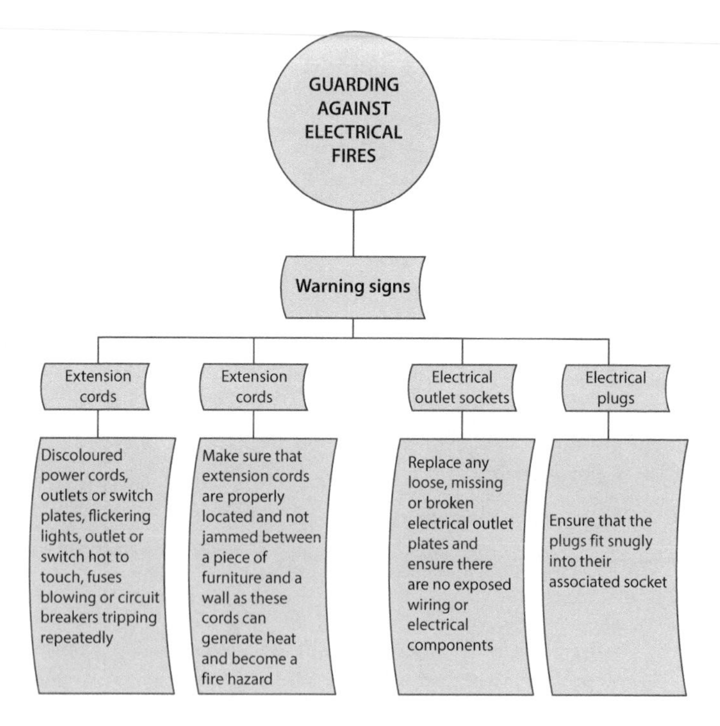

Figure 7.13 Basic fire precautions

The following are a few more of the most relevant stated requirements to be found in BS 7671:2018:

- every circuit shall be provided with a means of isolation from all live supply conductors by a linked switch or a linked circuit-breaker;
- preventive protection measures against the risk of fire occurring must be investigated at the earliest opportunity;
- some form of safety precaution should be installed at the origin of the circuit to be protected. These could include:
    - an RCD (for overall protection against the risk of fire; or for protection against the risk of fire in IT systems);
    - an RCM; or
    - an IMD; or
    - an AFDD.
- where practicable, the passage of circuits through locations presenting a fire risk should be avoided;
- locations where a fire risk does exist, conductors of circuits supplied at extra-low voltage should be protected either by barriers or enclosures;
- luminaires and floodlights should be fixed and protected so that a focusing or concentration of heat is not likely to cause ignition of any material;
- a wiring system which passes through the location but is not intended to supply electrical equipment within that location shall:
    - have no connection or joints in the location; and
    - is protected against overcurrent; and
    - does not use bare live conductors.
- a PEN conductor shall not be used unless it is a circuit traversing the location.

7.3.7.1 Fire precautions

Note: Within England and Wales, fire precautions are covered by the Regulatory Reform (Fire Safety) Order 2005. In Scotland, general fire safety requirements are covered in Part 3 of the Fire (Scotland) Act 2005.

The aim of these two Acts is to:

- reduce the likelihood of a fire occurring in a building;
- preventing its spread of a fire if an incident does occur; more specifically; however
- they are aimed at protecting both people, buildings and their contents from damage.

Other chapters in this book have explained in detail how BS 7671:2018 addresses the necessary precautions that need to be taken for different aspects of the Building and Wiring Regulations, but for the installation of cables and conductors the following situations also need to be considered:

- installing cables with improved fire-resisting characteristics in case of a fire hazard;
- installation of cables in non-combustible solid walls, ceilings and floors;
- installing cables in areas with constructional partitions which have a fire-resisting time capability of 30 minutes or 90 minutes;
- installing mineral insulated cables according to BS EN 60702.

7.3.7.2 *Precautions within a fire-segregated compartment*

- Cables that meet the requirements of BS EN 60332-1-2 can be installed without special precautions.

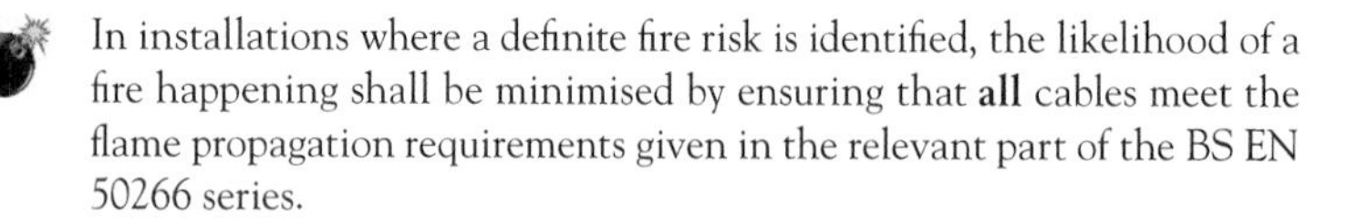

In installations where a definite fire risk is identified, the likelihood of a fire happening shall be minimised by ensuring that **all** cables meet the flame propagation requirements given in the relevant part of the BS EN 50266 series.

7.3.7.3 *Locations with risks of fire due to the nature of processed or stored materials*

If the possibility of fire owing to the manufacture, processing and or storage of flammable materials (usually referred to as BE2 conditions) exists in areas such as barns, woodworking facilities, paper mills and textile factories then cables shall, as a minimum, meet the flame propagation characteristics specified in BS EN 60332-1-2 and:

- a cable trunking system or cable ducting system needs to satisfy the fire conditions specified in BS EN 50085;
- precautions need to be taken to ensure that a cable or wiring system cannot propagate flame;
- a cable tray system or cable ladder must meet the requirements of BSEN 61537.

Except for mineral insulated cables, a wiring system shall be protected against insulation faults:

- in a TN or TT system, by an RCD having a rated residual operating current (IA) not exceeding 300 mA;
- in an IT system, by an insulation monitoring device with audible and visual signals.

Flexible cables shall either be:

- a heavy-duty type; or
- suitably protected against mechanical damage.

7.4 Safety precautions

7.4.1 Fuses

A fuse is an electrical safety device that operates to provide overcurrent protection of an electrical circuit. Its essential component is a metal wire or strip that melts when too much current flows through it, thereby interrupting the current. The Regulations clearly state that:

- fuse links shall preferably be of the cartridge type;
- a fuse carrier shall be arranged so that there is no contact between conductive parts belonging to two adjacent fuse bases; and
- a fuse base using screw-in fuses shall be connected so that the centre contact is connected to the conductor from the supply and the shell contact is connected to the conductor to the load.

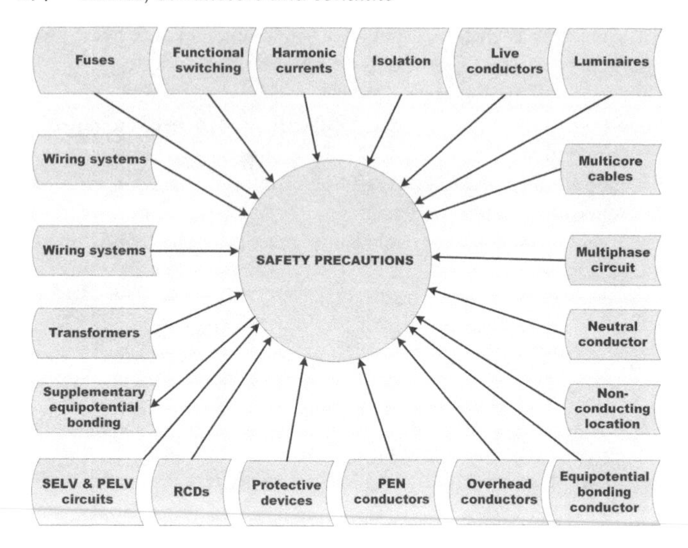

Figure 7.14 Safety precautions

7.4.2 *Functional switching*

Functional switching and functional switching devices are used in the normal operation of circuits to switch them on and off and:

- shall be provided for every part of a circuit that needs to be controlled independently from another part or parts of the installation; but
- may control several items of current-using equipment intended to operate simultaneously;
- need not necessarily be designed to control or switch off **all** live conductors of a circuit;
- may control the current without necessarily opening the corresponding poles.

Note: Equipment intended for protection only shall not be provided for functional switching of circuits.

7.4.3 *Harmonic currents*

Harmonic currents, generated by non-linear electronic loads, reduce system efficiency, cause apparatus overheating, and can have a significant impact on electrical distribution systems and the facilities they feed. To guard against this happening:

- overcurrent detection devices need to be provided for the neutral conductor in a multiphase circuit.

7.4.4 *Isolation*

Isolation is intended (for safety reasons) to make dead a circuit by separating an installation or section from every source of electric energy. For this reason:

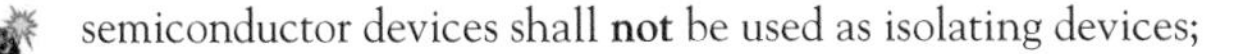

semiconductor devices shall **not** be used as isolating devices;

- every circuit shall be capable of being isolated from each live supply conductor and for disconnecting the neutral conductor;
- where an isolating device for a particular circuit is remote from the equipment to be isolated, the means of isolation should always be secured in the open position;
- if a switch is provided for this purpose:
    - it shall be capable of cutting off the full load current of the installation; and
    - In temporary electrical installations (marinas and similar locations) this switching device shall disconnect all live conductors including the neutral conductor.

7.4.4.1 Devices for isolation

In accordance with the Building Regulations, *the electrical installation of each building or part of a building shall be isolated by a single isolation device* and the Wiring Regulations state that, *every circuit shall be provided with a means of isolation from all live supply conductors by a linked switch or a linked circuit-breaker.* These requirements shall be achieved as follows:

- a means of isolation shall be installed on both sides of a static convertor;
- an isolation device shall isolate all live supply conductors from the circuit concerned;
- all fixed electric motors shall be provided with a readily accessible, efficient, means of switching off;

- isolation devices shall be designed and installed so as to prevent unintentional or inadvertent closure;
- isolation devices shall not be erected in agricultural and horticultural;
- means of isolation shall preferably be provided by a multipole switching device; however
- in TN-C-S and TN-S systems, isolation or switching of the neutral conductor is not required if protective equipotential bonding is installed.

Semiconductor devices shall **not** be used as isolating devices and no means of isolation or switching shall be inserted in the outer conductor of a concentric cable.

BS 7671:2018 Table 537.4 provides detailed guidance on the selection of protective, isolation and switching devices.

7.4.5 *Live conductors*

7.4.5.1 Some of BS 7671:2018's requirements and advice

The Regulations are full of advice about live wire conductors and the following are (i.e. in the Author's opinion) probably the most important ones:

- a main switch that is intended to be operated by ordinary persons shall interrupt both live conductors of a single-phase supply;
- bare live conductors shall be installed on insulators;
- conductors shall be able to carry fault current without overheating;
- in agricultural and horticultural premises:
  - the electrical installation of each building shall be isolated by a single isolation device;
  - means of isolation of all live conductors should also be provided.
- in solar PV power supply systems, earthing of one of the live conductors of the d.c. side is permitted;
- in Zone 1 of swimming pools (and other basins) the supply circuit of the equipment shall be protected by:
  - SELV; or
  - an RCD; or
  - electrical separation.

- live supply conductors **shall** be capable of being isolated from circuits;
- the supply to all live conductors **shall** be automatically interrupted if there is an overload current and fault current;
- where an installation is supplied from more than one source of energy, a switch may be inserted in the connection between the neutral point and the means of;
- where an IT system is designed **not** to disconnect in the event of a first fault, the occurrence of the first fault shall be indicated by either:
    - an IMD combined with an insulation fault location system (IFLS); or
    - an RCM.

Where no neutral point or midpoint exists, a line conductor may be connected to the Earth.

7.4.6 *Luminaires*

- In a TN or TT system (less E14 and E27 lampholders) the outer contact of every Edison screw or single centre bayonet cap type lampholder shall be connected to the neutral conductor.
- Groups of luminaires with only one common neutral conductor shall be provided with a device that simultaneously disconnects all line conductors.
- Extra-low voltage luminaires suspension devices and supporting conductors shall be capable of carrying not less than 5 kg.

7.4.7 *Multicore cables, conduits, ducting systems, franking systems or tray or ladder systems*

- If multicore cables are installed in parallel, each cable shall contain one conductor of each line.
- The line and neutral conductors of each final circuit shall be electrically separate from those of every other final circuit.
- Each part of a circuit shall be arranged so that the conductors are not distributed over different multicore cables, conduits, ducting, trunking, tray or ladder systems.

7.4.8 *Multiphase circuit*

- In multiphase circuits [a single-pole switching or protective device] shall **not** be inserted in the neutral conductor.

7.4.9 Neutral conductor

Consideration should be given to the fact that:

- if the neutral conductor in a three-phase TN or TT system is interrupted, the reinforced insulation can be temporarily stressed with the line-to-line voltage;
- if a line conductor of an IT system is accidentally earthed, insulation or components can be temporarily stressed with the line-to-line voltage;
- if a short circuit occurs in the low voltage installation between a line conductor and the neutral conductor, the voltage between the other line conductors and the neutral conductor will be affected.

To guard against the above problems, the following Regulations should be observed:

- the neutral conductor shall **not** be smaller than the line conductors;
- in an IT system, the neutral conductor shall **not** be distributed unless:
  - overcurrent detection is provided;
  - the neutral conductor is effectively protected against short-circuit by a protective device;
  - the circuit is protected by an RCD;
- in a TN-S or TN-C-S system the neutral conductor need not be isolated or switched;
- in temporary electrical installations, the neutral conductor of the star-point of the generator shall, except for an IT system, be connected to the exposed conductive-parts of the generator;
- if an autotransformer is connected to a circuit having a neutral conductor, the common terminal of the winding shall be connected to the neutral conductor.

7.4.10 Non-conducting location

In a non-conducting location, there is no need for a protective conductor.

7.4.11 Non-earthed equipotential bonding conductor

- Source supplies may supply more than one item of equipment provided that all exposed-conductive-parts of the separated circuit are connected together by an insulated and non-earthed equipotential bonding conductor.

7.4.12 *Overhead conductors*

- In caravan and camping parks; marinas and similar locations, all overhead conductors:
  - shall be insulated;
  - shall be at a height above ground of not less than 6 m in all areas subject to vehicle movement and 3.5 m in all other areas;
  - poles and other supports for overhead wiring shall be located so that they are unlikely to be damaged by vehicle movement.

See BS 7671:2018 Chapter 709 for more detailed requirements relating to marinas and similar locations.

7.4.13 *PEN conductors*

The following conditions are applicable to all PEN conductors:

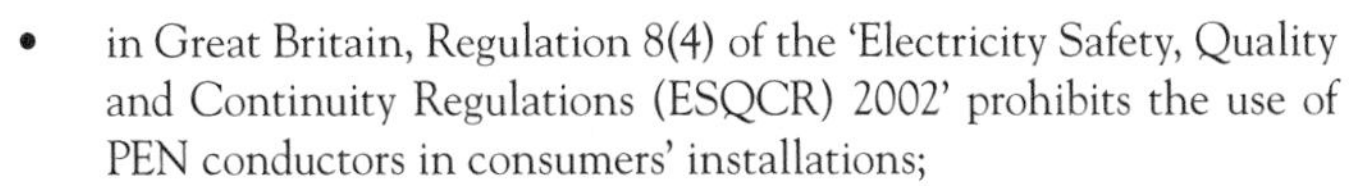

- in Great Britain, Regulation 8(4) of the 'Electricity Safety, Quality and Continuity Regulations (ESQCR) 2002' prohibits the use of PEN conductors in consumers' installations;
- PEN conductors may only be used within an installation where the supply is obtained from a private generating plant;
- automatic disconnection using an RCD shall **not** be applied to a circuit incorporating a PEN conductor;
- a separate metal cable enclosure shall **not** be used as a PEN conductor;
- PEN conductors shall **not** be isolated or switched;
- socket-outlet protective conductors in caravan and camping parks shall **not** be connected to any PEN conductor of the electricity supply;
- PEN conductors shall **not** be used in medical locations and medical buildings downstream of the main distribution board;
- in temporary electrical installations (where the type of system earthing is TN) a PEN conductor shall **not** be used downstream of the origin of the temporary electrical installation.

7.4.14 *Protection against overcurrent*

- A protective device shall be provided to break any overcurrent in the circuit conductors.
- Overload protection devices shall be installed at any point where a reduction would affect the value current-carrying capacity of the conductors of the installation.

- Overload protection devices may be installed along the run of that conductor provided:
    - it is protected against fault current; or
    - its length does not exceed 3 m;
    - it is installed so as to reduce the risk of fault, fire or danger to persons to a minimum; and

overload protection devices need not be provided:

- for a conductor situated on the load side;
- where the conductor is effectively protected against overload by a protective device;
- for a conductor which, is not likely to carry overload current;
- at the origin of an installation (where the distributor provides an overload device).

Where a single protective device protects two or more conductors in parallel there shall be no branch circuits or devices for isolation or switching in the parallel conductors.

7.4.15 Protection by Earth-free local equipotential bonding

- Earth-free, local equipotential bonding shall connect together every simultaneously accessible exposed conductive-part.
- Unless protection by automatic disconnection of supply can be applied, local protective bonding conductors shall **not:**
    - be in direct electrical contact with Earth; nor
    - through exposed-conductive-parts; nor
    - through extraneous-conductive-parts.
- protective equipotential bonding shall be applied to any metallic sheath of a telecommunication cable;
- socket-outlets shall be provided with a protective conductor contact that is connected to the equipotential bonding system;
- all flexible cables shall include a protective bonding conductor.

If two faults affect two exposed-conductive-parts (fed by conductors with a different polarity) at the same time, a protective device shall disconnect the supply as described in BS 7671:2018 Table 41.1.

7.4.16 Protective conductors

All exposed-conductive-parts of the installation shall be connected by a protective conductor to the main earthing terminal of the installation, which shall then be connected to the earthed point of the power supply system:

- for an outdoor lighting installation, a protective conductor need not be provided;
- in caravans and motor caravans, all circuit protective conductors shall be incorporated in a multicore cable or in a conduit together with the live conductors;
- in a fixed installation, a single conductor may-serve both as a protective conductor as well as a neutral PEN conductor;
- if an RCD is used for fault protection in a TN system, the circuit should also incorporate a PEN conductor (in the case of a TN-C-S system, the PEN conductor shall be on the source side of the RCD.

7.4.16.1 Preservation of electrical continuity of protective conductors

- A protective conductor shall be suitably protected against mechanical, chemical deterioration and electrodynamic effects.
- The sheath of a cable incorporating an uninsulated protective conductor shall be protected by insulating sleeving that complies with the BS EN 60684 series.
- A switching device shall not be inserted in a protective conductor unless:
    - the switch has been inserted in the connection between the neutral point and the means of earthing;
    - the switch is a linked switch arranged to disconnect and connect the earthing conductor.

Every connection and joint shall be accessible for inspection, testing and maintenance.

7.4.17 Protective devices and switches

- Single-pole fuses, switches or circuit-breakers shall only be inserted in the line conductor.
- No circuit-breaker, (except where linked) or fuse shall be inserted in an earthed neutral conductor.

Table 7.4 Requirements for protective bonding conductors in transportable units

Location	*Requirements*
Exhibitions and show grounds	• structural metallic parts which are accessible from within the stand, vehicle, wagon, caravan or container shall be connected through the main protective bonding conductors to the main earthing terminal within the unit
Mobile or transportable unit	• accessible conductive parts of the unit, shall be connected through the main protective bonding conductors to the main earthing terminal within the unit • the main protective bonding conductors shall be finely stranded
Caravans and motor caravans	In caravans and motor caravans, where protection by automatic disconnection of supply is used, an RCD and the wiring system must include a circuit protective conductor which shall be connected to: • the protective contact of the inlet; • the exposed-conductive-parts of the electrical equipment; and • the protective contacts of the socket-outlets Structural metallic parts which are accessible from within the caravan shall be connected through main protective bonding conductors to the main earthing terminal within the caravan

7.4.18 *Protective earthing*

- Exposed-conductive-parts shall be connected to a protective conductor which shall comply with Chapter 54 of BS 7671:2008 and run to and be terminated at each point in wiring and at each accessory.

7.4.19 *Protective equipotential bonding*

For each installation, a main protective bonding conductor will be used to connect all extraneous-conductive-parts to the main earthing terminal which include the following:

- central heating and air conditioning systems;
- exposed metallic structural parts of the building;

- gas installation pipes;
- water installation pipes;
- other installation pipework and ducting.

7.4.20 RCDs

Where it is necessary to limit the consequence of fault currents in a wiring system from the point of view of fire risk, the circuit shall be protected by an RCD that is installed at the origin of the circuit and which will switch all live conductors.

7.4.21 SELV and PELV circuits

- Protective separation of the wiring systems of Separated Extra Low Voltage (SELV) or Protective Extra Low Voltage (PELV) circuits from the live parts of other circuits shall be achieved by one of the following arrangements:
    - SELV and PELV circuit conductors;
    - using an earthed metallic sheath or earthed metallic screen.
- Plugs and socket-outlets in a SELV system shall not have a protective conductor contact.
- Exposed-conductive-parts of a SELV circuit shall not be connected to protective conductors.

7.4.22 Supplementary equipotential bonding

A supplementary equipotential bonding system:

- is provided by a supplementary conductor;
- is connected to the protective conductors of **all** equipment including those of socket-outlets.

7.4.23 Transformers

- Where an autotransformer is connected to a circuit that has a neutral conductor, the common terminal of the winding will be connected to the neutral conductor.
- Where a step-up transformer is used, a linked switch shall be provided for disconnecting the transformer from all live conductors of the supply.

A step-up autotransformer shall **not** be connected to an IT system.

7.4.24 Wiring systems

Whilst the only economic method of transmitting power from a grid station is by means of lines suspended from pylons, at lower voltages there is a choice between running them overhead or underground. The supply to most domestic buildings (particularly in towns) is predominantly underground but for electrical installations such as in agricultural buildings, the most cost-effective way is via overhead cables.

The downside of this, of course is the potential for the overhead cable to become a safety hazard (as shown in Figures 7.15 and 7.16) and protective methods must be used to guard against this possibility.

Examples of wiring systems are shown in Table 4A2 of the Wiring Regulations.

Figure 7.15 Overhead wiring systems (power lines)

Source: Courtesy Stingray

Figure 7.16 Overhead wiring systems (urban)

Source: Courtesy Stingray

- A voltage Band I circuit shall **not** be part of the same wiring system as a Band II circuit unless:
    - every cable or conductor is insulated for the highest voltage present;
    - each conductor of a multicore cable is insulated for the highest voltage present in the cable.
- Metallic structural parts of buildings shall not be used as live conductors.
- The minimum cross-sectional area of the extra-low voltage conductors shall normally be 1.5 mm^2 copper.

7.4.24.1 Marinas and similar locations

In marinas and similar locations, the following wiring systems are suitable for distribution circuits:

- overhead cables or overhead insulated conductors;
- cables with copper conductors and thermoplastic or elastomeric insulation and its sheath installed within an appropriate cable management system;

- cables with armouring and sleeving of thermoplastic or elastomeric material;
- mineral-insulated cables with a PVC protective covering.

7.4.24.2 Caravans and motor caravans

In caravans and motor caravans, the wiring systems shall be installed using one or more of the following:

- insulated single-core cables, with flexible class 5 conductors, in a non-metallic conduit;
- insulated single-core cables, with stranded class 2 conductors (minimum of 7 strands), in a non-metallic conduit;
- sheathed flexible cables.

7.5 Inspection and testing

Note: Inspection shall precede testing and shall normally be done with that part of the installation under inspection disconnected from the supply.

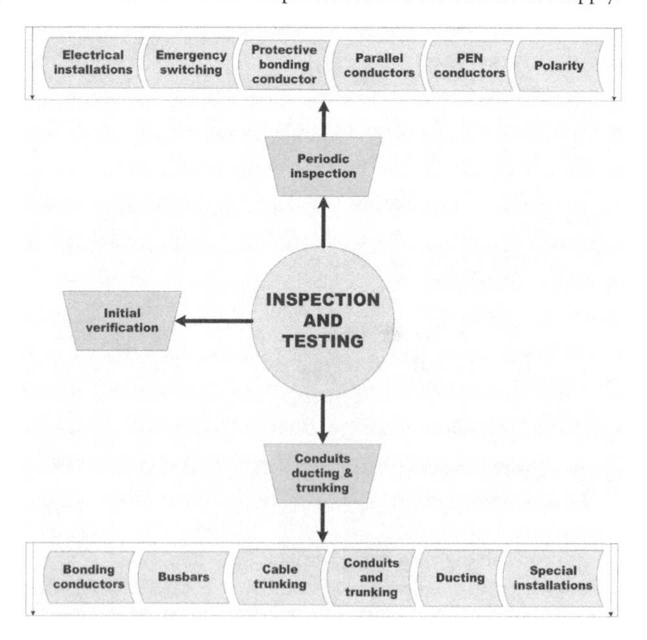

Figure 7.17 Inspection and testing

Every electrical installation should be subjected to periodic inspection and testing (combined with any maintenance and repair that is required) and a notice (see Figure 7.18) to that effect shall be fixed on or nearby to that installation as shown below:

During erection and on completion of an installation and before it is put into service, appropriate inspection and testing shall be carried out by skilled persons competent to verify that the requirements of this Standard have been met.

The inspection shall include at least the checking of the following items, where relevant to the installation and, where necessary, during erection:

- identification of conductors;
- connection and continuity of conductors;

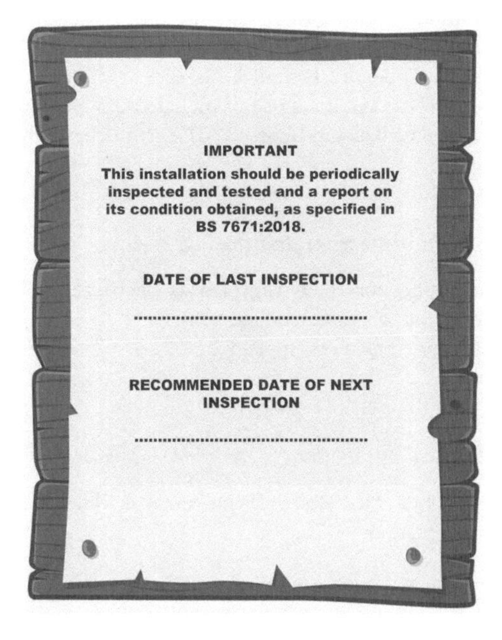

Figure 7.18 Warning notice for periodic electrical installation inspections and testing

- connection of single-pole devices for protection or switching in line conductors only;
- selection of conductors for current-carrying capacity and voltage drop, in accordance with the design;
- insulation resistance between live conductors and between live conductors and the protective conductor connected to the earthing arrangement (The insulation resistance measured against test voltages is shown in BS 7671:2018 Table 61).

For more stringent requirements required for fire alarm systems, see BS 5839-1.

- measurement of the Earth fault loop impedance.

7.5.1 Certification for initial verification

- On completion of the verification of a new installation (or an addition or alteration to an existing installation) an Electrical Installation Certificate based on the example contained in BS 7671:2018 Appendix 6 shall be issued to the person ordering the work. This certificate shall include details of any defect or omission revealed (and corrected) during the inspection and testing.

7.5.2 Periodic inspection and testing

- Periodic inspection and testing of every electrical installation shall be carried out, on an as required basis.
- A full report (based on Appendix 6 of BS 7671:2018) needs to be produced.

7.5.2.1 Electrical installations

The characteristics of the electrical equipment shall **not** be impaired by the process of erection.

- Every installation shall be divided into circuits, as necessary, to reduce the possibility of unwanted tripping of RCDs due to excessive PE currents that are **not** due to a fault.

7.5.2.2 Emergency switching

Emergency switching off is an emergency operation that is intended to dislocate the supply of electrical energy to all or part of an installation where a risk of electric shock or another risk of electrical origin is involved.

Selection and erection of devices for emergency switching off shall be in accordance with the following regulations and shall comply with Regulation 537.2.

Means for emergency switching off may consist of:

- an isolating device;
- one switching device; or
- a combination of devices;
- hand-operated switching devices;
- remote control switching of circuit-breakers, control and protective switching devices or RCDs.

Plugs and socket-outlets shall **not** be provided for use as means for emergency switching off.

The means of operating devices for emergency switching off shall be:

- clearly identified, preferably by colour;

If a colour is used for identification, this shall be RED with a contrasting background (e.g. yellow);

  - readily accessible;
  - capable of latching in the 'OFF' position.

- The operation of the emergency switching device shall have priority over any other function relative to safety and shall not be inhibited by any other operation of the installation.
- The release of an emergency switching device operated remotely shall not re-energise the relevant part of the installation.

7.5.2.3 Parallel conductors

- Where the currents in parallel conductors are unequal, the design current and requirements for overload protection for each conductor shall be considered individually.

7.5.2.4 PEN conductors

Where any necessary authorisation for use of a PEN conductor has been obtained:

- the outer conductor of a concentric cable shall not be common to more than one circuit;
- the conductance of the outer conductor of a concentric cable shall:
    - for a single-core cable, not be less than that of the internal conductor;
    - for a multicore cable not be less than that of the internal conductors connected in parallel;
- the continuity of every joint in the outer conductor of a concentric cable (and at a termination of that joint) may be supplemented by an additional conductor used for sealing and clamping the outer conductor.

Where two or more PEN conductors are connected in parallel in a system there should be equal load current sharing between them;

7.5.2.5 Polarity

A test of polarity shall be made and it shall be verified that:

- every fuse and single-pole control and protective device is connected in the line conductor only;
- circuits (except for E14 and E27 lampholders to BS EN 60238) have the outer or screwed contacts connected to the neutral conductor; and
- wiring has been correctly connected to socket-outlets and similar accessories;
- all polarity-sensitive appliances have terminals clearly marked '–' and '+', or that have two conductors, indicating polarity by colour or by identification tags or sleeves marked '–' or '+'.

7.5.2.6 Protective equipotential bonding conductors

Equipotential bonding is basically an electrical connection maintaining various exposed conductive parts and extraneous conductive parts at substantially the same potential and is very important for ensuring that:

- all socket-outlets are provided with a protective conductor contact connected to the equipotential bonding conductor;
- all flexible equipment cables (other than Class II equipment) shall have a protective conductor for use as an equipotential bonding conductor.

Such parts are bonded to the electrical service Earth point of the building to ensure safety of a building occupant.

7.5.3 Conduits, cable ducting, cable trunking, busbar or busbar trunking

7.5.3.1 Bonding conductors

- In each installation, main protective bonding conductors shall connect extraneous-conductive-parts to the main earthing terminal.

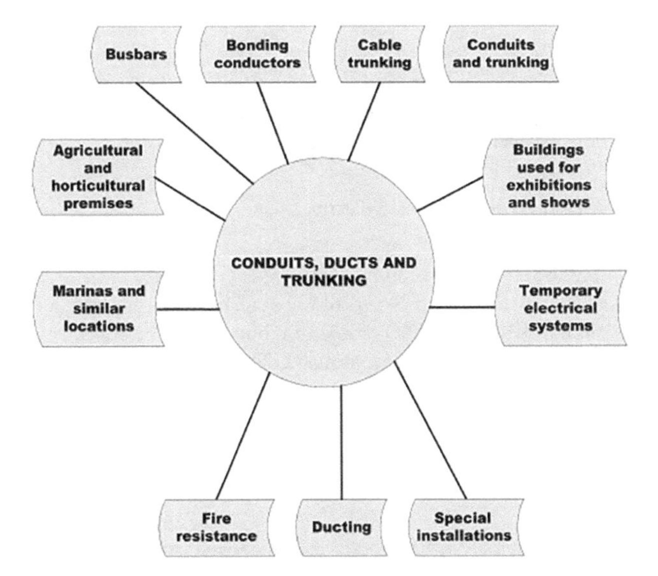

Figure 7.19 Conduits, ducting and trunking

7.5.3.2 Busbars

A busbar trunking system shall comply with BS EN 60439-2 (power-track with the BS EN 61534 series) and should always be installed taking account of external influences, particularly with respect to:

- busbars passing across expansion joints;
- any busbar trunking system used as a protective:
  - its cross-sectional area is in accordance with BS EN 60439-1;
  - the connection of other protective conductors at every predetermined tap-off point is always possible.

7.5.3.3 Cable trunking

If the cables of an installation are concealed in a wall or partition (the internal construction of which includes metallic parts) they shall:

- incorporate an earthed metallic covering; or
- be enclosed in earthed conduit; or
- be enclosed in earthed trunking or ducting; or
- be mechanically protected sufficiently to avoid damage to the cable during construction of the wall or partition and during installation of the cable.

Non-sheathed cables for fixed wiring:

- shall be enclosed in conduit, ducting or trunking;
- are permitted in a cable trunking system which provides a minimum of IP4X or IPXXD protection, **but** only if the cover can only be removed by means of a tool or a deliberate action.

The metal covering (including the sheath - bare or insulated) of a cable, trunking, ducting or metal conduit, may be used as a protective conductor for the associated circuit.

Wherever equipment is fixed on or in cable trunking, skirting trunking or in mouldings it should **not** be fixed on covers which can be inadvertently removed.

7.5.3.4 Conduits and trunking

- Conduit and trunking **shall** comply with the resistance to flame propagation requirements of BS EN 50085 or BS EN 50086.
- Conduits and conduit fittings **shall** comply with the appropriate British Standard shown in Table 7.5.

7.5.3.5 Ducting

A cable trunking system or cable ducting system shall satisfy the test under fire conditions specified in BS EN 50085.

A cable concealed in a wall or partition at a depth of less than 50 mm from a surface of the wall or partition shall:

- incorporate an earthed metallic covering; or
- be enclosed in earthed conduit; or
- be enclosed in earthed trunking or ducting; or
- be mechanically protected against damage sufficient to prevent penetration of the cable by nails, screws, etc.; or
- form part of a SELV or PELV circuit.

Note: Two or more circuits are allowed in the same conduit, ducting or trunking.

Conduit and trunking systems shall be in accordance with BS EN 61386-1 and BS EN 50085-1 respectively and shall meet the fire-resistance tests within these standards.

Table 7.5 Conduits and conduit fittings

Conduit fitting	*BS standard*	*BS EN standard*
Steel conduit and fittings	BS 31	BS EN 60423 and BS EN 50086-1)
Flexible steel conduit	BS 731-1	BS EN 60423 and BSEN50086-1
Steel conduit fittings with metric threads	BS 4568	BS EN 60423 and BS EN 50086-1
Non-metallic conduits and fittings	BS 4607	BS EN 60423 and BS EN 50086-2-1

7.5.3.6 Special installations

A cable installed under a floor or above a ceiling shall be:

- enclosed in earthed conduit, earthed trunking or ducting; or
- mechanically protected against damage.

A conduit system, cable trunking system or cable ducting system:

- which is going to be buried in the structure, shall be completely erected between access points before any cable is drawn in;
- be classified as non-flame propagating according to the relevant product standard and that:
    - the system satisfies the test of BS EN 60529 for IP33; and
    - any termination of the system in one of the compartments, separated by the building construction being penetrated, satisfies the test of BS EN 60529 for IP33.

Note: A voltage Band I circuit shall **not** be contained in the same wiring system as a Band II circuit.

Author's End Note

Although the Wiring Regulations are primarily aimed at electrical installations in and around commercial and residential buildings, BS 7671:2018 also covers the often-overlooked need for electrical safety in a number of special installations and locations that are subject to additional requirements due to the extra dangers they pose; for example, agricultural premises and mobile homes.

Chapter 8 of this book looks at these alternative locations and installations and their additional wiring and equipment requirements.

8 Special installations and locations

Author's Start Note

Whilst the Wiring Regulations apply to electrical installations in all types of buildings, there are also some indoor and out-of-doors special installations and locations (such as agricultural buildings) that are subject to additional requirements due to the extra dangers they pose. This chapter considers the requirements for these special locations and installations and whilst perhaps not being a complete list, represents the most important requirements.

The majority of these particular Regulations are ***additional*** *to all of the other requirements contained in BS 7671:2018 – and not meant as alternatives to them.*

8.1 General requirements

The following are intended to act as a reminder of the general requirements that are applicable to special installations and locations conductors.

8.1.1 Outdoor lighting installation

Outdoor lighting installations comprise one or more luminaires, a wiring system and lighting installations for:

- roads, parks, car parks, gardens, places open to the public, sporting areas, illumination of monuments and floodlighting;
- places such as telephone kiosks, bus shelters, advertising panels and town plans;
- road signs;
- temporary festoon lighting.

DOI: 10.1201/9781003165170-8

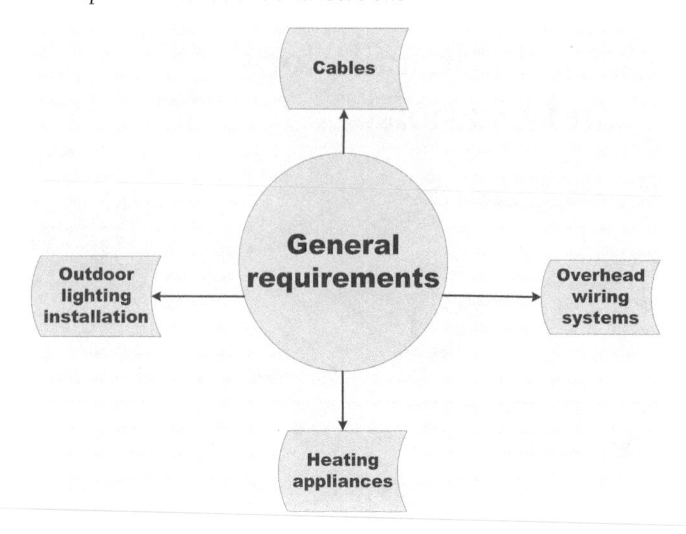

Figure 8.1 Special installations and locations – general requirements

The protective measures for outdoor lighting installations primarily include:

- access to a luminaire which is less than 2.80 m above ground level, shall only be possible after removing a barrier or an enclosure requiring the use of a tool.

Note: Placing out of reach and obstacles shall **only** be used where items of street furniture are closer than 1.5 m of a low voltage overhead line:

- lighting arrangements in telephone kiosks and bus shelters, etc. shall be protected by an RCD;
- enclosures for live parts shall only be accessible with a key or a tool;
- non-conducting locations and Earth-free local equipotential bonding shall **not** be used;
- a maximum disconnection time of 5 s applies to all circuits feeding fixed equipment used in highway power supplies;
- the earthing conductor of a street electrical fixture shall have a minimum copper equivalent cross-sectional area not less than 6 mm^2.

Where the protective measure is by double or reinforced insulation:

- no protective conductor shall be provided; and
- the conductive parts of the lighting column shall not be intentionally connected to the earthing system.

A device providing protection against the risk of fire shall:

- continuously monitor the power demand of the luminaires;
- automatically disconnect the supply circuit within 0.3 s in the case of a short-circuit (or failure);
- provide automatic disconnection while the supply circuit is operating with reduced power;
- provide automatic disconnection from the supply circuit if there is a failure which causes a power increase of more than 60 W;
- be fail-safe.

Suspension devices for extra-low voltage luminaires, shall be capable of carrying not be less than 5 kg.

8.1.2 *Cables*

A cable passing through a joist within a floor or ceiling construction or through a ceiling support (e.g. under floorboards), shall include:

- earthed trunking, or ducting;
- be mechanically protected against damage sufficient to prevent penetration of the cable by nails, screws, etc.; or
- be at least 50 mm, away from the joist or batten.

A cable concealed in a wall or partition at a depth of less than 50 mm from a surface of the wall or partition shall:

- be installed in an area within 150 mm from the top of the wall or partition.

If the wall or partition includes metallic parts (other than metallic fixings such as nails and screws) then it shall: (in addition to the requirements listed above).

- incorporate an earthed metallic covering; or
- be enclosed in earthed trunking or ducting; or
- be provided with additional protection by means of an RCD.

If a cable is not put in in a conduit or duct but is directly buried in the ground, then it **must** include a suitable protective conductor such as earthed armour or metal sheath or both.

Underground distribution cables should be buried at a minimum depth of 0.5 m so as to avoid being damaged by heavy vehicle movement.

8.1.3 Overhead wiring systems

Whilst the only economic method of transmitting power from a grid station is by means of lines suspended from pylons, at lower voltages there is a choice between running them overhead or underground. The supply to most domestic buildings (particularly in towns) is predominantly underground but for electrical installations such as in agricultural buildings, the most cost-effective way is via overhead cables.

The downside of this, of course is the potential for the overhead cable to become a safety hazard and protective methods must be used to guard against this possibility.

- All overhead conductors shall be insulated.
- Bare or insulated overhead distribution lines between buildings and structures need to be installed in accordance with the Electricity Safety, Quality and Continuity Regulations.

Figure 8.2 Overhead wiring systems to a farm

Source: Courtesy Artem Maltsev on Unsplash

- Poles and other supports for overhead wiring shall be located or protected so that they are unlikely to be damaged by any foreseeable vehicle movement.
- Overhead conductors shall be at a height above ground of not less than 6 m in all areas subjected to vehicle movement and 3.5 m in all other areas.

From a safety management point of view overhead cables shall not be used over waterways.

As previously mentioned, similar safety rules and Regulations exist for urban areas – but in some countries these rules are not entirely satisfactory (as you can see from Figure 8.3).

Figure 8.3 Overhead wiring in a town

Source: Courtesy Kael Bloom on Unsplash

8.1.4 Heating appliances

BS 7671:2018 Requirement

The equipment, system design, installation and testing of an electric surface heating system must meet the requirements of BS 6351.

Heating appliances must be fixed so as to minimise the risks of burns to livestock and of fire from combustible material.

Building Regulations Requirement

Any location that contains a forced air heating system or space heating appliance, must comply with the appropriate parts of the Building Regulations.

Both of these Regulations contain the general requirements for heating appliances (e.g. water heaters, boilers, heating units, heating conductors and cables, surface and underfloor heating systems, etc.) for special installations and locations and the following, whilst perhaps not being a complete list, represents the most important requirements:

- heating appliances must always be fixed;
- any circuit that supplies a heating unit needs to use an RCD to provide additional protection;
- precautions shall be taken to prevent heating elements creating high temperatures to adjacent material;
- electrical separation shall not be used for wall heating system;
- heating units shall be connected to the electrical installation via cold tails or suitable terminals;
- heating units shall not cross any expansion joints of the building or structure;
- electrical heating appliances used for the breeding and rearing of livestock shall comply with BS EN 60335-2-71;
- electric heating units shall not use the protective measure 'protection by electrical separation'.

8.1.4.1 Electric floor heating systems

Electric heating units that are embedded in the floor may be installed provided that they are:

- protected by SELV;
- covered by an earthed metallic grid or sheath that is connected to local supplementary equipotential bonding.

See BS EN 60335-2-96 for more detailed requirements relating to electric surface heating systems.

8.1.4.2 Electrode water heater

When an electrode water heater or boiler:

- is directly connected to a supply exceeding low voltage, the installation shall include an RCD;
- is directly connected to a three-phase low voltage supply, it shall be connected to the neutral of the supply as well as to the earthing conductor;
- is not piped to a water supply (or is in physical contact with any earthed metal) a fuse in the line conductor may be substituted for the circuit-breaker and the shell of the electrode water heater or electrode boiler need not be connected to the neutral of the supply.

Other installations and usage requirements contained in BS 7671:2018 and The Building Regulations include:

electric appliances producing hot water or steam must be protected against overheating;

- if located in a bathroom, electrical appliances must be fixed and permanently wired into the wall and controlled by a pull-cord inside the bathroom or by a switch located outside;
- heaters used for liquid or other substances must have an automatic device to prevent a dangerous rise in temperature;
- if the supply to a boiler/heater is single-phase and one electrode is connected to a neutral conductor earthed by the distributor, then the shell of that boiler/heater shall be connected to the neutral of the supply as well as to the earthing conductor;

- metal parts (other than the current-carrying parts of single-phase water heaters and boilers) that are in contact with the water shall be solidly and metallically connected to the metal water pipe supplying that heater/boiler;
- the protective conductor has to be connected to the shell of the water heater or electrode boiler;
- the heater/boiler must be permanently connected to the electricity supply via a double-pole linked switch that is either:
  - separate from and within easy reach of the heater/boiler; or
  - part of the boiler/heater (provided that the wiring from the heater or boiler is directly connected to the switch without use of a plug and socket-outlet).
- the shell of the heater/boiler must be bonded to the metallic sheath and armour (if any) of the incoming supply cable.

8.1.4.3 Forced air heating systems

Building Regulations Requirement

Locations that contain forced air heating systems and appliances that produce hot water or steam, and space heating appliances, must all comply with the appropriate parts of the Building Regulations.

There are various requirements for forced air heating systems to be found in BS 7671:2018 and the Building Regulations, but the main ones concern:

- that two, independent, temperature limiting devices must be made available;
- electric heating elements of forced air heating systems (other than those of central-storage heaters) should:
  - not be capable of being activated until the prescribed air flow has been established;
  - deactivate when the air flow is reduced or stopped.
- frames and enclosures of electric heating elements must be of non-ignitable material.

8.1.4.4 Heating conductors and cables

Heating cables passing through (or in close proximity to) a fire hazard:

- must be enclosed in material with an ignitability characteristic '13' as specified in BS 476 Part 12;
- must be protected from any mechanical damage.

Heating cables that are going to be laid (directly) in soil, a roadway, or the **structure** of a building must be installed so that they are:

- completely embedded in the substance it is intended to heat;
- not damaged by movement (either by it, or the substance in which it is embedded;
- compliant (in all respects) with the makers instructions and recommendations.

8.1.4.5 Locations with risks of fire due to heat storage

Equipment enclosures for heaters must not attain higher surface temperatures than:

- 90 C under normal conditions; and
- 115 C under fault conditions.

Where heating and ventilation systems containing heating elements are installed:

- the dust or fibre content and the temperature of the air must not present a fire hazard;
- temperature limiting devices must have a manual reset.

8.2 The requirements for special installations and locations

In addition to the normal safety protection methods against direct and indirect contact listed in other parts of the Regulations, special installations and locations such as:

- Agricultural and horticultural premises;
- Conducting locations with restricted movement;

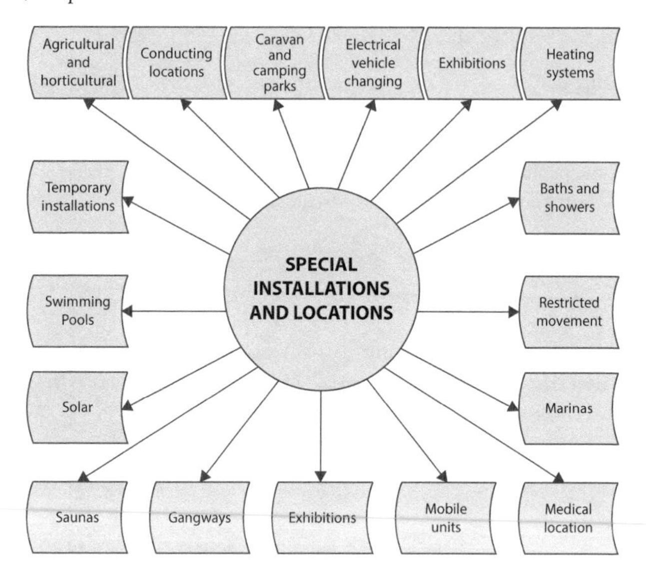

Figure 8.4 Special installations and locations

- Electrical installations in caravan/camping parks and similar locations;
- Electrical vehicle changing installations;
- Exhibitions, shows and stands;
- Floor and ceiling heating systems;
- Locations containing a bath or shower;
- Locations with restricted movement;
- Marinas and similar locations;
- Medical locations;
- Mobile and transportable units;
- Exhibitions, shows and sands;
- Operating and maintenance gangways;
- Rooms and cabins containing saunas;
- Solar, photovoltaic power supply systems;
- Swimming pools and other basins;
- Temporary electrical installations.

must also comply with the requirements for safety protection in respect of:

- electric shock;
- thermal effects;

- overcurrent;
- undervoltage;
- isolation and switching.

These requirements are described in more detail below.

8.2.1 Agricultural and horticultural premises

8.2.1.1 General

In contrast to normal domestic installations, an agricultural installation is usually prone to damp conditions and so contact with Earth will be better and people and animals are more liable to electric shock. For animals (whose body resistance is much lower than humans) this situation is worsened as their contact with Earth will be greater and so even a small voltage could prove lethal to them. Animals can also cause a lot of physical damage to electrical installations and animal effluents present a greater risk of corrosion.

Note: Horticultural installations are also subject to the same wet/high Earth contact conditions and for these reasons; special requirements have been introduced for all agricultural or horticultural installation.

8.2.1.2 Protective measures

In agricultural and horticultural premises, a TN-C system shall **not** be used.

In addition, placing obstacles out of reach shall **not** be used; instead bonding needs to connect all exposed-conductive-parts and extraneous conductive-parts that can be touched by livestock.

However, the Regulations specifically state that the protective measures of non-conducting location and Earth-free local equipotential bonding are **not** permitted.

8.2.1.2.1 PROTECTION AGAINST EXTERNAL INFLUENCES

In agricultural and/or horticultural premises:

- only water proofed electrical equipment shall be used;
- socket-outlets shall be installed so that they do not to come into contact with combustible material;

This does not apply to residential locations, offices, shops, etc. belonging to agricultural and horticultural premises and where (socket-outlets) need to comply with BS 1363-2 or BS 546.

- electrical equipment shall be protected against corrosive substances (particularly in dairies or cattle sheds).

8.2.1.2.2 PROTECTION AGAINST FIRE

Fire is a particular hazard in agricultural premises as there are normally large quantities of straw and other flammable material stored in these locations and so strict precautions must be taken. For example:

- electrical heating appliances used for the breeding and rearing of livestock shall be installed (in accordance with BS EN 60335-2-71) in order to minimise any risks of burns to livestock and of fire.

For fire protection purposes, RCDs:

- shall be installed which shall disconnect all live conductors.

8.2.1.2.3 SUPPLEMENTARY EQUIPOTENTIAL BONDING

- in locations intended for livestock, supplementary bonding shall connect all exposed-conductive-parts and extraneous conductive-parts that can be touched by livestock;
- nonessential conductive-parts in, or on, the floor (such as concrete reinforcement in cellars used to hold liquid manure) shall be connected to the supplementary equipotential bonding;
- protective bonding conductors shall be protected against mechanical damage and corrosion, and electrolytic effects.

8.2.1.3 Safety services

Isolator switches are purely mechanical devices which are designed to isolate an electrical circuit from its power source and the Regulations specifically state that:

- a single isolation device shall be provided for each building and in addition (e.g. during harvest time) for all live conductors - including the neutral conductor;
- isolation devices shall be clearly marked according to the part of the installation to which they belong.

Note: Isolation, switching and devices for emergency stopping or emergency switching shall not be erected where they are accessible to livestock or in any position where access may be impeded by livestock.

8.2.1.3.1 SAFETY PRECAUTIONS

Farmers and farm workers can easily be injured by livestock such as cattle, pigs, horses, sheep, dogs and other farm animals as they can be highly disordered and so, in a nutshell:

- animals are unpredictable, especially during the joining (mating) season and should be treated with caution at all times;
- make sure yards, sheds and equipment are in good repair;
- keep the walkways and laneways dry and non-slip wherever possible;
- ensure that workers are appropriately trained and familiar with the temperament of the animals on your farm;
- always wear suitable protective clothing and use appropriate animal-handling facilities and aids;
- take care when visiting other people's farms or sale yards.

With regard to electrical safety:

- where the supply of food, water, air and/or lighting to livestock is not guaranteed in the event of power supply failure:
  - a secure source of supply should always be provided (such as an alternative or back-up supply); and
  - separate final circuits for ventilation and lighting units shall be provided.
- where electrically powered ventilation is necessary in an installation one of the following needs to be provided:
  - a standby electrical source ensuring sufficient supply for ventilation equipment; or
  - temperature and supply voltage monitoring.

Note: A notice should be placed adjacent to the standby electrical source, indicating that it should be tested periodically according to the manufacturer's instructions.

8.2.1.3.2 ELECTRIC FENCE CONTROLLERS

Electric fencing systems have been developed to stop the free movement of animals across pastures.

- They are semi-permanent solutions that can be extended, altered, or removed to allow grazing.
- The system consists of plastic or wooden posts, insulators, conductive wire/rope/tape and an electrical energiser unit.
- This unit can be mains, battery or solar powered and it will send short electrical impulses along a conductive wire (tape, rope, etc.) so that when the wire or fence is touched by an animal the small current passes through it to the ground and causes the animal to feel a shock.
- The shock should be sufficient to alarm an animal but not to harm it – and in time the animal will learn to stay away from it.

For specific requirements regarding electric fence installations see BS EN 60335-2, BS 7671:2008 and BS EN 6100-1.

8.2.1.4 Selection and erection of electrical equipment

The selection of equipment is usually the responsibility of the designer, but most of the time, it is the installer who actually selects the equipment and the installer must ensure that:

- the electrical equipment fully complies with the Wiring Instructions;
- on completion of all electrical installations, an initial inspection of the equipment and installation is carried out;
- periodic checks are completed to an agreed Work Instruction and time schedule.

BS 7671:2018 Requirement

Only electrical heating appliances with visual indication of the operating position shall be used.

8.2.1.5 Supplies

In special installations and locations, whenever SELV or PELV is used protection **must** be provided by either:

- basic insulation; or
- barriers.

A TN-C system shall **not** be used.

8.2.1.5.1 AUTOMATIC DISCONNECTION OF SUPPLY

ADS (previously known as Earthed Equipotential Bonding and Automatic Disconnection of Supply or EEBADS) is the most widely used method for achieving protection against electric shock in an installation. It works by limiting the size and duration of any voltages between exposed-conductive parts and extraneous-conductive parts or Earth.

In circuits, whatever the type of earthing system, the following disconnection device shall be provided:

- in final circuits supplying socket-outlets with rated current not exceeding 32 A, an RCD (with a rated residual operating current not exceeding 30 mA);
- In final circuits supplying socket-outlets with rated current more than 32 A, an RCD with a rated residual operating current not exceeding 100 mA;
- in all other circuits, RCDs with a rated residual operating current not exceeding 300 mA.

8.2.1.6 Wiring systems

- All wiring systems must be selected and erected so that they are inaccessible to livestock or suitably protected against mechanical damaged by animal movement.
- Overhead lines should **always** be insulated.
- In areas of agricultural premises where vehicles and mobile agricultural machines are being used, the following methods of installation need to be used:
    - cables shall be buried in the ground at a depth of at least 0.6 m with added mechanical protection;

- cables in arable or cultivated ground shall be buried at a depth of at least 1 m;
- self-supporting suspension cables shall be installed at a height of at least 6 m.

8.2.1.6.1 CONDUIT SYSTEMS, CABLE TRUNKING SYSTEMS AND CABLE DUCTING SYSTEMS

- For locations where livestock is kept, all conduits shall have protection against corrosion of at least Class 2 (medium) for indoor use and Class 4 (high protection) outdoors according to BS EN 61386-21.
- For locations where the wiring system may be exposed to impact and mechanical shock due to vehicles and mobile agricultural machines, etc.:
  - conduits shall provide a degree of protection against impact of 5 J according to BS EN 61386-2;
  - cable trunking and ducting systems shall also provide a degree of protection against impact of 5 J according to BS EN 50085-2-1.

8.2.1.6.2 SOCKET-OUTLETS

Socket-outlets used in agricultural and horticultural premises shall comply with:

- BS EN 60309-1; or
- BS EN 60309-2 (when interchangeability is required); or
- BS 1363, BS 546 (provided the rated current does not exceed 20 A).

8.2.1.7 Identification and notices

The following documentation shall be provided to the user of the installation:

- a plan indicating the location of all electrical equipment;
- a single-line distribution diagram;
- the routing of all concealed cables;
- an equipotential bonding diagram showing the locations of all bonding connections.

A notice should be placed adjacent to the standby electrical source, indicating that it should be tested periodically according to the manufacturer's instructions.

> **Author's Hint**
>
> *See BS 7671:2018 Section 705 for more detailed requirements relating to agricultural or horticultural installations).*

8.2.2 *Conducting locations with restricted movement*

8.2.2.1 *General*

A restrictive conductive location is one in which the surroundings consist mainly of metallic or conductive parts such as a large metal container or boiler. People employed inside these locations (e.g. a person working inside the boiler whilst using an electric drill or grinder) would have their freedom of movement physically restrained and a large proportion of their body would be in contact with the sides of that location and, therefore, prone to shock hazards.

8.2.2.1.1 SELECTION AND ERECTION OF ELECTRICAL EQUIPMENT

- All equipment used for the distribution of electricity on construction and demolition sites shall meet the requirements of BS EN 61439-4.
- A plug or socket-outlet with a rated current equal to or greater than 16 A shall comply with the requirements of BS EN 60309-2.

8.2.2.1.2 PROTECTION AGAINST ELECTRIC SHOCK

If a functional Earth is required for certain equipment (for example measuring and control equipment):

- live parts shall be completely covered with insulation;
- supplementary equipotential bonding shall be provided between all exposed-conductive-and extraneous-conductive-parts inside a conducting location and the functional Earth;
- the unearthed source shall have simple separation and shall be positioned outside any conducting location with restricted movement;

- the protective measures of obstacles, placing out of reach, non-conducting location, Earth-free local equipotential bonding (or electrical separation for the supply to more than one item of current-using equipment) are not permitted and shall **not** be used;
- where more than one voltage is in use, plugs and sockets must be non-interchangeable to prevent misconnection.

These requirements do **not** apply to:

- construction site offices, cloakrooms, meeting rooms, canteens, restaurants, dormitories and toilets, etc.;
- installations covered by BS 6907.

For any circuit supplying one or more socket-outlets with a rated current exceeding 32 A, an RCD having a rated residual operating current not exceeding 500 mA shall be provided.

8.2.2.2 Supplies

Supplies will normally be obtained from the local electrical supply company but remote sites could need an IT supply (such as a generator).

- If this is the case, then care must be taken in complying with the safety requirements for this particular source – particularly with respect to disconnection and isolation.

8.2.2.2.1 AUTOMATIC DISCONNECTION OF SUPPLY

- Although an RCD should be used for Circuits supplying one or more socket-outlets with a rated current exceeding 32A, a PME earthing facility should **not** be used for the supply to a construction site, unless, all extraneous parts are reliably connected to the main earthing terminal.

8.2.2.2.2 ISOLATION DEVICES

Each Assembly for Construction Sites (ACS) must include suitable devices for the switching and isolating the incoming supply.

- These devices shall be capable of being permanently secured in the OFF position (e.g. via a padlock or locating the device inside a lockable enclosure).

Current-using equipment shall be supplied by ACSs comprising:

- overcurrent protective devices; and
- devices affording fault protection; and
- socket-outlets, if required.
- Whatever the nominal voltage, where SELV or PELV is used:
    - live parts shall be completely covered with insulation; or
    - basic protection shall be provided by barriers or enclosures.

8.2.2.3 *Wiring systems*

The Regulation's requirements for wiring systems include:

- cables must **not** be installed across a site road or a walkway unless they are adequately protected against mechanical damage;
- for reduced low voltage systems, low temperature 300/500 V thermoplastic or equivalent flexible cables should be used;
- for applications exceeding reduced low voltage, flexible cable [(such as the HO7RN-F (BS 7919) type or equivalent] which has a 450/750 V rating and is resistant to abrasion and water;
- a wiring system buried in a floor shall be sufficiently protected to prevent damage caused by the intended use of the floor.

Author's Hint

See BS 7671:2018 Section 706 for more detailed requirements relating to conducting locations with restricted movement.

8.2.3 *Electrical installations in caravan/camping parks and similar locations*

In order not to mix regulations on different subjects, such as those for the electrical installation of caravan and camping parks with those for the actual electrical installation inside a caravan or campervan, two sections have been created within BS 7671:2018. Namely:

- **Section 708** – electrical installations in caravan/camping parks and similar locations; and
- **Section 721** – electrical installations **inside** caravans and motor caravans.

8.2.3.1 General

In a caravan park or camping park, special consideration must be given to the:

- protection of people – owing to the fact that the human body may be in contact with Earth potential;
- protection of wiring – due to tent pegs or ground anchors; and
- movement of heavy or high vehicles.

The actual requirements for supplying electricity to leisure accommodation vehicles in caravan/camping parks and similar locations, are as follows.

8.2.3.2 Protective measures

- all socket-outlets shall be protected by an RCD complying with BS 4293, BS EN 61008-1 or BS EN 61009-1 with a 30-mA rating;
- all equipment that is installed in a caravan pitch or campsite shall be protected against mechanical damage (See BS EN 62262);
- the Regulations disallows protective measures such as obstacles, placing out of reach, non-conducting location, and Earth-free local equipotential bonding being used.

8.2.3.3 Selection and erection of equipment

The Regulations clearly state that all electrical equipment that is installed in an outside area:

- must to be protected against mechanical damage (See BS EN 62262);
- must ensure that adequate precautions have been taken against:
  - mechanical stress;
  - the presence of water;
  - the presence of foreign solid bodies.
- should ideally be via underground distribution circuits, must be buried at a depth of at least 0.6 m in;
- cables that are run below caravan pitches must be provided with additional protection. As shown in Figure 8.5.

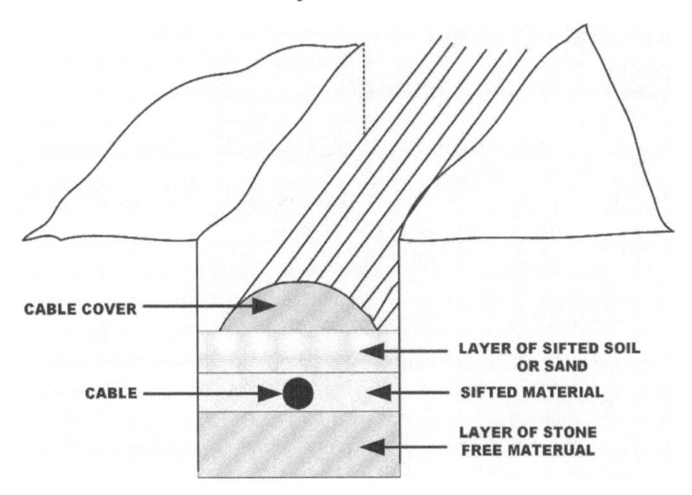

Figure 8.5 Cable covers

For conduit systems buried underground, see BS EN 61386-24.

8.2.3.4 *Supplies*

The nominal supply voltage to leisure accommodation vehicles shall not exceed 230 V a.c. single-phase, 400 V a.c. three-phase or 48 V d.c.

As the requirements of the Electricity Supply Regulations do not allow the supply neutral to be connected to any metalwork from (or in) a caravan **only** TT or TN-S systems may be used.

in general, the supply of electrical energy to caravan and tent sites must ensure that:

- the supply is via underground cables;
- overhead supplies use insulated as opposed to bare cables;
- cables installed outside the area of the caravan pitch are at least 3.5 m above ground level;
- each socket has its own fuse or circuit breaker;
- no more than four socket-outlets should be grouped in one location.

8.2.3.5 *Wiring systems*

Equipment supplying a caravan pitch:

- shall be adjacent to that pitch and be no more than 20 m from the connection facility;
- busbar trunking systems shall comply with BS EN 60439-6; and
- a powertrack system shall comply with the appropriate part of the BS EN 61534 series.

8.2.3.5.1 OVERHEAD CONDUCTORS

All overhead conductors shall:

- be insulated;
- be at least 6 m above ground level in all areas subject to vehicle movement and 3.5 m in all other areas.

8.2.3.5.2 SOCKET-OUTLETS

Socket-outlets shall:

- be individually protected by:
  - an RCD;
  - an overcurrent protective device.
- be placed at a height of 0.5 m to 1.5 m from the ground to the lowest part of the socket-outlet;
- have a current rating not less than 16 A; and
- at least one socket-outlet shall be provided for each caravan pitch.

8.2.3.5.3 UNDERGROUND DISTRIBUTION CIRCUITS

- Underground cables (unless provided with additional mechanical protection) shall be buried at a depth of at least 0.6 m to avoid being damaged by tent pegs, ground anchors or by the movement of vehicles.

Author's Hint

See BS 7671:2018 Section 708 for more detailed requirements relating to electrical installations in caravan/camping parks and similar locations.

8.2.4 Electrical installations in caravans and motor caravans

Author's Note

Some of the information contained in this particular Section are a repeat of what is already in the previous Section (i.e. Electrical installations in caravan/camping parks and similar locations).

I have done this deliberately so you that you won't have to constantly flick back and forth through the book to find the correct information.

8.2.4.1 General

A 'mobile home' is defined as a 'transportable leisure accommodation vehicle' that does not meet the requirements for use as a road vehicle and is usually a permanent fixture on a caravan park.

Note: Caravans that are used as mobile workshops will also be subject to the requirements of the Electricity at Work Regulations 1989 and locations containing baths or showers for medical treatment, or for disabled persons, may have additional special requirements.

8.2.4.2 Selection and erection of equipment

- The connection to the caravan pitch socket-outlet to the caravan shall comprise the following shall be via:
  - a plug complying with BS EN 60309-2; and
  - a flexible cord or cable of 25 m (±2 m) length incorporating a protective conductor.
- Accessories that are exposed to the effects of moisture shall be constructed (or enclosed) so as to provide protection from water splashing coming from all directions.

8.2.4.3 Protective measures

The protective measures of:

- electrical separation;

- obstacles and placing out of reach;
- non-conducting location and Earth-free local equipotential bonding;

are **not** permitted.

8.2.4.3.1 PROTECTIVE EQUIPOTENTIAL BONDING

- Structural metallic parts which are accessible from within the caravan shall be connected through main protective bonding conductors to the main earthing terminal within the caravan.

8.2.4.3.2 PROTECTIVE MEASURE – AUTOMATIC DISCONNECTION OF SUPPLY

- Where protection by automatic disconnection of supply is used, an RCD shall be provided and the wiring system shall include a circuit protective conductor.

8.2.4.3.3 PROTECTION AGAINST OVERCURRENT – FINAL CIRCUITS

Each final circuit shall be protected by an overcurrent protective device which disconnects all live conductors of that circuit.

8.2.4.4 Supplies

- the a.c. supply voltage shall not exceed 230 V single-phase, or 400 V three-phase;
- the d.c. supply voltage shall not exceed 48 V.

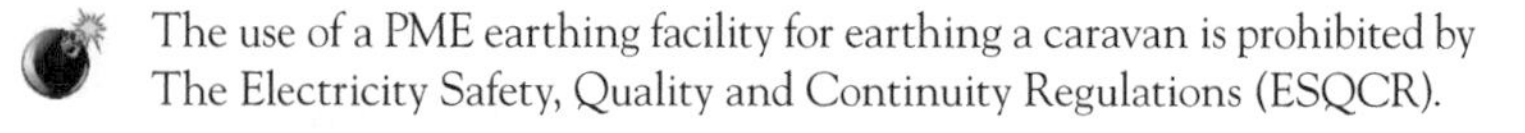

The use of a PME earthing facility for earthing a caravan is prohibited by The Electricity Safety, Quality and Continuity Regulations (ESQCR).

Each socket-outlet shall:

- be provided with individual overcurrent protection;
- be protected individually by an RCD;
- have a current rating of all socket-outlets not less than 16 A.

Socket-outlet protective conductors shall **not** be connected to any PEN conductor of the electricity supply.

8.2.4.4.1 ELECTRICAL INLETS

The a.c. electrical inlet on a caravan must comply with BS EN 60309-1 and shall:

- be installed not more than 1.8 m above ground level;
- be located in a readily accessible position; and
- shall not protrude significantly beyond the body of the caravan.

8.2.4.4.2 LUMINAIRES IN CARAVANS AND MOTOR CARAVANS

- Each luminaire in a caravan shall preferably be fixed directly to the structure or lining of the caravan.
- Where a pendant luminaire is installed in a caravan, it shall be made secure so as to prevent damage whilst the caravan is in motion.

8.2.4.4.3 ISOLATION

Each installation shall be provided with a main disconnector that will disconnect all live conductors.

Annex A721 to BS 7671:2018 provides guidance for extra-low voltage d.c. installations.

8.2.4.4.4 SOCKET-OUTLETS

Socket-outlets supplied at extra-low voltage shall have their voltage visibly marked.

8.2.4.5 *Wiring systems*

- Wiring shall be protected against mechanical damage.
- Wiring passing through metalwork shall be protected by bushes or grommets, securely fixed in position.
- Structural metal parts within the caravan shall be connected via bonding conductors to the main earthing terminal within the caravan.
- All cables (unless enclosed in rigid conduit) shall be supported at intervals not exceeding 0.4 m for vertical runs and 0.25 m for horizontal runs.

8.2.4.5.1 TYPES OF WIRING SYSTEM

- All cables shall meet the requirements of BS EN 60332-1-2.
- Non-metallic conduits shall comply with BS EN 61386-21.
- Cable management systems shall comply with BS EN 61386.

8.2.4.6 *Identification and notices*

- A notice shall be permanently fixed near the main isolating switch inside the caravan showing what checks have to be completed before connecting and disconnecting the caravan installation from the mains supply so that the caravan can be used safely.

see Figure 721 on page 308 of BS 7671:2018.

Author's Hint

See BS 7671:2018 Section 721 for more detailed requirements relating to electrical installations inside caravans and motor caravans.

8.2.5 Electric vehicle charging installations

8.2.5.1 General

With the increased number of all-electric or hybrid-electric cars currently being manufactured, the 2018 edition of BS 7671 now contains a new Section (722) to cover the requirements for circuits that are intended to supply electric vehicles for charging purposes either from an EV (Electric Vehicle) Charging Point or a dedicated Home Charging Point.

Plug-in hybrid electric vehicles have both an electric motor and a conventional petrol or diesel engine. Compared to a battery driven electric vehicle, this extends the total driving range but obviously lowers the all-electric range.

The electric car market is growing more quickly than most people had previously assumed it would! With more than 26,000 pure-electric cars on UK roads at the end of May 2021 and over 535,000 plug-in models including Plug-in Hybrids (PHEVs).

The number of electric vehicles as a proportion of all new vehicles showed a significant increase in 2020. In 2015 just 1.1% of new vehicles registered had a plug compared to 3.2% in 2019; by the end of December 2020, this figure has accelerated to 10.7% as an average for the year (6.6% BEV and 4.1% PHEV). These numbers reflect both the increase in demand for electric vehicles and the decline in demand for traditional, particularly diesel, vehicles.

8.2.5.2 *Protective measures*

Each EV charging point shall be supplied individually by a final circuit protected by an overcurrent protective device complying with:

- BS EN 60947-2, BS EN 60947-6-2 or BS EN 61009-1; or
- with the relevant parts of the BS EN 60898 series; or
- the BS EN 60269 series.

8.2.5.2.1 SAFETY SERVICES

- The protective measures of obstacles and placing out of reach, non-conducting location and Earth-free local equipotential bonding shall **not** be used.
- Where emergency switching off is required, devices shall be capable of breaking the full load current of the relevant parts of the installation and disconnecting all live conductors, including the neutral conductor.
- Each charging point shall be protected by its own RCD of at least Type A, having a rated residual operating current not exceeding 30 mA.

RCDs must comply with one of the following standards: BS EN 61008-1, BS EN 61009-1, BS EN 60947-2 or BS EN 62423.

8.2.5.3 *Selection and erection of equipment*

A dedicated final circuit must be provided for the connection to electric vehicles. At EV charging points, this is normally restricted to a single charging point used at its rated current, but others are equipped with a dedicated distribution circuit capable of supplying multiple electric vehicle charging points if load control is available.

When installed outdoors, the equipment shall be selected with a degree of protection against:

- the presence of water;
- the presence of solid foreign bodies;
- mechanical damage (impact of medium severity from other vehicles or foreign bodies).

8.2.5.4 *Supplies*

Each a.c. charging point shall incorporate one socket-outlet complying with BS 1363-2 marked 'EV' on its rear with a label on the front face or adjacent to the socket-outlet or its enclosure stating: '*suitable for electric vehicle charging*':

- each socket-outlet shall be mounted in a fixed position.

There are three main types of EV charging – rapid, fast, and slow. These represent the power outputs [measured in kilowatts (kW)], and therefore charging speeds, available to charge an EV will vary.

- **rapid chargers** are the fastest way to charge an EV, and predominantly cover d.c. charging. This can be split into two categories, ultra-rapid and rapid:
  - **ultra-rapid points** can charge at 100+ kW – often 150 kW and up to 350 kW - and are d.c. only;
  - **rapid chargers** include those which provide power from 7 kW to 22 kW, which typically fully charge an EV in 3–4 hours.

Conventional rapid points make up the majority of the UK's rapid charging infrastructure and charge at 50 kW d.c., with 43 kW a.c. rapid charging also often available.

Note: The most common public charge point found in the UK is a 7 kW untethered Type 2 inlet, though tethered connectors are available for both Type 1 and Type 2.

- Slow units (up to 3 kW) are best used for overnight charging and usually take between 6 and 12 hours for a pure-EV, or 2-4 hours for a PHEV (Plug-In Hybrid Electric Vehicle). EVs charge on slow devices using a cable which connects the vehicle to a 3-pin or Type 2 socket.

- In EV charging modes 3 and 4, an electrical or mechanical system shall be provided to prevent the plugging/unplugging of the plug unless the socket-outlet or the vehicle connector has been switched off from the supply.

8.2.5.4.1 HOME CHARGING POINT

Note: A home charging point is a compact weatherproof unit that mounts to a wall with a connected charging cable or a socket for plugging in a portable charging cable.

It's useful to have a standard 3 pin plug with an EVSE (Electric Vehicle Supply Equipment) as a backup charging option, but they are not designed to withstand these loads and should not be used long term.

Charging speed for electric cars is measured in kilowatts (kW) and a home charging point can charge your car at 3.7 kW or 7 kW giving about 15–30 miles of range per hour of charge (compared to 2.3 kW from a 3-pin plug which provides up to 8 miles of range per hour).

Author's Hint

See BS 7671:2018 Section 722 for more detailed requirements relating to electric vehicle charging installations.

8.2.6 Exhibitions, shows and stands

8.2.6.1 General

Whilst Chapter 51 of BS 7671:2018 still contains the requirements for external influences (see below), the 18th Edition of the Wiring Regulations has now been further developed in accordance with new ISO standards and the BS EN 60721 and BS EN 61000 series on environmental conditions.

8.2.6.2 Protective measures

The following requirements are intended for temporary electrical installations at exhibitions, shows, stands and mobile (or portable) displays and equipment:

- any cable supplying these temporary structures shall be protected by an RCD;

- accessible structural metallic parts shall be connected through the main protective bonding conductors to the main earthing terminal within the unit;
- accessible socket-outlet circuits (other than those for emergency lighting) shall also be protected by and RCD;
- all live parts shall be covered by insulation and located inside barriers and shelters;
- PME earthing shall **not** be used unless the installation is continuously monitored and a suitable means of Earth has been confirmed before the connection.

8.2.6.2.1 PROTECTION AGAINST ELECTRIC SHOCK

- the cable supplying temporary structures shall be protected at its origin by an RCD.
- protective measures for a non-conducting location such as Earth-free local equipotential bonding, obstacles and placing out of reach are **not** permitted.

8.2.6.2.2 PROTECTION AGAINST FIRE AND HEAT GENERATION

- lighting equipment with high temperature surfaces (such as bright lamps), shall be suitably guarded, installed, located and adequately ventilated.

8.2.6.2.3 PROTECTION AGAINST THERMAL EFFECTS

- if a luminaire is mounted within arm's reach it shall be securely fixed, sited and guarded in order to prevent any risk of injury to persons or ignition of materials.

8.2.6.2.4 PROTECTIVE EQUIPOTENTIAL BONDING

- all structural, metallic, parts (which are accessible from within the stand, vehicle, wagon, caravan or container) should be connected via the main protective bonding conductors to the main earthing terminal within the unit.

8.2.6.3 *Selection and erection of equipment*

- designated switchgear and controlgear must be placed in closed cabinets which can only be opened by the use of a key or a tool.

8.2.6.3.1 ELECTRICAL CONNECTIONS

- joints shall not be made in cables as a connection into a circuit.

8.2.6.3.2 ELECTRIC DISCHARGE LAMP INSTALLATIONS

All luminous tubes, signs or lamps that are used to illuminate a stand or as exhibit, shall comply with the following:

- **Location** – the sign or lamp shall be installed out of arm's reach;
- **Installation** – the facia or material behind luminous tubes, signs or lamps shall be non-ignitable;
- **Emergency switching devices** – a separate circuit (controlled by an easily visible, accessible and clearly marked emergency switching device) shall be used to supply signs, lamps or exhibits.

8.2.6.3.3 LUMINAIRES AND LIGHTING INSTALLATIONS

Insulation piercing lampholders shall **not** be used.

- ELV lighting systems for filament lamps shall comply with BS EN 60598-2-23.

8.2.6.3.4 SOCKET-OUTLETS AND PLUGS

- Each socket-outlet circuit not exceeding 32 A and final circuits (other than for emergency lighting) shall be protected by an RCD.
- Floor mounted socket-outlets shall be protected from accidental ingress of water.

8.2.6.4 *Supplies*

- The supply voltage for a temporary electrical installation in an exhibition, show or stand shall not exceed 230/400 V a.c. or 500 V d.c.
- Electronic converters shall conform to BS EN 61347-1.

8.2.6.4.1 EXTRA-LOW VOLTAGE PROVIDED BY SELV OR PELV

Where SELV or PELV is used, whatever the nominal voltage, basic protection shall be provided by:

- covering all live parts with an insulation that can only be removed by destruction; or
- by protective barriers or enclosures.

8.2.6.5 *Wiring systems*

Flexible cables shall **not** be laid in areas accessible to the public unless they are protected against mechanical damage.

8.2.6.5.1 TYPES OF WIRING SYSTEM

If there isn't a fire alarm system installed in a building that is used for exhibitions, etc., cable systems shall be either:

- flame retardant and low smoke resistant; or
- unarmoured single or multicore cables enclosed in metallic or non-metallic conduit or trunking.

8.2.6.5.2 LUMINAIRES AND LIGHTING INSTALLATIONS IN EXHIBITIONS SHOWS AND STANDS

- Exhibitions that contain a concentration of electrical equipment, luminaires and lamps that are liable to generate excessive heat, shall not be installed unless adequate ventilation is available.
- Insulation piercing lampholders shall **not** be used.
- Luminaires mounted within arm's reach – shall be firmly fixed and located or guarded so as to prevent possible risk of injuring persons or igniting of materials.

8.2.7 *Floor and ceiling heating systems*

8.2.7.1 *General*

This Section of BS 7671 has been completely revised since it was last published in 2008 and now applies to the installation of embedded electric floor and ceiling heating systems.

8.2.7.2 Protective measures

The following protective measures are **not** permitted:

- electrical separation for wall heating systems;
- non-conducting location and Earth-free local equipotential bonding;
- a circuit supplying heating equipment of Class II construction or equivalent insulation shall be provided with additional protection by the use of an RCD.

8.2.7.3 Selection and erection of equipment

8.2.7.3.1 HEATING UNITS

- Heating units that are being installed in ceilings must be capable of resisting water that might hit the product at a 15° angle or less (i.e. IPX1).
- Heating units that are intended to be installed in a floor of concrete or similar material shall be protected so that accidental submersion in up to 1 meter of water for up to half an hour will not affect the heating unit and it will still be able to operate safely.

Flexible sheet heating elements **must** comply with the requirements of BS EN 60335-2-96.

8.2.7.4 Identification and notices

The designer of the installation/heating system or installer shall provide a plan for each heating system which shall be fixed to, or located adjacent to, the distribution board of the heating system.

The installation plan shall contain the following details:

- cables, earthed conductive shields, etc.;
- heated area;
- insulation resistance of the heating installation and the test voltage used layout of the heating units in the form of a sketch, a drawing, or a picture;
- length/area of heating units;
- manufacturer and type of heating units;
- number of heating units installed;
- position/depth of heating units;
- position of junction boxes;

- rated power;
- surface power density;
- rated current of overcurrent protective device;
- rated residual operating current of RCD;
- rated resistance (cold) of heating units;
- rated voltage;
- the leakage capacitance;
- Installation and maintenance instructions.

See BS 7671:2018 Section 753 for more detailed requirements relating to heating cables and embedded heating systems.

8.2.8 Locations containing a bath or shower

8.2.8.1 General

When people use bathrooms and showers, most of the time they are naturally unclothed and wet and thus very vulnerable to electric shock due to their reduced body resistance. Special measures are, therefore, required to ensure that the possibility of direct and/or indirect contact is reduced.

The requirements are based on three Zones which take into account the limitations of walls, doors, fixed partitions, ceilings and floors. The three Zones are as shown in Figure 8.6 and Table 8.1.

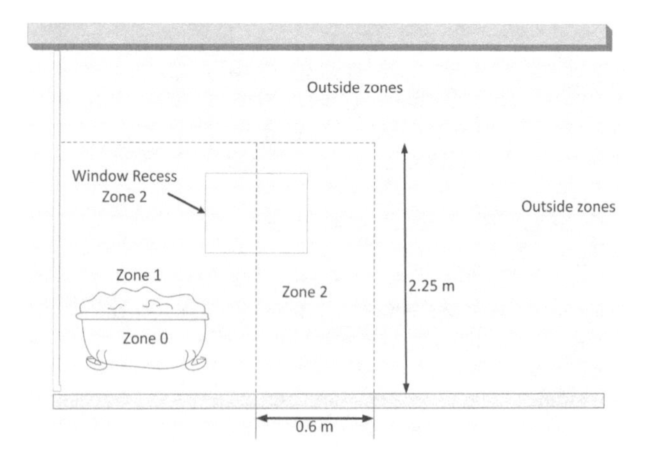

Figure 8.6 Zone limitations

Table 8.1 Limitations of zones

Zone	*Description*
Zone 0	Is the interior of the bath tub or shower basin  **Note:** for showers without a basin, the height of Zone 0 is 0.10 m and the same horizontal extent as Zone 1

Zone 1	Zone 1 does not include zone 0 and is limited by
	• the finished floor level and the highest fixed shower head or water outlet or 2.25 m above the finished floor level
	• the vertical surface
	  • around the bath tub or shower basin and
	  • 1.20 m from the centre point of the fixed water outlet on the wall (or ceiling for showers without a basin)
	 **Note:** The space under the bath tub or shower basin is considered to be Zone 1

Zone 2	Zone 2 is limited by • the finished floor level and the highest fixed shower head or water outlet or 2.25 m above the finished floor level • the vertical surface at the boundary of Zone 1 and 0.60 m from the border of Zone 1  **Note:** For showers without a basin, there is no Zone 2 but the horizontal dimension of Zone 1 is increased to 1.20 m

Locations containing baths or showers for medical treatment, or for disabled persons, may have special requirements.

8.2.8.2 Protective measures

The protective measures of obstacles, placing out of reach, non-conducting location and Earth-free local equipotential bonding are **not** permitted.

Where SELV or PELV is used, basic protection for equipment in zones 0, 1 and 2 shall be provided by:

- barriers or enclosures;
- or protection by RCDs.

8.2.8.2.1 ELECTRICAL SEPARATION

Protection by electrical separation shall only be used for circuits supplying:

- one item of current-using equipment; or
- one single socket-outlet.

8.2.8.2.2 SUPPLEMENTARY EQUIPOTENTIAL BONDING

- Local supplementary protective equipotential bonding shall include all simultaneously accessible exposed-conductive-parts of fixed equipment and extraneous-conductive-parts.
- The equipotential bonding system shall be connected to the protective conductors of all equipment including those of socket-outlets.
- Special requirements exist for locations containing baths and showers for medical use or for disabled persons (see Building Regulations Part M) and Section 710 of BS 7671:2018).

8.2.8.2.3 ELECTRIC FLOOR HEATING SYSTEMS

- The so-called protective measure, '*Protection by electrical separation*' is **not** permitted for electric floor heating systems and only heating cables or thin sheet flexible heating elements may be installed – **provided** that they have either a metal sheath or a metal enclosure or a fine mesh metallic grid which is connected to the protective conductor of the supply circuit.

Note: Compliance with the latter requirement is not required if the protective measure SELV is provided for the floor heating system.

8.2.8.3 *Selection and erection of equipment*

The following requirements do not apply to switches and controls which are part of fixed current-using equipment suitable for use in a particular zone or to insulating pull cords of cord operated switches.

8.2.8.3.1 IN ZONE 0

- Switchgear or accessories shall **not** be installed.

8.2.8.3.2 IN ZONE 1

- Only the switches of a SELV circuit may be installed with the safety source installed outside zones 0, 1 and 2.

8.2.8.3.3 IN ZONE 2

Switchgear, controlgear and accessories that include switches or socket-outlets shall **not** be installed with the exception of:

- switches and socket-outlets of SELV;
- shaver supply units complying with BS EN 61558-2-5.

All other socket-outlets are prohibited within a distance of 3 m horizontally from the boundary of zone 1.

8.2.8.4 *Current-using equipment*

8.2.8.4.1 ZONE 0

Current-using equipment may only be installed provided that:

- the equipment is fixed and permanently connected;
- the equipment is protected by SELV.

8.2.8.4.2 ZONE 1

Only the following fixed and permanently connected current-using equipment may be installed:

- electric showers;
- equipment protected by SELV or PELV (provided that the source is installed outside zones 0, 1 and 2);
- luminaires;
- shower pumps;
- towel rails;
- ventilation equipment;
- water heating appliances;
- whirlpool units.

8.2.8.4.3 EXTERNAL INFLUENCES

Installed electrical equipment shall have at least the following degrees of protection:

- in zone 0: protection from accidental submersion in 1 m of water for up to 30 minutes;
- in zones 1 and 2: resistance to water splashes from any direction.

Author's Hint

See BS 7671:2018 Section 701 for more detailed requirements relating to Locations containing a bathroom of shower.

8.2.9 *Marinas and similar locations*

The particular requirements of this section are applicable only to circuits intended to supply pleasure craft or houseboats in marinas and similar locations. They do **not** apply to the supply for houseboats (if they are normally supplied directly from the public network) **or** to the **internal** electrical installations of pleasure craft or houseboats.

8.2.9.1 *General*

- The nominal supply voltage to pleasure craft or houseboats is 230 V a.c. single-phase, or 400 V a.c. three-phase.

In the UK, the ESQCRs prohibit the connection of a PME earthing device to any metalwork of a boat.

8.2.9.2 *Protective measures*

For marinas, particular attention has to be given to the likelihood of corrosive elements, movement of structures, mechanical damage, presence of flammable fuel and the increased risk of electric shock.

The protective measures of obstacles, placing out of reach, non-conducting location and Earth-free local equipotential bonding are **not** permitted.

8.2.9.2.1 PROTECTION AGAINST OVERCURRENT

- Each socket-outlet shall be protected by an individual overcurrent protective device.
- A fixed connection for the supply to each houseboat shall be protected by an overcurrent protective device.

8.2.9.2.2 PROTECTION BY RCDS

- Socket-outlets shall be protected individually by an RCD.
- Final circuits intended for a fixed connection of supply to houseboats shall be protected individually by an RCD which shall disconnect all poles, including the neutral.

8.2.9.3 Selection and erection of equipment

8.2.9.3.1 ISOLATION

- At least one means of isolation (with a maximum of four outlet-sockets) shall be installed in each distribution cabinet.

This switching device shall disconnect all live conductors including the neutral conductor.

8.2.9.4 Supplies

Socket-outlets shall:

- comply with BS EN 60309-1 (above 63 A) and BS EN 60309-2 (up to 63 A).

Every socket-outlet shall be:

- located as close as practicable to the berth to be supplied;
- installed in the distribution board or in separate enclosures which can hold a maximum of four socket-outlets;
- placed at a height of not less than 1 m above the highest water level;
- In general, single-phase socket-outlets (with a rated voltage of 200 V–250 V and rated current 16 A) shall be provided.

One socket-outlet shall supply one pleasure craft or houseboat – only!

Author's Hint

See BS 7671:2018 Section 709 for more detailed requirements relating to marinas and similar locations.

8.2.10 Medical locations

BS 7671:2018 Section 710 is a relatively new Section specifically aimed at electrical installations in medical locations which was originally produced for the 2008 edition of this standard. It has now been brought into alignment with the Medical Devices Directive (MDD) and other International, European and British standards; plus, HTM 06-01, published by the Department of Health.

Note: Unfortunately, the numbering system of this particular sub-section in BS 7671:2018 rather seems to have a mind of its own which makes it difficult to cross refer – so be prepared!

8.2.10.1 General

BS 7671:2018 Section 710 is aimed at patient healthcare facilities and although the requirements mainly refer to hospitals, private clinics, medical and dental practices, healthcare centres and dedicated medical rooms in the workplace, this section will also apply to electrical installations in locations designed for medical research and (where applicable) to veterinary clinics.

The requirements do not, however, apply to the actual medical electrical equipment itself, as this is fully covered in ISO 13485 (The Medical Devices Directive) and the BS EN 60601 series on Medical Electrical Equipment and Systems.

8.2.10.1.1 OFFICIAL REQUIREMENTS FOR HOSPITALS

HTM 06-01(a Government guidance document entitled *Electrical services supply and distribution*) provides the legal requirements, design applications, operation and maintenance of the electrical infrastructure within healthcare premises.

The approved layout of a typical Hospital Operating Theatre is shown in Figure 710.2 together with a typical medical IT in Figure 710.3 of BS 7671:2018.

To save you having to hunt through this Standard and other Government and official documents for the meaning of some of the medical terms mentioned within this section, and elsewhere in my book, Annex 8.1 to Section 8.3.10 provides an aide memoire to the most important ones.

8.2.10.2 *Protective measures*

Within patient healthcare facilities, the risk to patients suffering through reduced body resistance is enhanced and it is extremely important that the following rules and regulations are strictly adhered to.

8.2.10.2.1 EXPOSED-CONDUCTIVE-PARTS OF EQUIPMENT

- In Group 2 medical locations, where PELV is used, exposed-conductive-parts of equipment, (e.g. operating theatre luminaires) shall be connected to the circuit protective conductor.

8.2.10.2.2 EQUIPOTENTIAL BONDING BUSBAR

- Equipotential bonding busbars must be located in or near the medical location using a protective conductor.

8.2.10.2.3 SUPPLEMENTARY EQUIPOTENTIAL BONDING

For Group 1 and Group 2 medical locations, supplementary equipotential bonding shall be installed as follows:

- Group 1 medical location – one per patient location;
- Group 2 medical location – a minimum of four connection points.

The equipotential bonding busbar needs to be located in or near the medical location and all connections must be accessible, labelled, clearly visible and capable of being easily being disconnected individually.

8.2.10.2.4 PROTECTION AGAINST ELECTRIC SHOCK

The protective measures:

- of obstacles and placing out of reach;
- non-conducting location;

- Earth-free local equipotential bonding; and
- electrical separation for the supply of more than one item of current-using equipment;

are **not** permitted in medical locations.

8.2.10.2.5 FUNCTIONAL EXTRA-LOW VOLTAGE

- FELV is **not** permitted as a method of protection against electric shock in medical locations.

8.2.10.2.6 EXTRA-LOW VOLTAGE PROVIDED BY SELV OR PELV

- When using SELV and/or PELV circuits in Group 1 and Group 2 medical locations, protection by basic insulation of live parts must be provided.
- Where PELV is used In Group 2 medical locations, all exposed-conductive-parts of equipment (such as operating theatre luminaires) must be connected to the circuit protective conductor.

8.2.10.3 *Safety services*

- Medical locations shall have safety standby power supply systems available to energise the installations that are required for continuous operation in case of failure of the general power system.
- This safety standby power supply system shall be capable of automatically taking over if the main power supply drops for more than 0.5 s and by more than 10%.

A list of examples with suggested reinstatement times is given in Table A710 of Annex A710 to BS 7671:2018.

8.2.10.3.1 EMERGENCY LIGHTING SYSTEMS

In the event of mains power failure, the changeover period to the standby the safety services sources such as:

- emergency lighting and exit signs;
- switchgear and control gear for emergency generating sets;
- essential service rooms;
- locations of central fire alarm and monitoring systems;

shall not exceed 15 s. In addition:

- Group 1 medical locations, rooms shall be supplied with at least one luminaire in case of emergency;
- Group 2 medical locations, rooms shall have at least of 90% of their normal lighting requirements supplied from the standby safety service.

Note: Escape route luminaires shall be arranged on alternate circuits.

8.2.10.3.2 OTHER SERVICES

Services such as:

- electrical equipment for medical gas supply;
- fire detection and fire alarms;
- fire extinguisher systems;
- firefighter lifts;
- paging systems;
- smoke extractions; and
- ventilation systems;

and other vitally important ME equipment used in Group 2 medical locations, may require a changeover system to a standby safety service within 15 s.

8.2.10.4 Earthing

In accordance with the requirements of the ESQCR, PEN conductors **shall not** be used in medical locations and medical buildings downstream of the main distribution board.

8.2.10.4.1 MEDICAL FACILITY EARTH FAULTS

- In the event of a first fault to Earth, a total loss of supply in medical Group 2 locations shall be prevented.
- Cables intended to supply temporary structures shall be protected at their origin by an RCD.
- Supplementary protective equipotential bonding shall be installed via the shortest route to the earthing conductor.

RCDs **should** be used for:

- lighting in places such as telephone kiosks, bus shelters, advertising panels;
- socket-outlet circuits not exceeding 32 A and all final circuits other than for emergency lighting;
- surgical applications; and
- where an electrical installation includes a PV power supply system without a simple separation between the a.c. side and the d.c. side.

RCDs should **not** be used:

- for final circuits supplying medical electrical equipment and systems intended for life support;
- where a medical IT system is functioning.

8.2.10.5 *Selection and erection of equipment*

Distribution boards shall meet the requirements of BS EN 61439 series.

8.2.10.5.1 RISK OF EXPLOSION

The gases used as anaesthetics in operating theatres are extremely flammable if present in high concentrations.

- For this reason, all electrical devices (such as socket-outlets and/or switches) should be installed at least 0.2 m below any medical-gas outlet, so as to minimise the risk of ignition of flammable gases.

8.2.10.6 *Supplies*

- In medical locations at least two different sources of supply shall be provided, one for the electrical supply system and one for safety services.
- Automatic changeover facilities from the main distribution network to an electric safety source shall comply with BS EN 60947-6-1.

8.2.10.6.1 TRANSFORMERS

- Transformers should always be installed in close proximity to a medical location and they shall comply with the requirements of BS-EN-61558-2-15 and BS EN 61439.

Capacitors **shall not** be used in transformers for medical IT systems.

8.2.10.7 Wiring systems

Owing to the growing number of medical implants [for example a Cardiac Resynchronisation Therapy Defibrillator (CRT-D)] or 'biventricular pacemaker'. to guard against heart failure and a cardiac arrest) special considerations have to be made concerning EMI and EMC as not all medical equipment is housed inside a Faraday Cage (but fortunately my CRT-D has one!).

8.2.10.7.1 ARC FAULT DETECTION DEVICES

- In Group O medical locations, AFDDs shall be used subject to a risk assessment.
- In Group 1 and Group 2 medical locations, AFDDs are **not** required to be installed.

8.2.10.7.2 RCDS

Care shall be taken to ensure that simultaneous use of many items of equipment connected to the same circuit cannot cause unwanted tripping of the associated RCD.

In Group 1 and Group 2 medical locations, the following shall apply:

- where RCDs are required, only type A (according to BS EN 61008 and BS EN 61009) or type B (according to IEC 62423) shall be selected, depending on the possible fault current arising.

Where a medical IT system is used, additional protection by means of an RCD shall **not** be used.

8.2.10.7.3 TN SYSTEMS

- In Group 1 final circuits, rated up to 63 A, RCDs shall be used.
- In Group 2 final circuits, medical locations, protection by automatic disconnection of supply by means of RCDs may only be used in a medical IT system.

8.2.10.7.4 IT SYSTEMS

- For each group of rooms serving the same function, at least one IT system is necessary and each IT system shall be equipped with an Insulation Monitoring Device (MED-IMD) in accordance with Annex A and Annex B of BS EN 61557-8.

In Group 2 ME locations, an IT system shall always be used for:

- final circuits supplying medical electrical equipment and systems intended for life support;
- surgical applications; and
- other electrical equipment located in the '*patient environment*'.

For each group of rooms serving the same function, at least one IT system is necessary and this shall be equipped with an IMD that has an acoustic and visual alarm system and this alarm system shall have:

- a green signal lamp to indicate normal operation;
- a yellow signal lamp which lights when the minimum value set for the insulation resistance is reached;
- an audible alarm which sounds when the minimum value set for the insulation resistance is reached.

Note: A record of all faults occurring in a medical location shall be maintained

8.2.10.7.5 IT SYSTEM ME SOCKET-OUTLETS

It is a mandatory requirement that socket-outlet circuits in the medical IT system for Group 2 medical locations:

- intended to supply ME equipment shall be unswitched;
- shall be coloured blue and clearly and permanently marked '*Medical Equipment Only*'.

At each patient's place of treatment (e.g. bedheads):

- each socket-outlet shall be supplied by an individually protected circuit; or
- several socket-outlets shall be separately supplied by a minimum of two circuits.

8.2.10.8 *Inspection and testing*

International Electrotechnical Commission (IEC) and the BSI manufacturers' standards for medical electrical equipment consist of two types of testing - Type Testing and Routine Testing.

8.2.10.8.1 TYPE TESTING

'*Type testing*' is carried out by an approved test on a single representative sample equipment for which certification of compliance with a standard is being sought.

8.2.10.8.2 ROUTINE TESTING

'*Routine testing*', on the other hand, is intended to provide an indication of the inherent safety of the equipment without subjecting it to undue stress that would be liable to cause deterioration.

8.2.10.9 Documentation related to the installation

Once the installation is completed, an overall plan of the electrical installation together with records, drawings, wiring diagrams and modifications relating to the medical location, shall be provided. These should consist of – but are not limited to:

- single line overview diagrams of the distribution system of the normal power supply and the power supply for safety services;
- distribution board block diagrams showing switchgear, controlgear and distribution boards;
- schematic diagrams of controls;
- functional description for the operation of the safety power supply system;
- verification of compliance with the requirements of standards.

8.2.10.9.1 INITIAL VERIFICATION

In addition to the requirements of BS 7671:2018 Chapter 64 and HTM 06-01 (Part A) the following tests **shall** be carried out, both prior to commissioning and after alteration or repairs and before re-commissioning:

- complete functional tests of all IMDs associated with the medical IT system;
- measurements of leakage current from the IT transformers of the output circuit and enclosure in no-load condition;
- measurements to verify that the resistance of the supplementary equipotential bonding is within stipulated limits.

8.2.10.9.2 PERIODIC INSPECTION AND TESTING

Periodic inspection and testing should be carried out in accordance with HTM 06-01 (Part B) and local Health Authority requirements as follows (and at the given intervals):

- **Annually** – complete functional tests of all IMDs associated with the medical IT system;
- **Annually** – measurements to verify that the resistance of the supplementary equipotential bonding is within the stipulated limits;
- **Every 3 years** – measurements of leakage current of the output circuit and of the enclosure of the medical IT transformers in no-load condition.

The dates and results of each verification **shall** be recorded.

Author's Hint

See BS 7671:2018 Section 710 for more detailed requirements relating to Medical locations.

8.2.10.10 Annex 8.1 to Section 8.3.10

This Annex is intended as an aide memoir to the meaning to some of the most important medical terms used within this section, and elsewhere in the book.

Applied part refers to that part of the medical electrical equipment that, whilst in normal use, comes into physical contact with the patient.

Intracardiac Procedure is a procedure whereby an electrical conductor is placed within the heart of a patient or is likely to come into contact with the heart, such conductor being accessible outside the patient's body. In this context, an electrical conductor includes insulated wires such as cardiac pacing electrodes or intracardiac Electrocardiogram (ECG) electrodes, or insulated tubes filled with conducting fluids.

Medical electrical equipment (ME equipment). Electrical equipment having an applied part for transferring energy to or from the patient or detecting such energy transfer to or from the patient and which is:

- provided with not more than one connection to a particular supply main; and
- intended by the manufacturer to be used;

- in the diagnosis, treatment or monitoring of a patient; or
- for compensation or alleviation of disease, injury or disability.

ME equipment also includes the accessories that have been defined by the manufacturer as being necessary for enabling normal use of the ME equipment.

Medical electrical system (ME system). Combination, as specified by the manufacturer, of items of equipment (and their accessories), at least one of which is medical electrical equipment to be interconnected by functional connection or by use of a multiple socket-outlet.

Medical IT system. IT electrical system fulfilling specific requirements for medical applications.

Note: These supplies are also known as '*isolated power supply systems*'.

Medical location. A location intended for purposes of diagnosis, treatment including cosmetic treatment, monitoring and care of patients.

Patient. Living being (person or animal) undergoing a medical, surgical or dental procedure.

Patient environment. Any volume in which intentional or unintentional contact can occur between a patient and parts of the medical electrical equipment or medical electrical system or between a patient and other person touching parts of the medical electrical equipment or system.

8.2.10.10.1 SAFETY

Electrical installations in medical locations [and also electrical installations in locations designed for medical research and (where applicable) veterinary clinics] must be capable of ensuring the continued safety of patients and medical staff. For convenience, these medical locations are classified into groups depending on the level of safety required. For example:

- the type of contact between applied parts and the patient;
- the threat to the safety of the patient that represents a discontinuity (failure) of the electrical supply; as well as
- the purpose for which the location is used.

To ensure the protection of patients from possible electrical hazards, additional protective measures need to be applied in medical locations.

Care should also be taken to ensure that other installations do not compromise the level of safety provided by installations meeting the requirements of this section.

8.2.10.10.2 MEDICAL LOCATIONS

Medical locations are split into Groups as follows:

Group 0. Medical location where no applied parts are intended to be used and where discontinuity (failure) of the supply cannot cause danger to life.

Group 1. Medical location where discontinuity of the electrical supply does not present a threat to the patient.

8.2.10.10.3 CLASSES AND TYPES OF MEDICAL ELECTRICAL EQUIPMENT

All medical electrical equipment is categorised into classes according to the type of protection against electric shock that it uses.

8.2.10.10.3.1 Class I equipment Class I equipment has a protective Earth where the basic means of protection is the insulation between live parts and exposed conductive parts such as the metal enclosure. In the event of a fault (that would otherwise cause an exposed conductive part to become live) the supplementary protection (i.e. the protective Earth) comes into effect.

Note: Large fault current flows from the mains part to Earth via the protective Earth conductor, which will cause a protective device (usually a fuse) in the mains circuit to disconnect the equipment from the supply.

8.2.10.10.3.2 Class II equipment Class 2 equipment uses either double insulation or reinforced insulation. In double insulated equipment the basic protection is provided by the first layer of insulation. If the basic protection fails then supplementary protection is provided by a second layer of insulation – thus preventing contact with live parts.

8.2.10.10.3.3 Class III equipment Class III equipment is either battery operated or supplied by a SELV transformer and as the voltages do not exceed 25 V a.c or 60 V d.c. protection against electric shock is a minimal requirement.

8.2.10.10.4 PATIENT SAFETY

Since the hazard to people will depend on the treatment being administered, hospital locations are divided into groups as follows.

Group Zero: where no treatment or diagnosis using medical electrical equipment is administered, e.g. consulting rooms.
Group One: where medical electrical equipment is in use, but not for treatment of heart (intracardiac) conditions.
Group Two: where medical electrical equipment is in use for heart (intracardiac) conditions.

8.2.10.11 Annex 8.2 to Section 8.3.10

This Annex contains examples for some of the Group numbers that are allocated for classification of the medical location safety services listed in Table 8.2.

8.2.11 Mobile and transportable units

8.2.11.1 General

For the purposes of this section, the term 'unit' is intended to mean a vehicle and/or mobile (self-propelled or towed) transportable structure (such as a container or cabin) in which all (or part of) an electrical installation is contained and which is provided with a temporary supply by means of, for example, a plug and socket-outlet.

8.2.11.2 Protective measures

The protective measures of obstacles, placing out of reach, non-conducting location and Earth-free local equipotential bonding are **not** permitted.

Automatic disconnection of the supply shall be provided by means of an RCD.

8.2.11.2.1 PROTECTIVE EQUIPOTENTIAL BONDING

- The chassis and other accessible conductive parts of the unit shall be connected through the main protective bonding conductors to the unit's main earthing terminal.

Table 8.2 Medical location services

	Group		
Medical location	*0*	*1*	*2*
Anaesthetic area			X
Angiographic examination room			X
Bedrooms		X	
Delivery room		X	
ECG, EEG, EHG room		X	
Endoscopic room		X	
Examination or treatment room		X	
Haemodialysis room		X	
Heart catheterisation room			X
Hydrotherapy room		X	
Intensive care room			X
Intermediate Care Unit (IMCU)			X
Magnetic resonance imaging (MRI) room		X	X
Massage room	X	X	
Nuclear medicine		X	
Operating plaster room			X
Operating preparation room			X
Operating recovery room			X
Operating theatre			X
Premature baby room			X
Physiotherapy room		X	
Radiological diagnostic and therapy room		X	X
Urology room		X	

8.2.11.2.2 SOCKET-OUTLETS

- socket-outlets that are intended to supply current-using equipment outside the unit will have an RCD protection device with the exception of socket-outlets which are supplied from circuits that are protected by:
    - SELV;
    - PELV; or
    - electrical separation.

8.2.11.2.3 PME EARTHING SYSTEM

A PME earthing facility shall **not** be used as a means of earthing.

8.2.11.3 Selection and erection of equipment

The connecting devices used to connect the unit to the supply shall be:

- housed within an enclosure of insulating material which will provide a degree of protection not less than IP44.

8.2.11.3.1 PROXIMITY TO NON-ELECTRICAL SERVICES

- Except for ELV equipment, no electrical equipment (or wiring system) shall be installed in a gas cylinder storage compartment.

If cables have to run through such a compartment, they shall be protected against mechanical damage by either:

- a conduit complying with the BS EN 61386 series; or
- a ducting system complying with the appropriate part of the BS EN 50085 series.

Both of which shall be able to withstand an impact equivalent to AG3 without visible physical damage.

- ELV cables and electrical equipment may only be installed within the LPG cylinder compartment – **if**, the installation indicates the operation of the gas cylinder (e.g. indication of empty gas cylinders) or is for use within the compartment.

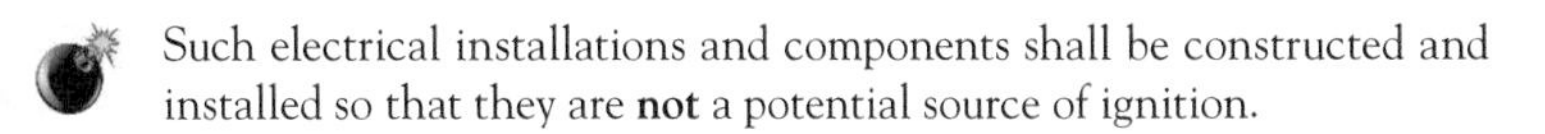

Such electrical installations and components shall be constructed and installed so that they are **not** a potential source of ignition.

8.2.11.3.2 SOCKET-OUTLETS

- Socket-outlets located outside the unit shall be provided with an enclosure affording a degree of protection not less than IP44.

8.2.11.3.3 SUPPLIES

Generating sets that are able to produce voltages other than SELV or PELV (and which are mounted in a mobile unit) **shall be** automatically switched off in case of an accident to the unit.

8.2.11.3.4 IT SYSTEMS

An IT system can be provided by:

- an isolating transformer or a low voltage generating set; or
- a transformer providing simple separation (see BS EN 61558-1) with an RCD and an Earth electrode installed to provide automatic disconnection in the in case the transformer fails.

8.2.11.4 Wiring systems

The wiring system shall be installed using one or more of the following:

- unsheathed flexible cable installed in either a conduit (complying with the BS EN 61386 series) or trunking or ducting; (that complies the BS EN 50085 series);
- sheathed flexible cable with thermoplastic or thermosetting insulation.

When installed, the conduit or duct shall be able to withstand an impact equivalent to AG3 without visible physical damage.

Flexible cables (used for connecting the unit to the supply) shall have a minimum cross-sectional area of 2.5 mm^2 copper.

8.2.11.5 Identification and notices

A permanent notice shall be fixed to the unit in a prominent position stating:

- the type of supplies which may be connected to the unit;
- the voltage rating of the unit;
- the number of supplies, phases and their configuration;
- the on-board earthing arrangement;
- the maximum power requirement of the unit.

Author's Hint

See BS 7671:2018 Section 717 for more detailed requirements relating to Mobile and transportable units.

8.2.12 Exhibitions, shows and stands

8.2.12.1 General

- Where mobile equipment is likely to be used, the socket-outlet, plugs and devices that are used to connect the unit to the supply shall comply with BS EN 60309-2.

The protective measures of obstacles, placing out of reach, non-conducting location and Earth-free local equipotential bonding are **not** permitted.

Automatic disconnection of the supply shall be provided by means of an RCD.

8.2.12.2 *Protective equipment*

- Accessible conductive parts of the unit, such as the chassis, shall be connected through the main protective bonding conductors to the main earthing terminal.

8.2.12.2.1 IT SYSTEM

- Every socket-outlet of an IT system that is intended to supply current-using equipment outside the unit, shall be protected by an RCD – with the exception of socket-outlets that are supplied from circuits already protected by:
    - SELV; or
    - PELV; or
    - electrical separation.

8.2.12.2.2 TN SYSTEM

- A PME earthing facility shall **not** be used with a TN system as a means of earthing.

8.2.12.2.3 ADDITIONAL RCD PROTECTION

- In a.c. systems, additional protection by means of an RCD shall be provided for mobile equipment with a current rating not exceeding 32 A when used outdoors.

8.2.12.3 *Selection and erection of equipment*

Connecting devices used to connect the unit to the supply shall:

- meet the requirements of the BS EN 60309-2 series;
- be protected against solid objects that are bigger than 1 mm and water splashing from all directions; and
- plugs shall be within an insulated enclosure.

8.2.12.3.1 PROXIMITY TO NON-ELECTRICAL SERVICES

Other than ELV equipment for gas supply control, no electrical equipment or wiring systems shall be installed in any gas cylinder storage compartment unless protected against solid objects and water.

- Cables running through a gas cylinder storage compartment shall be protected from mechanical damage by storing them inside a conduit system (see BS EN 61386) or within a ducting system (see BS EN 50085).
- ELV cables and electrical equipment may only be installed within the LPG cylinder compartment **if**, the installation is for use within the compartment.

8.2.12.4 *Supplies*

Mobile generating sets which produce voltages other than SELV or **shall be** automatically switched off in case of an accident to the unit.

8.2.12.4.1 IT SYSTEM

An IT system can be provided by:

- an isolating transformer or a low voltage generating set which includes an insulation monitoring and fault location system; or
- by a transformer that provides simple separation (in accordance with BS EN 61558-1) and automatic disconnection in case of transformer failure.

8.2.12.5 *Wiring systems*

- Flexible cables (having a minimum cross-sectional area of 2.5 mm^2 copper and mechanically protected by enclosing them within a conduit or ducting system) shall enter the unit via an insulated inlet so as to avoid the possibility of energising exposed-conductive-parts of the unit.

8.2.12.6 *Identification and notices*

A permanent notice shall be fixed adjacent to each supply inlet connector stating:

- the type of supplies which may be connected to the unit;
- the number of supplies, phases and their configuration;
- the voltage rating of the unit;
- the on-board earthing arrangement;
- the maximum power requirement of the unit.

Author's Hint

See BS 7671:2018 Section 711 for more detailed requirements relating to exhibitions, shows and stands.

8.2.13 *Operating and maintenance gangways*

Section 729 of BS 7671:2018 is a comparatively small section which centres on the operation and safe maintenance of switchgear and controlgear within areas that include gangways and where access is restricted to skilled or instructed persons.

8.2.13.1 *Requirements for operating and maintenance gangways*

Gangways shall permit at least a 90° opening of equipment doors or hinged panels.

For further details, please refer to Annex A 729 of BS 7671:2018.

- Gangways longer than 10 m shall be accessible from both ends.

See Figure 729.3 of BS 7671:2018 for examples of positioning of doors in closed restricted access.

8.2.13.2 Restricted access areas

Restricted access areas shall:

- be clearly and visibly marked by appropriate signs;
- not provide access to unauthorised persons; and
- provide closed restricted access areas which will allow easy evacuation in case of danger, without any form of opening mechanism.

Where basic protection is provided by barriers or enclosures, the following minimum dimensions shown in Table 8.3 apply after barriers and enclosures are in a fixed position and circuit-breakers and switch handles are in the most difficult position, including 'isolation'. (Also see Figure 729.1 of BS 7671:2018).

Table 8.3 Minimum gangway dimensions

Gangway	*Dimensions*
Gangway width between barriers or enclosures and switch handles or circuit-breakers • in the most onerous position, or : and • in the most onerous position and the wall	700 mm
Gangway width between barriers or enclosures or other barriers or enclosures and the wall	700 mm
Height of gangway to barrier or enclosure above floor.	2000 m
Live parts that have been placed out of reach	2500 mm

Author's Hint

See BS 7671:2018 Section 729 for more detailed requirements relating to Operating and maintenance gangways.

8.2.14 Rooms and cabins containing saunas

Saunas, similar to swimming pools, bathrooms and showers, etc., are primarily used by people who are unclothed and wet and thus very vulnerable to electric shock due to their reduced body resistance. Special measures are, therefore, required to ensure that the possibility of direct and/or indirect contact is reduced.

8.2.14.1 General

BS 7671:2018's requirements for hot air saunas are based on three zones which take into account the limitations of walls, doors, fixed partitions, ceilings and floors and the electric heater itself. The three zones are shown in Figure 8.7 and description of the zone in Table 8.4.

8.2.14.2 Protective measures

The protective measures of obstacles, placing out of reach, non-conducting location and Earth-free local equipotential bonding are **not** permitted.

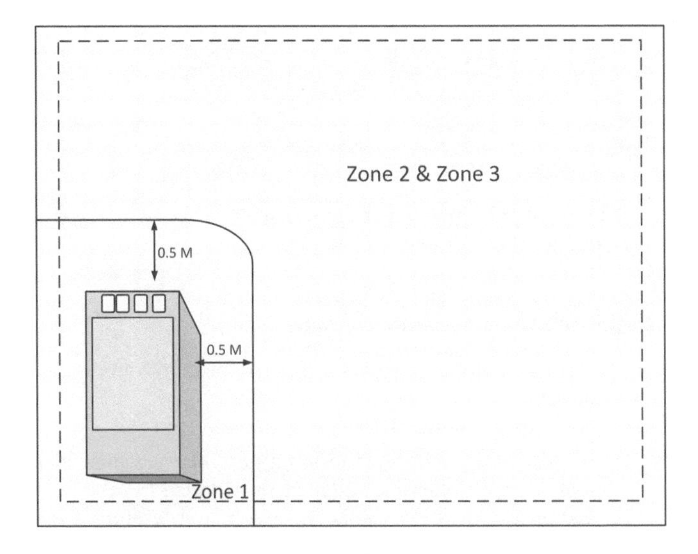

Figure 8.7 Zones of ambient temperature

Table 8.4 Ambient temperature zones

Zone 1	In zone 1, only the sauna heater and equipment belonging to the sauna heater shall be installed The section containing the sauna heater is limited by the floor, the cold side of the thermal insulation of the ceiling and a vertical surface around the sauna heater, 0.5 m from its surface
Zone 2	Zone 2 is the volume outside zone 1, limited by the floor, the cold side of the thermal insulation of the walls and a horizontal surface located 1.0 m above the floor In zone 2: there is no special requirement concerning heat-resistance of equipment
Zone 3	Zone 3 is the volume outside zone 1, limited by the cold side of the thermal insulation of the ceiling and walls and a horizontal surface located 1.0 m above the floor Equipment shall withstand a minimum temperature of 125°. Insulation and sheaths of cables shall withstand a minimum temperature of 170°C

8.2.14.2.1 EXTERNAL INFLUENCES

- Equipment shall have a degree of protection against water splashes and a low-pressure water jet spray.

8.2.14.2.2 PROTECTION BY RCDS

- Additional protection shall be provided for all circuits of the sauna, by the use of one or more RCDs.

8.2.14.3 *Selection and erection of equipment*

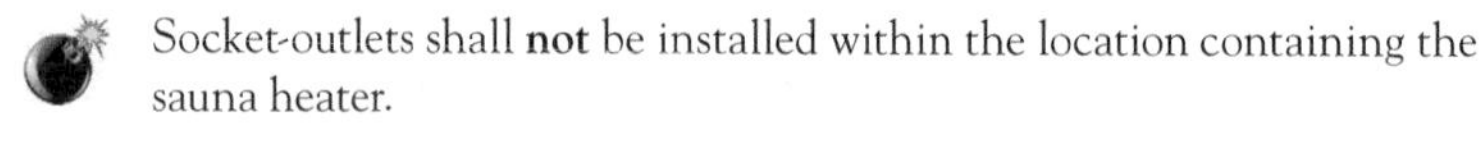

Socket-outlets shall **not** be installed within the location containing the sauna heater.

- Switchgear and controlgear forming **part of** the sauna heater equipment or of other fixed equipment installed in zone 2, may be installed within the sauna room or cabin whilst all other switchgear and controlgear (e.g. for lighting) must be placed outside.

8.2.14.4 *Supplies*

Where SELV or PELV is used, basic protection shall be provided by:

- *covering*, all live parts with an insulation that can only be removed by destruction; or
- *barriers or enclosures* that provide protection to at least IP4X or IPXXD.

8.2.14.5 *Wiring systems*

Metallic sheaths and metallic conduits shall **not** be accessible during normal use.

- The wiring system should preferably be installed outside the zones (i.e. on the cold side of the thermal insulation).
- If the wiring system is installed on the warm side of the thermal insulation in zones 1 or 3, then it must be heat-resisting.

Author's Hint

See BS 7671:2018 Section 703 for more detailed requirements relating to rooms and cabins containing sauna heaters.

8.2.15 *Solar, photovoltaic power supply systems*

Photovoltaics is a technology that converts light directly into electricity by using photons from sunlight to knock electrons into a higher state of energy, thereby creating electricity. Solar cells are connected in multiples as solar photovoltaic arrays which convert energy from the sun into a usable form of electricity.

8.2.15.1 *Protective measures*

- Where protective bonding conductors are installed, they shall be set parallel to the d.c. cables, a.c. cables and accessories.
- The protective measures of non-conducting location, however, and Earth-free local equipotential bonding shall **not** be used on the d.c. side.

8.2.15.1.1 DOUBLE OR REINFORCED INSULATION

- Protection by the use of Class II or equivalent insulation is the preferred option on the d.c. side.

8.2.15.1.2 EXTRA-LOW VOLTAGE PROVIDED BY SELV OR PELV

- For SELV and PELV systems, the uninterrupted operating control must not exceed 120 V d.c.

8.2.15.1.3 PROTECTION AGAINST ELECTROMAGNETIC INTERFERENCE IN BUILDINGS

- Wiring loops shall be kept as small as possible in order to minimise the effects of lightening causing EMI.

8.2.15.2 Supplies

- Electrical equipment on the d.c. side must all be suitable for direct voltage and direct current.
- On the a.c. side, the PV supply cable shall be connected to the supply side of the overcurrent protective device.
- To allow maintenance of the PV convertor, the equipment should be capable of being isolated both from the d.c. side and the a.c. side. If this is not possible, then an RCD (type B) shall be installed to provide fault protection by automatically disconnecting the supply.

The protective measures of non-conducting location and Earth-free local equipotential bonding are **not** permitted on the d.c. side.

8.2.15.3 Devices for isolation

- A switch-disconnector shall be provided on the d.c. side of the PV convertor.

8.2.15.3.1 EARTHING ARRANGEMENTS

- Earthing of one of the live conductors of the d.c. side is permitted, provided that there is some form of simple separation between the a.c. side and the d.c. side.

Note: Any connections with Earth on the d.c. side should be electrically connected so as to avoid corrosion (see BS EN 13636 and BS EN 15112).

8.2.15.3.2 PROTECTION AGAINST FAULT CURRENT

- The PV supply cable on the a.c. side shall be protected against fault current by an overcurrent protective device installed at the connection point to the a.c. mains.

8.2.15.4 *Wiring systems*

- Wiring systems shall withstand all expected external influences such as wind, ice formation, temperature and solar radiation.

8.2.15.4.1 PV MODULES

- may be connected in series up to the maximum allowed operating voltage of the modules and the PV convertor;
- shall be installed to ensure that there is sufficient heat dissipation during the greatest solar radiation for the site.

8.2.15.5 *Identification and notices*

- All junction boxes (PV generator and PV array boxes) shall carry a warning label indicating that parts inside the boxes may still be live after isolation from the PV convertor.

Author's Hint

See BS 7671:2018 Section 712 for more detailed requirements relating to Solar, photovoltaic (PV) power supply systems.

8.2.16 *Swimming pools and other basins*

The particular requirements of this section apply to the basins of swimming pools, fountains and of paddling pools together with the surrounding Zones of these areas. Swimming pools are, by design, '*wet areas*' as people using them are normally damp (or in most cases soaking wet!)

which will increase their vulnerability to electric shock. Special measures are, therefore, required to ensure that the possibility of direct and/or indirect contact is reduced.

8.2.16.1 General

As shown in Figure 8.8 (and detailed in Table 8.5), the following requirements concern the dimensions of three Zones.

Zones 1 and 2 may be limited by fixed partitions having a minimum height of 2.5 m.

8.2.16.2 Protective measures

In swimming pools and other basins, the protective measures of obstacles, placing out of reach, non-conducting location; and Earth-free local equipotential bonding are **not** permitted.

8.2.16.2.1 ZONES 0 AND 1 (SWIMMING POOLS AND OTHER BASINS)

Except for fountains, in Zone 0 and Zone 1, **only** protection by SELV (installed outside of Zones 0, 1 and 2) is permitted.

- Extraneous-conductive-parts in Zones 0, 1 and 2 must always be connected via supplementary protective bonding conductors to the protective conductors of exposed-conductive-parts of equipment situated in these Zones.

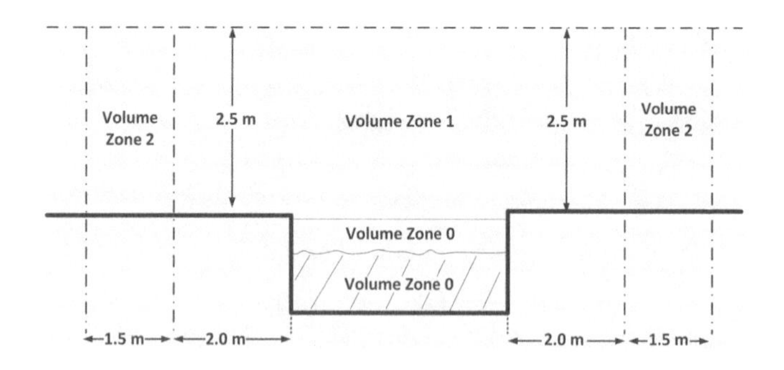

Figure 8.8 Swimming and paddling pool zone dimensions

Table 8.5 Swimming poll zones

Zone 0	This Zone is the interior of the basin of the swimming pool or fountain including any recesses in its walls or floors, basins for foot cleaning and waterjets or waterfalls and spaces below them
Zone 1	This Zone is limited by • Zone 0 • a vertical plane 2 m from the rim of the basin • the floor or surface expected to be occupied by persons • the horizontal plane 2.5 m above the floor or the surface expected to be occupied by persons Where the swimming pool or fountain contains diving boards, springboards, starting blocks, chutes or other components expected to be occupied by persons, Zone 1 comprises the area limited by • a vertical plane situated 1.5 m from the periphery of the diving boards, springboards, starting blocks, chutes and other components such as accessible sculptures, viewing bays and decorative basins • the horizontal plane 2.5 m above the highest surface expected to be occupied by persons
Zone 2	This Zone is limited by: • the vertical plane external to Zone 1 and a parallel plane 1.5 m from the former • the floor or surface expected to be occupied by persons • the horizontal plane 2.5 m above the floor or surface expected to be occupied by persons

Note: There is no Zone 2 for fountains

- Any equipment that is located inside the basin that is **only** intended to be in operation when people are **not** inside Zone 0, shall be supplied by a circuit protected by:
  - SELV;
  - automatic disconnection of the supply (using an RCD); or
  - electrical separation.
- The socket-outlet of any circuit supplying such equipment (and similarly the control device of such equipment) shall have a notice warning users that this equipment maybe used **only** when the swimming pool is **not** occupied by persons.

8.2.16.2.2 ZONES 0 AND 1 (FOUNTAINS)

- In Zone 0, a cable for electrical equipment shall be installed as far outside the basin rim as is reasonably practicable.
- In Zone 1, a cable shall be selected, installed and provided with mechanical protection to medium severity (AG2) and be suitable for submersion in water up to 10 m in depth.
- In Zones 0 and 1 of fountains, one or more of the following protective measures shall be employed:
    - SELV;
    - automatic disconnection of supply (using an RCD);
    - electrical separation.

There is no zone 2 for fountains.

8.2.16.2.3 REQUIREMENTS FOR SELV CIRCUITS

Where SELV is used, basic protection shall be provided by:

- covering all live parts with insulation; or
- barriers or enclosures.

8.2.16.3 *Selection and erection of equipment*

8.2.16.3.1 CURRENT-USING EQUIPMENT OF SWIMMING POOLS

- In Zones 0 and 1, it is **only** permitted to install fixed current-using equipment that has been specifically designed for use in a swimming pool.
- Equipment that is intended to be in operation **only** when people are outside Zone 0, may be used in all Zones, provided that it is supplied via a protected circuit.

8.2.16.3.2 ZONAL ERECTIONS

- In Zones 0, 1 and 2, any metallic sheath (or metallic covering) of a wiring system shall be connected to the supplementary equipotential bonding.

Note: Cables should preferably be installed in conduits made of insulating material.

8.2.16.3.3 EXTERNAL INFLUENCES

Electrical equipment shall have at least the following degree of water proof protection in accordance BS EN 60529:

- Zone 0: IPX8;
- Zone 1: IPX4, IPX5 (where water jets are likely to occur for cleaning purposes);
- Zone 2:
    - IPX2 for indoor locations;
    - IPX4 for outdoor locations;
    - IPX5 where water jets are likely to occur for cleaning purposes.

8.2.16.3.4 JUNCTION BOXES

A junction box shall **not**:

- be installed in Zones 0;
- be installed in Zones 1 unless it is a SELV circuit.

8.2.16.3.5 INSTALLATION OF ELECTRICAL EQUIPMENT IN ZONE 1 OF SWIMMING POOLS AND OTHER BASINS

Fixed equipment (such as filtration systems, jet stream pumps, etc.) that are designed for use in swimming pools and other basins and which are supplied at low voltage, are permitted in Zone 1, subject to all the following requirements:

- the equipment shall be located inside an insulating enclosure providing at least Class II or equivalent insulation and protection against mechanical impact of medium severity (AG2);
- the equipment shall only be accessible via a locked door which shall disconnect all live conductors and the supply cable;
- the equipment's supply circuit shall be protected by:
    - SELV; or
    - an RCD; or
    - electrical separation.

Switchgear, controlgear and socket-outlets shall **not** be installed In Zones 0.

In Zone 2, a socket-outlet or a switch is permitted **only** if the supply circuit is protected by:

- SELV;
- an RCD;
- electrical separation.

8.2.16.3.6 UNDERWATER LUMINAIRES FOR SWIMMING POOLS

- A luminaire for use in the water or in contact with the water shall be fixed and shall comply with BS EN 60598-2-18.
- Underwater lighting located behind watertight portholes, and serviced from behind, shall comply with BS EN 60598 Wiring systems.

8.2.16.3.7 WIRING FOUNTAINS

For a fountain, the following additional requirements shall be met:

- electrical equipment in Zones 0 or 1 shall be provided with mechanical protection;
- a luminaire installed in zones 0 or 1 shall be fixed and shall comply with BS EN 60598-2-18;
- an electric pump shall comply with the requirements of BS EN 60335-2-41.

8.2.16.3.8 PROTECTION, ISOLATION, SWITCHING, CONTROL AND MONITORING

- In Zones 0 or 1, switchgear, controlgear and socket-outlets shall **not** be installed.
- In Zone 2, a socket-outlet or a switch is permitted, but **only** if the supply circuit is protected by an RCD or SELV - or by electrical separation.

8.2.16.4 *Identification and notices*

Any equipment that is located inside the basin that is **only** intended to be in operation when people are **not** inside Zone 0, shall be supplied by a circuit protected by:

- SELV;
- automatic disconnection of the supply (using an RCD); or
- electrical separation.

The socket-outlet of any circuit supplying such equipment (and similarly the control device of such equipment) shall have a notice warning users that this equipment maybe used **only** when the swimming pool is **not** occupied by persons.

Author's Hint

See BS 7671:2018 Section 702 for more detailed requirements relating to Swimming pools and other basins.

8.2.17 *Temporary electrical installations*

Section 740 of BS 7671:2018 specifies the minimum electrical installation requirements for the safe design, installation and operation of temporarily erected mobile or transportable electrical machines and structures which incorporate electrical equipment.

One of the main requirements is that an adequate number of socket-outlets shall be installed to allow the user's requirements to be safely met.

8.2.17.1 General

The design and installation of temporary electrical installations should always take into account that:

- socket-outlets dedicated to lighting circuits which should be placed out of arm's reach;
- when used outdoor, plugs, socket-outlets and couplers must comply with BS EN 60309;
- all plug and socket-outlets (except for SELV) must be of the non-reversible type and capable of being connected to a protective conductor;
- all a.c. socket-outlets deigned for household and similar use should be of the shuttered type (preferably complying with BS 1363).

In certain instances, additional requirements may be necessary for isolation and switching and BS 7671:2018 Table 537.4 summarises the functions provided by the devices for isolation and switching, together with indication of the relevant product standards.

- All low voltage plug and socket-outlets shall conform to BS EN 60309.

A plug and socket-outlet must **not** be used as a device for connecting a water heater and/or boiler to the supply.

Plugs and socket-outlets in a SELV system must **not** have a protective conductor contact.

8.2.17.2 Protective measures

- Switchgear and controlgear needs to be placed in cabinets which can be opened only by the use of a key or a tool.
- The protective measures of obstacles, placing out of reach, non-conducting location and Earth-free local equipotential bonding are **not** permitted.
- In addition, PME earthing facility shall also **not** be used.

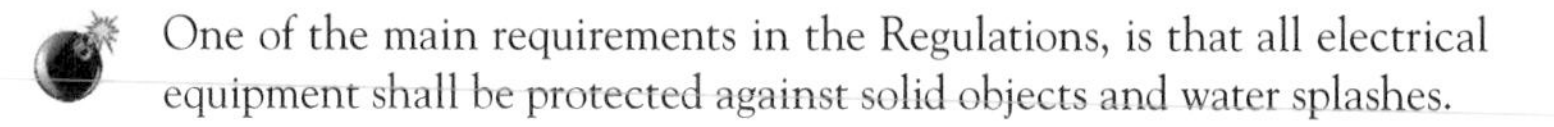

One of the main requirements in the Regulations, is that all electrical equipment shall be protected against solid objects and water splashes.

8.2.17.2.1 PROTECTION AGAINST ELECTRIC SHOCK

- Automatic disconnection of supply to a temporary electrical installation has to be provided at the origin of the installation by one, or more, RCDs.

8.2.17.2.2 RCDS

Owing to the increased risk of damage to cables in temporary installations, all final circuits for:

- lighting;
- socket-outlets; and
- mobile equipment connected by means of a flexible cable or cord;

must be protected by RCDs.

The supply for a battery-operated emergency lighting circuit should be connected to the same RCD protecting the lighting circuit unless the:

- circuits are protected by SELV or PELV; or
- circuits protected by electrical separation; or
- lighting circuits are placed out of arm's reach.

8.2.17.2.3 SUPPLEMENTARY EQUIPOTENTIAL BONDING

- Extraneous-conductive-parts in, or on, the floor (such as reinforcement of cellars for liquid manure) shall be connected to the supplementary equipotential bonding.
- In locations intended for livestock, supplementary bonding shall connect all exposed-conductive-parts and extraneous-conductive-parts that can be touched by livestock.
- Where a metal grid is laid in the floor, it shall be included within the supplementary bonding of the location.

8.2.17.3 Safety services

8.2.17.3.1 SAFETY ISOLATING TRANSFORMERS AND ELECTRONIC CONVERTORS

- A manually reset protective device shall shield the secondary circuit of each transformer or electronic convertor.
- Enclosures containing rectifiers and transformers shall be adequately ventilated and the vents shall **not** be obstructed when in use.
- Electronic converters shall conform to BS EN 61347-2-2.
- Safety isolating transformers shall comply with BS EN 61558-2-6.

8.2.17.4 Selection and erection of equipment

- Switchgear and controlgear should be placed in locked cabinets which can only be opened by the use of a key or a tool, except for those parts designed and intended to be operated by ordinary persons.
- Electrical equipment needs to be protected against solid objects and water splashing.

8.2.17.4.1 JOINTS

- Joints shall **not** be made in cables except where necessary as a connection into a circuit.
- If joints are required, they shall either use connectors in accordance with the relevant British or Harmonized Standard or be made in an enclosure that is dust resistant.

The electrical installation between its origin and the electrical equipment must be inspected and tested after it has been assembled on site.

8.2.17.4.2 ELECTRIC DISCHARGE LAMP INSTALLATIONS

- All luminous tubes, signs or lamps must be installed out of arm's reach or adequately protected to reduce the risk of injury to persons.

8.2.17.4.3 EMERGENCY SWITCHING DEVICE

- A separate circuit (controlled by an emergency switch) **shall** be used to supply luminous tubes, signs or lamps.
- The switch **shall** be easily visible, accessible and marked in accordance with the requirements of the local authority.

8.2.17.4.4 FLOODLIGHTS

- Where transportable floodlights are used, they should be mounted so that the luminaire is inaccessible.
- Supply cables must be flexible and have adequate protection against mechanical damage.
- Luminaires and floodlights need to be fixed and protected against a possible concentration of heat.

8.2.17.4.5 LAMPHOLDERS

- Insulation-piercing lampholders shall **not** be used unless the lampholders are non-removable, once fitted to the cable.

8.2.17.4.6 LAMPS IN SHOOTING GALLERIES

It is essential that all lamps in shooting galleries and other sideshows where projectiles are used are suitably protected against accidental damage.

8.2.17.4.7 LUMINAIRES

- Every luminaire and decorative lighting chain shall:
    - have a suitable IP rating;
    - be installed so as not to spoil its ingress protection; and
    - be securely attached to the structure, or support, intended to carry it.
- Its weight shall not be carried by the supply cable, unless it has been specifically selected for this purpose.

- Access to the fixed light source should only be possible after removing a barrier or an enclosure.
- Lighting chains shall use HO5RN-F (BS 7919) cable or equivalent.

8.2.17.5 *Selection and erection of electrical equipment*

- Electrical installations (including those intended as temporary installations) in booths, stands and amusement devices must have their own means of isolation, switching and overcurrent protection.
- Isolation devices must disconnect all live conductors (i.e. line and neutral conductors).

8.2.17.6 *Supplies*

8.2.17.6.1 AUTOMATIC DISCONNECTION OF SUPPLY

- Supplies to a.c. motors and RCDs (where used) should be of the time-delayed type (BS EN 60947-2) or of the S-type (BS EN 61008-1 or BS EN 61009-1) to prevent unwanted tripping.

8.2.17.6.2 ELECTRIC DODGEMS

- Electric dodgems must be electrically separated from the supply mains by means of a transformer or a motor-generator set and operate at voltages not exceeding 50 V a.c. or 120 V d.c.

Note: Placing out of arm's reach is also acceptable for electric dodgems.

- Each amusement device must have a connection point that is readily accessible and permanently marked to indicate its rated voltage, current and frequency.

8.2.17.6.3 GENERATORS

All generators should be located or protected so as to prevent any danger and injury to people by unintended contact with hot surfaces and dangerous parts.

- Where a generator supplies a temporary installation that is part of a TN, TT or IT system, care shall be taken to ensure that the earthing arrangements are in accordance with the Regulations.

- The neutral conductor of the star-point of the generator shall (except for an IT system) be connected to the exposed conductive-parts of the generator.

8.2.17.6.4 SUPPLY FROM THE PUBLIC NETWORK

- The line and neutral conductors from different sources of supply shall **not** be interconnected downstream from the origin of the temporary electrical installation.

8.2.17.6.5 TN SYSTEM

In accordance with ESQCR, a PME earthing facility shall **not** be used as a means of earthing in a TN system.

8.2.17.7 Wiring systems

8.2.17.7.1 CABLES AND CABLE MANAGEMENT SYSTEMS

- All cables shall meet the requirements of BS EN 60332-1-2.
- Armoured cables or cables that are protected against mechanical damage shall be used wherever there is a possible mechanical risk.
- Buried cables shall be protected against mechanical damage and (if buried in the ground) be marked by electrical tape (as shown by the example in Figure 8.9).
- Cables shall have a minimum rated voltage of 450/750 V.
- Cable trunking systems and cable ducting systems shall comply with BS EN 50085 regarding protection against impact.
- Conduit systems shall comply with the relevant part of the BS EN 61386 series.
- If flexible conduit systems are provided, they shall comply with BS EN 61386-23.
- Mechanical protection shall be used in public areas and in areas where wiring systems are crossing roads or walkways.
- Tray and ladder systems shall comply with BS EN 61537.
- Where subjected to movement, wiring systems shall be of flexible.

Figure 8.9 Warning tape – cable below

Author's Hint

See BS 7671:2018 Section 740 for more detailed requirements relating to Temporary electrical installations for structures, amusement devices and booths at fairgrounds, amusement parks and circuses.

Author's End Note

Having thoroughly reviewed the various basic requirements of the Wiring Regulations, its relationship with the Building Regulations, the need for pay special attention to electrical safety, earthing, external influences and some of the special installations and locations in which electric and electromagnetic equipment have to be controlled, the next chapter looks at the methods to install, maintain and (when required) repair these electrical installations.

9 Installation, maintenance and repair

Author's Start Note

Although this is a comparatively small chapter (compared with some of the others in this book!), it is nevertheless an extremely important chapter as it provides some guidance on the requirements for installation, maintenance and repair. It also lists the Regulations' requirements for maintenance, etc. with respect to electrical installations.

Finally, in Appendix 9.1, there is a set of check lists for the quality, safety and environmental control of electrical equipment and electrical installations.

According to studies recently completed by CEN/CENELEC, the installation and maintenance engineer is the primary cause of reliability degradations during the in-service stage of most electrical installations! The problems associated with poorly trained, poorly supported and/or poorly motivated personnel with respect to reliability and dependability, therefore, requires careful assessment and quantification.

Because of this requirement, quality standards for the installation, maintenance, repair and inspection of in-service products have had to be laid down in engineering standards, handbooks and operating manuals.

Note: Schedules for installing wiring systems (together with guidance for selecting the appropriate size of cable, current ratings and so on) are shown in Appendix 4 Tables 4Aa and 4Ab of BS 7671:2018.

One Part of the Regulations is devoted entirely to inspection and testing (i.e. Part 6) and emphasises the need for continual improvement by stating:

DOI: 10.1201/9781003165170-9

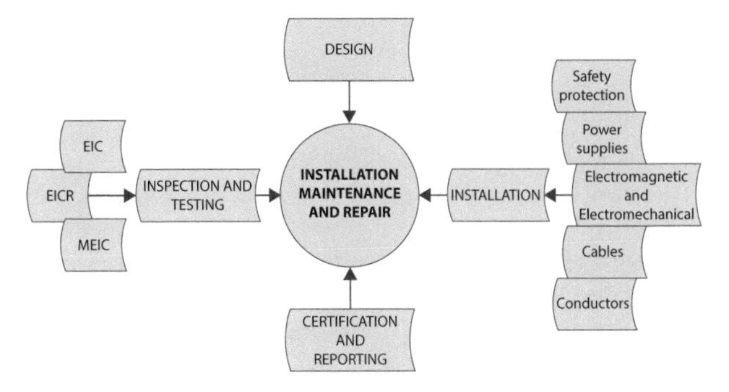

Figure 9.1 Installation, repair and maintenance

> *Every installation (or alteration to an existing installation) shall, during erection and on completion before being put into service, be inspected and tested to verify, so far as is reasonably practicable, that the requirements of the Regulations have been met.*
>
> *The verification shall be made by a competent person and on completion of the verification, a certificate shall be prepared.*

9.1 General

As previously described in Chapter 8, the requirements of BS 7671:2018 apply to the design, erection and verification of virtually all domestic and no-domestic electrical installations **except for**:

- aircraft equipment (e.g. BS EN 6049-003:2018);
- mobile and fixed offshore installations (e.g. IEC 60092-360:2021);
- equipment for motor vehicles (e.g. was BS EN 55012:2002);
- equipment on board ships (e.g. BS 8450:2006);
- lightning protection systems for buildings and structures (e.g. the BS EN 62305 series);
- systems for the distribution of electricity to the public (e.g. the BS EN 61439 series);
- railway rolling stock and signalling equipment (e.g. the BS EN 50152 series);
- lift installations (covered by BS 5655 and the BS EN 81 series);
- mines and quarries specifically covered by Statutory Governmental Regulations.

9.2 Installation

Electrical equipment must always be installed so that:

- design temperatures are not exceeded;
- electrical equipment is not damaged or weakened during the installation process;
- electrical joints and connections are properly constructed with regard to conductance, insulation, mechanical strength and protection;
- equipment is fully accessible for operation, inspection, testing, fault detection, maintenance and repair;
- high temperatures or electric arcs are minimised;
- installation boxes and distribution are enclosed in non-flammable material;
- live parts are inside enclosures or behind barriers identified by a suitable warning label;
- precautions have been taken to prevent persons or livestock from unintentionally touching live parts;
- the risk of ignition of flammable materials has been reduced;
- there is sufficient space for the initial installation (and later replacement of) individual items of electrical equipment.

Figure 9.2 Installation

Author's Hint

There are also some system-specific instructions for special installations and locations contained in Chapter 8 of this book and within Part 7 of BS 7671:2018.

9.2.1 *Restricted access areas*

Section 729 of BS 7671:2018 is a fairly new (and relatively small addition) to the standard which centres on the operation and safe maintenance of switchgear and controlgear within areas that include gangways and where access is restricted to skilled or instructed persons and lists the following requirements:

- *Restricted access areas shall be clearly and visibly marked and not provide access to unauthorised persons;*
- *The width and access of gangways and access areas shall be suitable for a normal working environment whilst taking into account, operational access, emergency access, emergency evacuation and for the transport of equipment.*

Where basic protection is provided by obstacles, barriers or enclosures, the following minimum dimensions shown in Table 9.1 will apply:

Table 9.1 Minimum dimensions of gangways

Gangway	*Dimensions*
Gangway width between • barriers or enclosures, and switch handles or circuit-breakers • obstacles, and switch handles or circuit-breakers	700 mm
Gangway width between • barriers or enclosures, and or other barriers or enclosures and the wall • between obstacles, or other obstacles and the enclosures and the wall	700 mm
Height of gangway to barrier or enclosure above floor	2000 mm
Live parts placed out of reach	2500 mm

9.2.2 Design

The most important requirements for designing the installation of electrical equipment are that:

- equipment is selected and installed with regard to the safety and intended use of the installation;
- equipment and their connections are accessible for operational, inspection and maintenance purposes (except for equipment complying with BS 5733 that is designed to be a maintenance-free accessory);
- and equipment is arranged to allow easy access for periodic inspection, testing and maintenance;
- the design of the electrical installation shall take into account the environmental conditions to which it will be subjected.

Every installation needs to be divided into circuits in order to:

- avoid danger and minimise inconvenience in the event of a fault;
- ensure safe inspection, testing and maintenance;
- prevent the indirect energising of a circuit that is otherwise intended to be isolated;
- reduce the possibility of unwanted tripping of RCDs that are not due to a fault;
- reduce the effects of EMI.

The electrical installation must be designed to ensure:

- a switching device has not been inserted in the neutral conductor, alone;
- installed equipment is appropriate for all predicted external influences;
- the installation of equipment does not affect the protection afforded by an enclosure;
- the protection of persons, livestock and property is guaranteed.

9.2.3 Power supplies

- During operation, inspection, testing, fault detection, maintenance and repair, a disconnection device must be installed.
- In a TN, TT or IT system, an RCD with a rated residual operating current of not more than 30 mA will need to be installed to protect every circuit.

- Stationary batteries need to be installed in a secure environment.
- The power output of a low power supply system should be inspected so as to ensure that it is limited to 500 W for 3-hour duration, or 1500 W for 1-hour duration.
- Batteries shall be of heavy-duty industrial design (such as those complying with BS EN 60623 or BS EN 60896) with a minimum design life of 5 years.

Note: For multiphase circuits, the phase sequence needs to be maintained and overcurrent detection provided for the neutral conductor.

Where an installation includes alternative or additional sources of supply, warning notices (see example in Figure 9.3) shall be affixed at the following locations in the installation:

- at the origin of the installation;
- at the meter position (if remote from the origin).

The warning notice shall have the following wording:

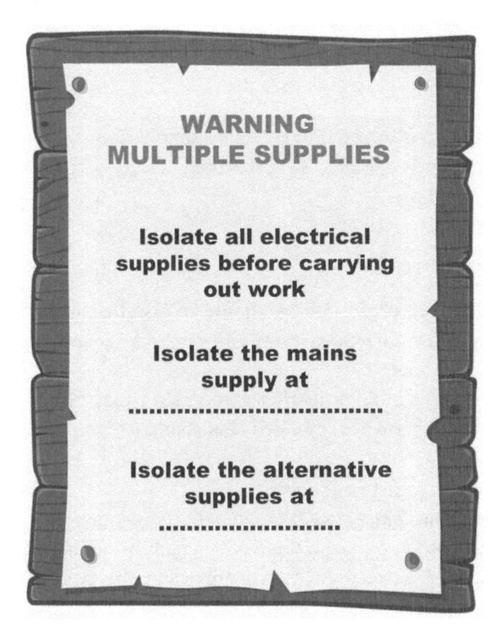

Figure 9.3 Warning notice – alternative or additional sources of supply

- at the consumer unit or distribution board to which the alternative or additional sources are connected;
- at all points of isolation of all sources of supply.

9.2.4 *Safety protection*

Disconnecting devices shall be provided so as to allow electrical installations, circuits or individual items of equipment to be switched off (or isolated) for the purposes of operation, inspection, fault detection, testing, maintenance and repair. In addition:

- measuring and monitoring equipment shall be chosen according to BS EN 61557;
- the protective measures of placing out of reach and obstacles **shall** be used unless the maintenance of equipment is restricted to specially trained skilled persons;
- safety isolating transformers shall comply with BS EN 61558-2-6.

If mechanical maintenance could involve a risk of physical injury, then either a:

- CPS device is inserted in the main supply circuit;
- or a manually operated switch is used.

9.2.5 *Protective devices*

Overload fault protective devices should be installed at any point where a reduction in the current-carrying capacity of an installation's conductors is likely to occur.

An overload device protecting a conductor may be installed along the run of that conductor provided that that part is:

- protected against fault current;
- shorter than 3 m;
- installed so as to reduce the risk of a fault occurring; and
- installed so as to reduce to a minimum the possibility of a fire occurring or danger to persons.

Unless an RCD or other protective device is installed to interrupt the supply in the event of the first Earth fault, an RCM should be provided to indicate that there is a possible fault from a live part to Earth.

9.2.6 *Electromagnetic comparability*

All electrical installations and equipment shall be in accordance with the current EMC regulations and standards for a particular installation (e.g. IEC 60601-1-2 for medical products).

9.2.7 *Electromechanical stresses*

Every conductor or cable shall have adequate strength and be installed so that it can withstand electromechanical forces caused by any current it may have to carry whilst in service.

9.2.8 *Thermal effects*

Electrical installations must be installed so:

- that the risk of igniting any flammable materials is minimised;
- there will be minimal risk of burns to persons or livestock.

9.2.9 *Wiring systems*

Wiring systems will meet requirements if:

- the rated voltage of the cable(s) is not less than the nominal voltage of the system;
- adequate mechanical protection (e.g. non-metallic trunking) of the basic insulation is provided.

Where an electrical service is located near to one or more **non-electrical** services, it shall meet the following conditions:

- the wiring system shall be suitably protected against the hazards likely to arise from other services;
- fault protection shall be provided by means of automatic disconnection of supply.

A wiring system which passes through the location (but is not intended to supply electrical equipment within that location) shall meet the tests on electric cables described in BS EN 60332-1-2.

In some cases, additions and alterations are made to existing installations which could mean that some of the wiring complies with the current 2018 edition of BS 7671 but there is also wiring to the previous 2008 version. If this occurs, a warning notice (see example in Figure 9.4) shall be placed at or near the appropriate distribution board with the following wording:

9.2.10 *Cables*

- buried cables, conduits and ducts shall be at a sufficient depth to avoid being damaged;
- cables should preferably be installed in conduits made with an insulating material;
- flexible cables used for fixed wiring must be of the heavy-duty type;

Figure 9.4 Warning notice – non-standard colours

- the location of buried cables must be marked by cable covers or marker tape;
- surface-run and overhead cables shall be protected against mechanical damage;
- cables installed under a floor or above a ceiling shall not risk being damaged by contact with other fixings;
- cables shall not pass from one fire-segregated compartment to another;
- cables should not be installed across a site road unless they are protected against mechanical damage;
- cables complying with the requirements of BS EN 60332-1-2 may be installed without special precautions, however;
- special requirements may apply to battery cables.

Cables shall **not** be run in a lift or hoist well unless they form part of the lift installation.

9.2.11 *Conductors*

A bare live conductor **shall** always be installed on insulators.

9.3 Inspection and testing

During erection, installation, completion, addition or alteration to an installation (and before being put into service) appropriate inspections and tests must be carried out by competent and skilled persons to verify that the requirements of BS 7671:2018 have been met.

By definition, this requirement also applies to alterations and/or additions to an existing installation, as well as entirely new installations.

Material changes of use

Where there is a material change of use of a building, any work carried out shall ensure that the building complies with the applicable requirements of the Building Act 1984.

9.3.1 *Frequency of inspection and testing*

The amount and timing of inspections and maintenance of an installation depends on the type of installation or equipment, its use and operation, the current frequency and quality of maintenance provided and any external influences to which it is subjected.

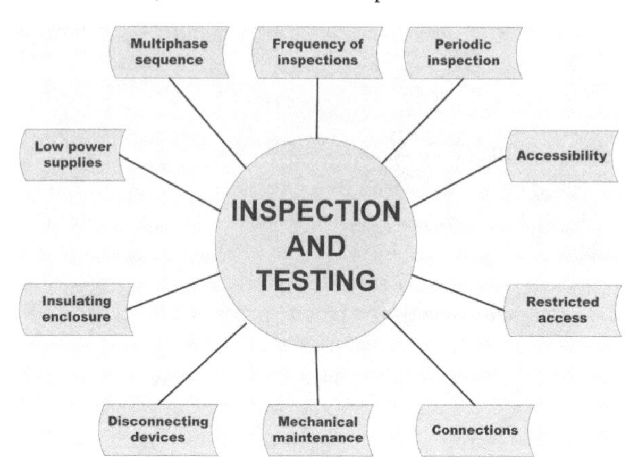

Figure 9.5 Inspection and testing

The results and recommendations of the previous report, if any, shall also be taken into account.

Full records of these inspections and tests **must** be retained.

9.3.2 Periodic inspection

The aim of periodic inspection and testing is to:

- confirm that the installation continues to work in a satisfactory condition;
- confirm that the safety of the installation has not deteriorated or been damaged;
- ensure the continued safety of persons and livestock against the effects of electric shock and burns;
- identify any installation defects and non-compliances with the requirements of the Regulations which may give rise to danger;
- protect property being damaged by fire and heat caused by a defective installation.

9.3.3 Maintenance inspection

The Regulations state that maintenance inspections **shall** consist of a careful scrutiny of the installation (dismantled or otherwise) using the appropriate tests described in Chapter 6 of BS 7671:2018.

The maintenance inspections are made to verify that the installed electrical equipment:

- complies with the requirements of the Regulations and the appropriate National Standards and European Harmonised Directives;
- is correctly selected and erected in accordance with the Regulations, etc.;
- is not visibly damaged or defective so as to impair safety.

Precautions have been taken to ensure that maintenance inspections do not cause:

- danger to persons or livestock;
- damage to property and equipment (even if the circuit is defective).

Equipment is normally expected to have been designed to have a useful life of not less than 20 years.

Note: '*Useful life*' normally means '*the period for which the equipment will continue to operate with the specified level of reliability and safety*'.

However, as previously mentioned, the frequency of maintenance inspections will depend on:

- the type of installation, its use and operation;
- the frequency and quality of maintenance; and
- the external influences to which it is subjected.

Generally speaking, the following CEN/CENELEC recommendations are relevant for all installed electrical and/or electronic equipment.

CEN/CENELEC Recommendations

For ease of maintenance, all equipment provided should have

- *easily accessible test points to facilitate fault location;*
- *modules have been constructed so as to facilitate the connection of test equipment (e.g. logic analysers, emulators and test ROMs, etc.);*
- *fault location provision to allow functional areas within each module or equipment to be isolated.*

9.4 Certification and reporting

As explained in Chapter 10 of this book, the certificates listed in Table 9.2 (and shown in Figures 9.3–9.5) shall be completed where appropriate:

Please note that all Electrical Installation Certificates, Electrical Installation Condition Reports and Minor Electrical Installation Condition Reports **MUST** be compiled and signed (or otherwise authenticated) by a competent, suitably **qualified** person or persons.

Table 9.2 Certificates and report required

Type of certificate	*Content*
Electrical Installation Certificate	An Electrical Installation Certificate (EIC) is a Safety Certificate issued by a qualified electrician as confirmation that an electrical installation (or project) fully complies with the requirements of the BS 7671:2018 It is used only for the initial certification of a new installation or for an alteration or addition to an existing installation where new circuits have been introduced
Electrical Installation Condition Report	An Electrical Installation Condition Report (EICR) is a periodic inspection report on a property's safety, relating to its fixed wiring. Its main purpose is to guarantee the safety of the residents and to ensure they are not susceptible to electrical shocks and/or fire
Minor Electrical Installation Works Certificate	A Minor Electrical Installation Works Certificate (MEIC) is used for all minor works (additions and alterations) to an existing electrical installation work that does not include the provision of a new circuit

Appendix 9.1 – Example stage audit checks

Design stage

Table 9.3 Example stage audit checks

	Item		*Related item*	*Remark*
1	**Requirements**	1.1	Information	Has the customer fully described his requirement?
				Has the customer any mandatory requirements?
				Are the customer's requirements fully understood by all members of the design team?
				Is there a need to have further discussions with the customer?
				Are other suppliers or subcontractors involved?
				If yes, who is the prime contractor?
		1.2	Standards	What international standards need to be observed?
				Are they available?
				What national standards need to be observed?
				Are they available?

(Continued)

Table 9.3 (Continued)

	Item	*Related item*		*Remark*
1				What other information and procedures are required? Are they available?
		1.3	Procedures	Are there any customer-supplied drawings, sketches or plans? Have they been registered?
2	**Quality Procedures**	2.1	Procedures Manual	Is one available? Does it contain detailed procedures and Instructions for the control of all drawings within the drawing office?
		2.2	Planning Implementation and Production	Is the project split into a number of Work Packages? If so: • are the various Work Packages listed? • have Work Package Leaders been nominated? • is their task clear? • is their task achievable? Is a time plan available? Is it up to date? Regularly maintained? Relevant to the task?
3	**Drawings**	3.1	Identification	Are all drawings identified by a unique number? Is the numbering system strictly controlled?

(Continued)

Table 9.3 (Continued)

	Item		*Related item*	*Remark*
		3.2	Cataloguing	Is a catalogue of drawings maintained? Is this catalogue regularly reviewed and up to date?
		3.3	Amendments and	Is there a procedure for authorising the issue of amendments, changes to drawings?
			Modifications	Is there a method for withdrawing and disposing of obsolete drawings?
4	**Components**	4.1	Availability	Are complete lists of all the relevant components available?
		4.2	Adequacy	Are the selected components currently available and adequate for the task? If not, how long will they take to procure? Is this acceptable?
		4.3	Acceptability	If alternative components have to be used are they acceptable to the task?
5	**Records**	5.1	Failure reports	Has the Design Office access to all records, failure reports and other relevant data?
		5.2	Reliability data	Is reliability data correctly stored, maintained and analysed?
5	**Records**	5.3	Graphs, diagrams, plans	In addition to drawings, is there a system for the control of all relevant graphs, tables, plans, etc.? Are CAD (Computer Aided Design) facilities available? (If so, go to 6.1)?

(Continued)

Table 9.3 (Continued)

	Item	Related item		Remark
6	**Reviews and Audits**	6.1	Computers	If a processor is being used: • are all the design office personnel trained in its use? • are regular backups taken? • is there an antivirus system in place?
		6.2	Manufacturing Division	Is a close relationship being maintained between the design office and the manufacturing division?
		6.3		Is notice being taken of the manufacturing division's exact requirements, their problems and their choices of components, etc.?

Installation stage

Table 9.4 Installation stage

	Item	Related item		Remark
1	**Degree of quality**	1.1	Quality Control procedures	Are quality control procedures available? Are they relevant to the task? Are they understood by all members of the manufacturing team? Are they regularly reviewed and up to date? Are they subject to control procedures?

(Continued)

Table 9.4 (Continued)

	Item		*Related item*	*Remark*
		1.2	Quality Control Checks	What quality checks are being observed? Are they relevant? Are there laid down procedures for carrying out these checks? Are they available? Are they regularly updated?
2	**Reliability of product design**	2.1	Statistical data	Is there a system for predicting the reliability of the product's design? Is sufficient statistical data available to be able to estimate the actual reliability of the design, before a product is manufactured? Is the appropriate engineering data available?
		2.2	Components and parts	Are the reliability ratings of recommended parts and components available? Are probability methods used to examine the reliability of a proposed design? If so, have these checks revealed design deficiencies such as: • assembly errors? • operator learning, motivational, or fatigue factors? • latent defects? • improper part selection?

Acceptance stage

Table 9.5 Acceptance stage

	Item	Related item		Remark
1	**Product performance**			Does the product perform to the required function? If not, what has been done about it?
2	**Quality level**	2.1	Workmanship	Does the workmanship of the product fully meet the level of quality required or stipulated by the user?
		2.2	Tests	Is the product subjected to environmental tests? If so, which ones? Is the product field tested as a complete system? If so, what were the results?
3	**Reliability**	3.1	Probability function	Are individual components and modules environmentally tested? If so, how?
		3.2	Failure rate	Is the product's reliability measured in terms of probability function? If so, what were the results? Is the product's reliability measured in terms of failure rate? If so, what were the results?
		3.3	Mean time between failures	Is the product's reliability measured in terms of Mean Time Between Failure (MTBF)? If so, what were the results?

In-service stage

Table 9.6 In-service stage

	Item	*Related item*		*Remark*
1	**System reliability**	1.1	Product basic design	Are statistical methods being used to prove the product's basic design? If so, are they adequate? Are the results recorded and available? What other methods are used to prove the product's basic design? Are these methods appropriate?
2	**Equipment reliability**	2.1	Personnel	Are there sufficient trained personnel to carry out the task? Are they sufficiently motivated? If not, what is the problem?
		2.1.1	Operators	Have individual job descriptions been developed? Are they readily available?
2	**Equipment reliability**	2.1.1	Operators	Are all operators capable of completing their duties?
		2.1.2	Training	Do all personnel receive appropriate training? Is a continuous On-the-Job Training (OJT) programme available to all personnel? If not, why not?
		2.2	Product dependability	What proof is there that the product is dependable?

(Continued)

Table 9.6 (Continued)

Item	Related item		Remark
			How is product dependability proved? Is this sufficient for the customer?
	2.3	Component reliability	Has the reliability of individual component been considered? Does the reliability of individual components exceed the overall system reliability?
	2.4	Faulty operating procedures	Are operating procedures available? Are they appropriate to the task? Are they regularly reviewed?
	2.5	Operational abuses	Are there any obvious operational abuses? If so, what are they? How can they be overcome?
	2.5.(1)	Extended duty cycle	Do the staff have to work shifts? If so, are they allowed regular breaks from their work? Is there a senior shift worker? If so, are his duties and responsibilities clearly defined? Are computers used? If so, are screen filters available? Do the operators have keyboard wrist rests?
	2.5.(2)	Training	Do the operational staff receive regular On-The-Job Training?

(Continued)

Table 9.6 (Continued)

	Item	*Related item*		*Remark*
				Is there any need for additional in-house or external Training?
3	**Design capability**	3.1	Faulty operating procedures	Are there any obvious faulty operating procedures? Can the existing procedures be improved upon?

Author's End Note

In accordance with BS 7671:2018, every installation (or alteration to an existing installation) shall, during erection and on completion before being put into service, be inspected and tested to verify, so far as is reasonably practicable, that the requirements of the Regulations have been met.

The final Chapter of this book provides an extensive list of the types of inspections and tests that should be used together with details of the sort of Certificates and Reports that will be required by the authorities – but, as per the previous chapters which contain similar lists, please remember that this is only the author's impression of the most important aspects of the Wiring Regulations and electricians should ***always*** *consult the latest edition of BS 7671 to satisfy compliance.*

10 Inspection and testing

Author's Start Note

In accordance with the Regulations, every installation (or alteration to an existing installation) during erection and on completion before being put into, or back into, service, has to be inspected and tested to verify, so far as is reasonably practicable, that the requirements of the Regulations have been met.

Part 6 of BS 7671:2018 has been completely restructured and now aligns with the CENELEC (European Committee for Electrotechnical Standardization) Standard in order to meet the latest requirements for the initial and periodic inspection and testing.

This final chapter of the book provides some guidance on the requirements for installation, maintenance, inspection, certification and repair of electrical installations. It lists the Regulations' primary requirements for these activities with respect to electrical installations and (in Appendix 10.1) provides an example Stage Audit Checklist for designers and engineers to use.

This chapter also contains (i.e. in Appendix 10.2) complete, bullet pointed check lists for the design, construction, inspection and testing of any new electrical installation, or new work associated with an alteration or addition to an existing installation.

To meet these requirements, it is essential for any electrician engaged in inspection, testing and certification of electrical installations has a **full** working knowledge of the Wiring Regulations contained in BS 7671:2018 (and its latest amendments).

DOI: 10.1201/9781003165170-10

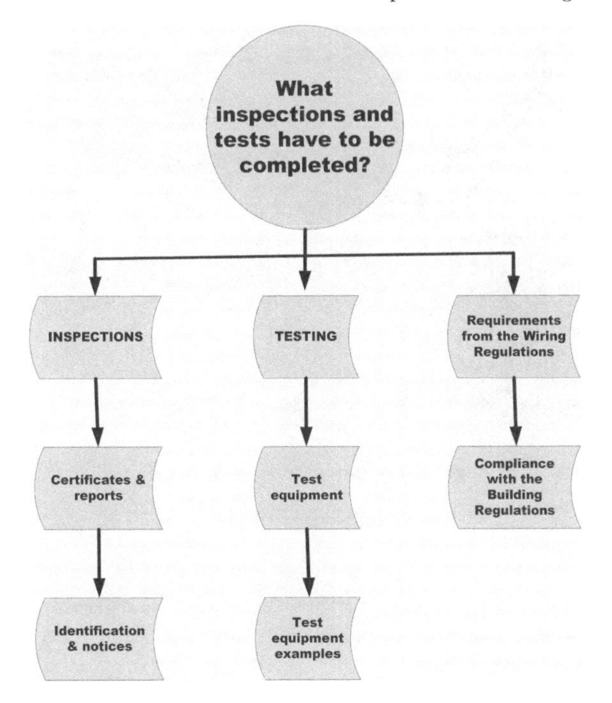

Figure 10.1 Inspections and tests

The electrician must also have above average experience and knowledge of the type of installation under test in order to carry out **any** Inspection and Test. Without this prerequisite, it could be quite dangerous – particularly concerning installations such as that shown in Figure 10.2!

10.1 What inspections and tests have to be completed and recorded?

Every installation must be inspected and tested during erection and on completion before being put into service to verify:

- the requirements of the Wiring and Building Regulations have been met; and
- the frequency and type of maintenance that the installation can be expected to receive during its intended life.

Figure 10.2 Example of a non-conforming electrical installation!

Source: Courtesy Herne European Consultancy Ltd

Precautions must, however, always be taken to avoid danger to persons and livestock, damage to property and installed equipment, during inspections and testing.

10.2 Inspections

An electrical inspection is an official evaluation of any new or existing electrical installation and involves a mixture of measurements, tests, and assessments.

The inspection (which should always precede testing and is normally done with that part of the installation under inspection disconnected from the supply) is made to verify that the installed electrical equipment is:

- correctly selected and erected in accordance with the Regulations;
- in compliance with the relevant requirements of the applicable British or Harmonized Standard, appropriate to the intended use of the equipment; and
- not visibly damaged or defective so as to impair safety.

An inspection check list is to be found at Appendix 10.2.

10.3 Testing

The following are précised details of the most important elements of the Wiring Regulations that an electrician must test for in order to confirm that the electrical installation meets the fundamental design requirements of the BS 7671:2018 and is installed in conformance with the requirements of that British Standard.

10.3.1 Design requirements

Check to confirm that the number and type of circuits required for lighting, heating, power, control, signalling, communication and information technology, etc. have taken consideration of:

- any special conditions;
- requirements for control, signalling, communication and information technology, etc.;
- the daily and yearly variation of demand;
- the loads to be expected on the various circuits;
- the location and points of power demand.

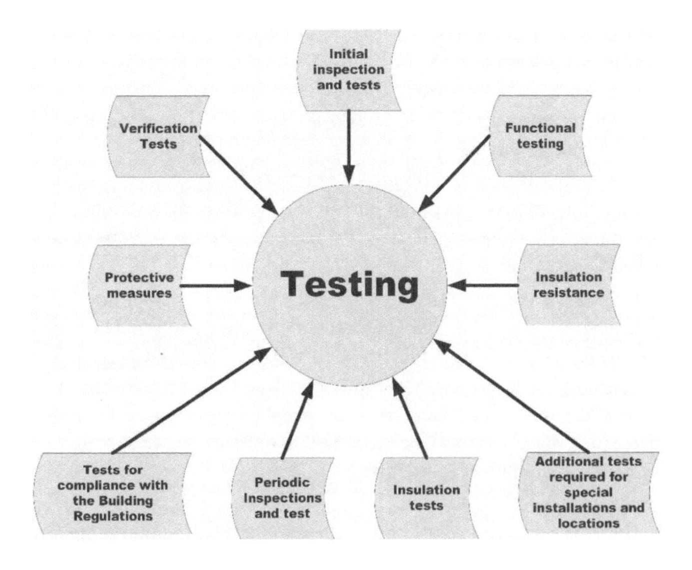

Figure 10.3 Testing

10.3.2 Electricity distributor

Check to confirm that the electricity distributor:

- has ensured that their equipment on consumers' premises:
  - is suitable for its purpose;
  - is safe in its particular environment;
  - clearly shows the polarity of the conductors;
- ensured that the cut-out and meter is mechanically protected and can be safely maintained;
- evaluated and agreed proposals for new installations or significant alterations to existing ones;
- maintained the supply within defined tolerance limits;
- provided an earthing facility for all new connections;
- provided certain technical and safety information to the consumer to enable them to design their installations.

10.3.3 Initial inspection and tests

In accordance with both the Wiring and the Building Regulations, all new installations (plus additions and/or alterations to existing circuits) need an initial verification to:

- ensure equipment and accessories meet the requirements of the relevant standard;
- comply with the requirements of BS 7671;
- comply with the requirements of the Building Regulations;
- ensure that the installation is not damaged so as to impair safety.

The contractor or other person responsible for the new work, or a person authorised to act on their behalf, shall record on the Electrical Installation Certificate or the Minor Electrical Installation Works Certificate, any defects found, so far as is reasonably practicable, in the existing installation.

The following tests **shall** be carried out (and in the following order) before the installation is energised:

- a continuity test of all protective conductors (including main and supplementary equipotential bonding);
- a continuity test of all ring final circuit conductors;

- a measurement of the insulation resistance:
  - between live conductors and between each live conductor and Earth;
  - of live parts from those of other circuits and those of other circuits and from Earth;
  - of the main switchboard and each distribution circuit.
- a polarity test to verify that, fuses, single-pole control and protective devices, lamp holders and wiring, meet requirements;
- confirmation that insulation for protection against direct and/or indirect contact meets requirements;
- confirmation that functional extra low voltage circuits meet all the test requirements for low voltage circuits;
- verification that the separation of circuits is protected by SELV, PELV and that this electrical separation meets requirements;
- verification (by measurement) that the amount of protection against indirect contact provided by a non-conducting location meets requirements.

The following tests **shall** then be carried out once the installation is energised:

- a measurement of the electrode resistance to Earth for earthing systems incorporating an Earth electrode;
- a measurement of Earth loop impedance;
- a measurement of prospective short-circuit and Earth fault;
- functional tests to verify the effectiveness of a RCDs and test assemblies (e.g. switchgear, controlgear, drives, controls and interlocks) to show that they are properly mounted, adjusted and installed in accordance with the Regulations.

10.3.4 Functional testing

Equipment, such as:

- switchgear and controlgear assemblies, drives, controls and interlocks;
- systems for emergency switching off and emergency stopping;
- insulation monitoring;

shall be subjected to a functional test to show that they are properly mounted, adjusted, installed and operate in accordance with the relevant requirements of the Wiring Regulations.

Note: Where fault protection and/or additional protection is to be provided by an RCD, the effectiveness of any test facility incorporated in the device shall be verified.

10.3.5 Insulation resistance

Check to confirm that:

- the insulation resistance (measured with all its final circuits connected but with current-using equipment disconnected) between live (phase and neutral) conductors, and between each live conductor and Earth, is not be less than that shown in Table 10.1;
- the separation of live parts from those of other circuits and from Earth is in accordance with the values show in Table 10.1, by measuring the insulation resistance.

More stringent requirements are applicable for the wiring of fire alarm systems in buildings, see BS 5839-1.

Where an SPD or other equipment is likely to influence the verification test, or be damaged, such equipment shall be disconnected before carrying out the insulation resistance test.

In locations exposed to a fire hazard, a measurement of the insulation resistance between the live conductors should be applied.

Table 10.1 Protective measures

Type of protection	*Description*
Protection against overload current	That the protective device is capable of breaking any overload current flowing in the circuit conductors before the current can damage the insulation of the conductors
Protection by Earth-free local equipotential bonding	That Earth-free local equipotential bonding prevents the appearance of a dangerous voltage between simultaneously accessible parts in the event of failure of the basic insulation
Protection by electrical separation	That equipment used as a fixed source of supply has been manufactured so that the output is separated from the input and from the enclosure by insulation for protection against indirect contact

(Continued)

Table 10.1 (continued)

Type of protection	*Description*
Protection by extra-low voltage systems (other than SELV)	That if an extra-low voltage system complies with the requirements for SELV, it is not be connected to a live part or a protective conductor forming part of another system and is not connected to • Earth; • an exposed-conductive-part of another system; • a protective conductor of any system; or • an extraneous-conductive-part. Protection against direct contact has been provided by either • insulation capable of withstanding 500 V a.c. rms for 60 seconds or • barriers or enclosures with a degree of protection of at least IP2X or IPXXB.
Protection by insulation of live parts	That the insulation protection has been designed to prevent contact with live parts
	Whilst, generally speaking, this method is for protection against direct contact, it also provides a degree of protection against indirect contact.
Protection by non-conducting location	That this form of protection prevents simultaneous contact with parts which may be at different potentials through failure of the basic insulation of live parts
	Protection by non-conducting location shall **not** be used in installations and locations subject to increased risk of shock such as agricultural and horticultural premises, caravans, swimming pools, etc.

(Continued)

Table 10.1 (Continued)

Type of protection	*Description*
Protection by residual current devices	That: • parts of a TT system that are protected by a single RCD have been placed at the origin of the installation, unless that part between the origin and the device complies with the requirements for protection by using Class II equipment or an equivalent insulation; Where there is more than one origin this requirement applies to each origin. • installations forming *part of* an IT system have been protected by an RCD supplied by the circuit concerned or made use of an insulation monitoring device.
Protection by SELV	That circuit conductors for each SELV system have been physically separated from those of any other system.
Protection by the use of Class II equipment or equivalent insulation	That this form of protection prevents a fault in the basic insulation causing a dangerous voltage to appear on the exposed metalwork of electrical equipment

10.3.5.1 Insulation resistance/impedance of floors and walls

In a non-conducting location at least three measurements shall be made in the same location and shall be repeated for each relevant surface of the location.

Any insulation or insulating arrangement of extraneous-conductive-parts when tested at 500 V d.c.:

- shall be greater than 1 MΩ;
- shall be capable of withstanding a test voltage of at least 2 kV a.c. rms; and
- shall not pass a leakage current exceeding 1 mA under normal conditions of use.

Further information on measurement of the insulation resistance/impedance of floors and walls can be found in Appendix 13 of BS 7671:2008.

10.3.6 *Site insulation*

Confirm that:

- insulation applied on site to protect against direct contact is capable of withstanding the applied British Standard test voltage;
- any supplementary equipment insulation designed to protect against indirect contact is tested; and
- the insulation is capable of withstanding, without breakdown or flashover, an applied test voltage.

10.3.7 *Protective measures*

The Wiring Regulations stipulate that a continuity test of all protective conductors (including main and supplementary equipotential bonding) shall be made by confirming, checking and testing:

10.3.8 *Protection by automatic disconnection of the supply*

Where RCDs are used for protection against fire, the conditions for protection by automatic disconnection of the supply shall be verified.

The effectiveness of fault protection measures for fault protection by automatic disconnection of supply depends on the type of system used and is as follows:

10.3.8.1 TN system

Compliance shall be verified by:

- measurement of the Earth fault loop impedance;
- verification of the characteristics and/or the effectiveness of the associated protective device.

10.3.8.2 TT system

Compliance shall be verified by:

- measurement of the resistance of the Earth electrode for exposed-conductive-parts of the installation;
- verification of the characteristics and/or effectiveness of the associated protective device.

10.3.8.3 IT system

Compliance shall be verified by calculation or measurement of the current (Id) in case of a first fault at the live conductor.

Where conditions that are similar to conditions of a TT system occur (and in the event of a second fault in another circuit) verification shall be made according to a TT system.

Where conditions that are similar to conditions of a TN system occur, in the event of a second fault in another circuit verification shall be made according to a TT system.

10.3.9 Protection against direct and indirect contact

The two methods for protecting against shock from both direct and indirect contact are:

- SELV– i.e. where an extra low voltage circuit is separated from other circuits that carry a higher voltage;
- where equipment is arranged so that current that could flow through the body (or livestock) is limited to a safe level (e.g. by using electric fences).

10.3.9.1 SELV

Protection by SELV is acceptable when:

- the nominal circuit voltage does not exceed extra-low voltage;
- the supply is from one of the following:
  - a safety isolating transformer complying with BS 3535;
  - a motor-generator with windings providing electrical separation equivalent to that of the safety isolating transformer specified above;

- a battery or other form of electrochemical source;
- a source independent of a higher voltage circuit (e.g. an engine driven generator);
- electronic devices which (even in the case of an internal fault) restrict the voltage at the output terminals so that they do not exceed extra low voltage.

Confirm and test that:

- a mobile source for SELV has been selected and erected in accordance with the requirements for protection by the use of Class II equipment or by an equivalent insulation where the user is protected by at least two layers of insulation;
- all live parts of a SELV system are:
  - not connected to Earth;
  - not connected to a live part or a protective conductor forming part of another system.
- circuit conductors for each SELV system are physically separated from those of any other system;
- conductors of systems with a higher voltage than SELV are separated from the SELV conductors by an earthed metallic screen or sheath;
- SELV circuit conductors that are contained in a multicore cable with other circuits having different voltages are insulated, individually or collectively, for the highest voltage present in the cable or grouping;
- electrical separation between live parts of a SELV system (including relays, contactors and auxiliary switches) and any other system are maintained;
- if the nominal voltage of a SELV system exceeds 25 V a.c. rms or 60 V ripple-free d.c., protection against direct contact can be provided by:
  - a barrier or enclosure;
  - insulation capable of withstanding a type-test voltage of 500 V a.c. rms for 60 seconds.
- the socket-outlet of a SELV system is:
  - incompatible with the plugs used for other systems in the same premises;
  - does not have a protective conductor contact.
- luminaire supporting couplers which have a protective conductor contact, are not be installed in a SELV system.

10.3.10 Protection against direct contact

If an electrical installation has been designed and installed correctly, then there shouldn't be at too much risk from direct contact – but carelessness (such as changing an electric light bulb without switching the mains off first) or overconfidence (such as working on a circuit with the power on) are the prime causes of injuries and death from electric shock.

The main protective methods against direct contact causing an electric shock are:

- barriers or an enclosure – greater than IP2X or IPXXB or IP4X;
- insulation of live parts;
- obstacles and placing out of reach – for installations and locations where an increased risk of shock exists, this protective measure should **not** be used;
- PELV and FELV.

The use of an RCD cannot prevent direct contact, but may be used to **supplement** other protective means that are used.

10.3.10.1 Protection by placing out of reach

Check to insure that:

- bare (or insulated) overhead lines being used for distribution between buildings and structures are installed in accordance with the Electricity Safety, Quality and Continuity Regulations 2002;
- bare live parts (other than overhead lines) are not within arm's reach or within 2.5 m of:
    - any exposed conductive part;
    - an extraneous conductive part;
    - a bare live part of any other circuit.
- if a bulky or long conducting object is normally handled in these areas, the distances required shall be increased accordingly.

No additional protection against **overvoltages of atmospheric origin** is necessary for;

- installations that are supplied by low voltage systems which do not contain overhead lines;
- installations that contain overhead lines and their location is subject to less than 25 thunderstorm days per year;

provided that they meet the required minimum equipment, impulse withstand, voltages shown in Table 10.2.

Table 10.2 Required minimum impulse withstand voltage kV

Required minimum impulse withstand voltage U_w				
Nominal voltage of the installation (V)	*Category I equipment with reduced impulse voltage*	*Category II equipment with normal impulse voltage*	*Category III equipment with high impulse voltage*	*Category IV equipment with very high impulse voltage*
230/240	1.5	2.5	4	6
277/480				
400/690	2.5	4	6	8
1000	4	6	8	12

Note: Suspended cables having insulated conductors with **earthed metallic coverings**, are considered to be an *'underground cables'*!

Check to ensure that installations that are supplied by (or include) low voltage overhead lines incorporate protection against overvoltages of atmospheric origin or (if the location is subject to more than 25 thunderstorm days per year) protection against overvoltages of atmospheric origin are provided by a surge protective device with a protection level not exceeding Category II; or;

- where protective measures against indirect contact **only** have been dispensed with, confirm that:
  - exposed-conductive-parts (including small isolated metal parts such as bolts, rivets, nameplates not exceeding 50 mm × 50 mm and cable clips) cannot be gripped or be contacted by a major surface of the human body;
  - inaccessible lengths of metal conduit do not exceeding 150 mm^2;
  - metal enclosures mechanically protecting equipment comply with the relevant British Standard;
  - overhead line insulator brackets (and metal parts connected to them) are not within arm's reach;

- the steel reinforcement of steel reinforced concrete poles is not accessible;
- there is no risk of fixing screws used for non-metallic accessories coming into contact with live parts;
- unearthed street furniture that is supplied from an overhead line is inaccessible whilst in normal use.

10.3.11 Protection against indirect contact

The main protective methods against indirect contact causing an electric shock are shown below and in Figure 10.4:

- automatic disconnection of supply;
- Earth free local equipotential bonding;
- earthing and protective conductors;
- earthing arrangements for combined protective and functional purposes;
- protection by electrical separation;
- main equipotential bonding conductors;
- non conducting location (absence of protective conductors);
- supplementary equipotential bonding conductors;
- locations with increased risk of shock;
- electrical separation;
- polarity;
- use of Class II equipment.

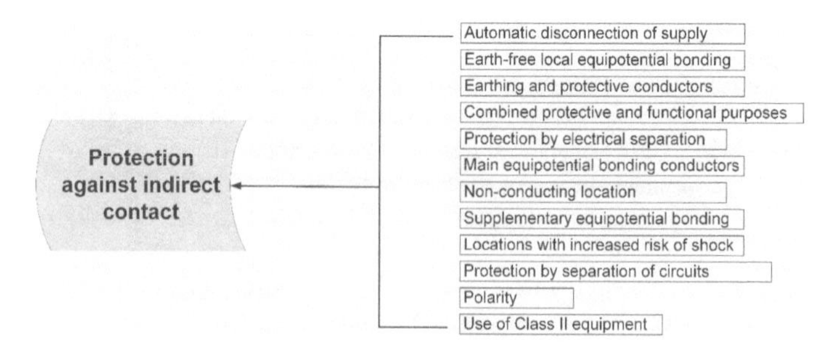

Figure 10.4 Protection against indirect contact

10.3.11.1 Automatic disconnection of supply

Confirm and test that:

- installations which are part of a TN system, meet the requirements for Earth fault loop impedance and for circuit protective conductor impedance as specified in the Wiring Regulations;
- for circuits supplying fixed equipment that are outside of the earthed equipotential zone (and which have exposed-conductive-parts that could be touched by a person who has direct contact directly with Earth) that the Earth fault loop impedance ensures that disconnection occurs within the time stated in Table 10.3 and as stated in BS 7671:2018;
- if the installation is part of a TT system, all socket-outlet circuits are protected by an RCD;
- automatic disconnection using an RCD is not be applied to a circuit incorporating a PEN conductor;
- installations that provide protection against indirect contact have a circuit protective conductor run to (and terminated at) each point in the wiring and at each accessory.

Except suspended lamp holders which have no exposed-conductive-parts.

10.3.11.2 Earth-free local equipotential bonding

Earth-free local equipotential bonding is effectively a Faraday cage.

Confirm that:

- Earth-free local equipotential bonding has only been used in special situations which are Earth-free;
- a cautionary notice has been fixed in a prominent position adjacent to every point of. Access to the location concerned.

Table 10.3 Maximum disconnection times for TN systems

Installation nominal voltage	*Maximum disconnection time (TN system)*
50 V	0.8 s
120 V	0.4 s
230 V	0.2 s
400 V	0.2 s
Greater than 400 V	0.1 s

10.3.11.3 Earthing and protective conductors

A protective conductor may consist of one or more of the following:

- a single-core cable;
- a conductor in a cable;
- an insulated or bare conductor in a common enclosure with insulated live conductors;
- a fixed bare or insulated conductor;
- a metal covering (for example, the sheath, screen or armouring of a cable);
- a metal conduit, enclosure or electrically continuous support system for conductors;
- an extraneous-conductive-part.

Verify and test that:

- the thickness of tape or strip conductors is capable of withstanding mechanical damage and corrosion (see BS 7430);
- the connection of earthing conductors to the Earth electrode are:
  - soundly made;
  - electrically and mechanically satisfactory;
  - labelled in accordance with the Regulations;
  - suitably protected against corrosion.
- all installations have a main earthing terminal to connect the following to the earthing conductor:
  - the circuit protective conductors;
  - the main bonding conductors;
  - functional earthing conductors (if required);
  - lightning protection system bonding conductor (if any).
- earthing conductors are capable of being disconnected to enable the resistance of the earthing arrangement to be measured;
- all joints:
  - are capable of disconnection only by means of a tool;
  - are mechanically strong;
  - ensure the maintenance of electrical continuity.
- confirm that unless a protective conductor forms part of a multicore cable that the cross-sectional area, (up to and including 6 mm^2) has been protected, throughout.

Table 10.4 Minimum cross-sectional area of the main equipotential bonding conductor in relation to the neutral

Copper equivalent cross-sectional area of the supply neutral conductor	*Minimum copper equivalent cross-sectional area of the main equipotential bonding conductor*
4 mm^2 up to 10 mm^2	6 mm^2
16 mm^2 up to 35 mm^2	10 mm^2
50 mm^2	16 mm^2
70 mm^2	25 mm^2

Where PME conditions apply, verify and test that:

- the main equipotential bonding conductor has been selected in accordance with the neutral conductor of the supply and Table 10.4.

Note: Local distributor's network conditions may require a larger conductor.

- buried earthing conductors have a cross-sectional area not less than that stated in Table 10.5.

Verify and test that:

- the cross-sectional area of all protective conductor (less equipotential bonding conductors) is not less than:

$$S = \sqrt{\frac{I^2 t}{k}}$$

Table 10.5 Minimum cross-sectional areas of a buried earthing conductor

Protection	*Protected against mechanical damage*	*Not protected against mechanical damage*
Protected against corrosion by a sheath		16 mm^2 copper 16 mm^2 coated steel
Not protected against corrosion	25 mm^2 copper 50 mm^2 steel	25 mm^2 copper 50 mm^2 steel

- if the protective conductor is not an integral part of a cable; or contained in a conduit, ducting, trunking or enclosure formed by a wiring system, then the cross-sectional area shall not be less than:
    - 2.5 mm^2 copper equivalent if protection against mechanical damage is provided; or
    - 4 mm^2 copper equivalent if mechanical protection is **not** provided.
- protective conductors buried in the ground, shall have a cross sectional area not less than that stated in Table 10.5.

10.3.11.4 Earthing arrangements for combined protective and functional purposes

The following conductors may serve as a PEN conductor provided that the part of the installation concerned is not supplied through an RCD:

- a conductor of a cable not subject to flexing and with a cross-sectional area not less than 10 mm^2 (for copper) or 16 mm^2 (for aluminium);
- the outer conductor of a concentric cable where that conductor has a cross-sectional area not less than 4 mm^2.

Verify and test that PEN conductors have only been used if:

- authorisation to use a PEN conductor has been obtained by the distributor; or
- the installation is supplied by a privately owned transformer or convertor and there is no metallic connection with the distributor's network; or
- the installation is supplied from a private generating plant.

Then you will need to check that:

- the outer conductor of a concentric cable is not common to more than one circuit;
- the conductance of the outer conductor of a concentric cable (and the terminal link or bar):
    - for a single-core (or multicore cable in a multiphase or multipole circuit cable) is not less than the internal conductor;
    - for a multicore cable (serving a number of points contained within one final circuit) is not be less than that of the internal conductors;

- the continuity of all joints in the outer conductor of a concentric cable (and at a termination of that joint) is supplemented by an additional conductor;
- isolation devices or switching have not been inserted in the outer conductor of a concentric cable;
- PEN conductors of all cables have been insulated or have an insulating covering suitable for the highest voltage to which it may be subjected;
- if neutral and protective functions are provided by separate conductors, those conductors are not then be re-connected together beyond that point;
- separate terminals (or bars) have been provided for the protective and neutral conductors at the point of separation;
- PEN conductors have been connected to the terminals or bar intended for the protective earthing conductor and the neutral conductor.

Note: Where earthing is required for protective as well as functional purposes, then the requirements for protective measures shall take precedence.

10.3.11.5 Protection by electrical separation

Verify that:

- equipment used as a fixed source of supply is either:
  - selected and/or installed with double insulated electrical protection (Class II) or equivalent protection; or
  - manufactured so that the output is separated from the input and from the enclosure by insulation satisfying the conditions for Class II;
- for circuits supplying a single piece of equipment, no exposed-conductive-part of the separated circuit is connected:
  - to the protective conductor of the source;
  - to any exposed-conductive-part of any other circuit.
- live parts of a separated circuit are not connected (at any point) to another circuit or to Earth;
- live parts of a separate circuit are electrically separated from all other circuits;

- mobile supply sources (fed from a fixed installation) are selected and/or installed with Class II or equivalent protection;
- protection by electrical separation has been applied to the supply of individual items of equipment by means of a transformer complying with BS 3535 (the secondary of which is not earthed) or a source affording equivalent safety;
- protection by electrical separation has been used to supply several items of equipment from a single separated source (but only for special situations);
- separated circuits, preferably, use a separate wiring system;

If this is not feasible, multicore cables (without a metallic sheath) or insulated conductors (in an insulating conduit) may be used;

- source supplies are only supplying more than one item of equipment provided that;
  - all exposed-conductive-parts of the separated circuit are connected together by an insulated and non-earthed equipotential bonding conductor;
  - the non-earthed equipotential bonding conductor is not connected to a protective conductor, or to exposed-conductive-part of any other circuit or to any extraneous-conductive-part;
  - all socket-outlets are provided with a protective conductor contact (that is connected to the equipotential bonding conductor);
  - all flexible equipment cables [other than reinforced or double insulated (Class H) equipment] have a protective conductor for use as an equipotential bonding conductor;
  - exposed-conductive-parts which are fed by conductors of different polarity (which are liable to a double fault occurring) are fitted with an associated protective device;
  - any exposed-conductive part of a separated circuit cannot come in contact with an exposed-conductive part of the source.
- the supply source to the circuit is either:
  - a motor-generator;
  - an isolating transformer complying with BS 3535; or
- the voltage of an electrically separated circuit does not exceed 500 V;

- all parts of a flexible cable (or cord) that is liable to mechanical damage is visible throughout its length;
- a warning notice (warning that protection by electrical separation is being used) is fixed in a prominent position adjacent to every point of access to the location concerned.

10.3.11.6 Main equipotential bonding conductors

All main equipotential (and supplementary) bonding conductors must be tested for continuity.

Confirm that main equipotential bonding conductors have (for each installation) been connected to the main earthing terminal of that installation. These can include the following:

- central heating and air conditioning systems;
- exposed metallic structural parts of the building;
- gas installation pipes;
- water service pipes;
- other service pipes and ducting;
- the lightning protective system.

In accordance with the Building Regulations, where an installation serves more than one building, the above requirement shall be applied to **each** separate building.

10.3.11.7 Non-conducting location

Test that the insulation of extraneous-conductive-parts:

- does not pass a leakage current exceeding 1 mA in normal use;
- is not be less than 0.5 MΩ (when tested at 500 V d.c.);
- is able to withstand a test voltage of at least 2 kV a.c. rms.

Check that the:

- degree of protection against indirect contact provided by a non-conducting location is verified by measuring the resistance of the location's floors and walls to the installation's main protective conductor at not less than three points on each relevant surface;

- the insulation of extraneous-conductive-parts are:
    - able to withstand a test voltage of at least 2 kV a.c. rms;
    - do not pass a leakage current exceeding I mA in normal use; and
    - not less than 0.5 MΩ when tested at 500 V d.c.

10.3.11.8 Supplementary equipotential bonding

Test all supplementary equipotential bonding conductors for continuity, particularly in locations intended for livestock to confirm that:

- any metallic grid that is laid in the floor for supplementary bonding is connected to the protective conductors of that installation;
- supplementary bonding connects all exposed and extraneous-conductive-parts which can be touched by livestock.

10.3.11.9 Locations with increased risk of shock

For installations and locations with increased risk of shock (e.g. saunas, bathrooms and agricultural/horticultural premises, etc.) certain additional measures may be required, such as:

- automatic disconnection of supply by means of an RCD with a rated residual operating current not exceeding 30 mA;
- supplementary equipotential bonding.

In these cases, test to confirm that circuits supplying fixed equipment which are outside of the earthed equipotential zone (and which have exposed-conductive-parts that could be touched by a person who has direct contact directly with Earth) that the Earth fault loop impedance ensures that disconnection occurs within the time stated in Table 10.4 and BS 7671:2018.

Test, measure and confirm that:

- automatic disconnection using an RCD has **not** been applied to a circuit incorporating a PEN conductor;
- if the installation is part of a TT system, all socket-outlet circuits have been protected by an RCD;
- installations that provide protection against indirect contact by automatically disconnecting the supply, have a circuit protective conductor run to (and terminated at) each point in the wiring and at each accessory;

Except suspended lamp holders which have no exposed-conductive-parts.

- one or more of the following types of protective device have been used:
  - an overcurrent protective device;
  - an RCD.
- where an RCD is used in a TN-C-S system, a PEN conductor has **not** been used on the load side;
- the protective conductor to the PEN conductor is on the source side of the RCD;
- the maximum disconnection times to a circuit supplying socket-outlets and to other final circuits which supply portable equipment intended for manual movement during use, or hand-held Class I equipment does not exceed those shown in Table 10.6;
- where a fuse is used to satisfy this disconnection requirement, maximum values of Earth fault loop impedance (Zs) corresponding to a disconnection time of 0.4 s are as stated in Table 10.7 for a nominal voltage to Earth (Uo) of 230 V;
- for a distribution circuit and a final circuit supplying only stationary equipment, the maximum disconnection time of 5 s is not exceeded:

Table 10.6 Maximum Earth fault loop impedance (Zs) for fuses, for 0.4 s disconnection time with Uo of 230 V (see Regulation 413-02-10)

General purpose (gG) fuses to BS 88-2.1 and BS 88-6								
Rating (Amperes)	6	10	16	20	25	32	40	50
Zs (ohms)	8.89	5.33	2.82	1.85	1.5	1.09	0.86	0.63

Table 10.7 Minimum values of insulation resistance

Circuit nominal voltage (V)	*Test voltage d.c. (V)*	*Minimum insulation resistance (MΩ)*
SELV and PELV	250	0.25
Up to and including 500 V (with the exception of the above systems)	500	0.5
Above 500 V	1000	1.0

Note: See appropriate part of BS 88 for types and rated currents of fuses other than those mentioned in Table 10.6.

10.3.11.10 Protection by separation of circuits

Ensure that:

- the separation of circuits is verified for protection by:
    - SELV;
    - PELV;
    - electrical separation; and
    - that the separation of live parts from those of other circuits and those of other circuits from Earth is verified by measuring that the insulation resistance in accordance with values show in Table 10.7.
- functional extra low voltage circuits meet all the test requirements for low voltage circuits.

10.3.11.11 Polarity

Complete a polarity test to verify that:

- circuits (other than BS EN 60238 E14 and E27 lamp holders) which have an earthed neutral conductor centre contact bayonet (and Edison screw lamp holders) have their outer or screwed contacts connected to the neutral conductor;
- fuses and single-pole control and protective devices are only connected in the line conductor;
- wiring has been correctly connected to socket-outlets and similar accessories.

10.3.11.12 Use of Class II equipment

Test to confirm that

- circuits supplying Class II equipment have a circuit protective conductor that is run to (and terminated at) each point in the wiring and at each accessory; (except suspended lamp holders which have no exposed-conductive-parts);

- the metal work of exposed Class II equipment is mounted so that it is not in electrical contact with any part of the installation that is connected to a protective conductor;
- when Class II equipment is used as the sole means of protection against indirect contact, the installation or circuit concerned is under effective supervision whilst in normal use.

This form of protection shall **not** be used for circuits that include socket-outlets or where a user can change items of equipment without authorisation.

10.3.12 Additional tests with the supply connected

Other than insulation tests the following tests are to be completed with the supply connected:

- Earth electrode resistance;
- Earth fault loop impedance;
- prospective fault current.
- voltage drop.

10.3.12.1 Earth electrode resistance

If the earthing system incorporates an Earth electrode as part of the installation, measure the electrode resistance to Earth (using test equipment similar to that shown in Appendix 10.1).

If the electrode under test is being used in conjunction with an RCD protecting an installation, test (prior to energising the remainder of the installation) between the phase conductor at the origin of the installation and the Earth electrode with the test link open.

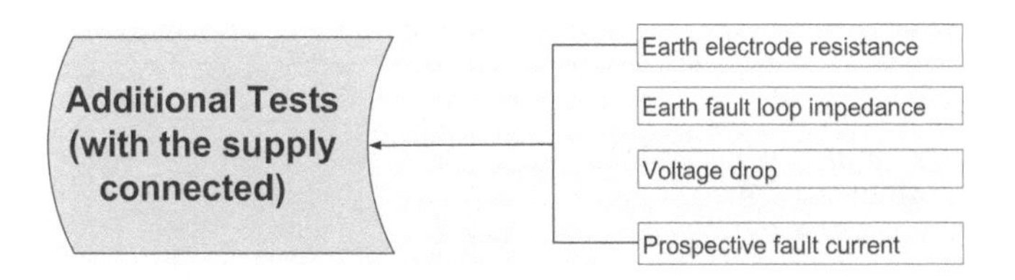

Figure 10.5 Additional tests with the supply connected

Note: The resulting impedance reading (i.e. the electrode resistance) should then be added to the resistance of the protective conductor for the protected circuits.

10.3.12.2 Earth fault loop impedance

This is an extremely important test to ensure that under Earth fault conditions, overcurrent devices disconnect fast enough to reduce the risk of electric shock. The Regulations stipulate that:

> *If the protective measures employed require knowledge of Earth fault loop impedance, then the relevant impedances must be measured.*

Where a fuse is used, maximum values of Earth fault loop impedance (Zs) corresponding to a disconnection time of 0.4 s are stated in Table 10.8 for a nominal voltage to Earth (Uo) of 230 V.

Note: the circuit loop impedances given in Table 10.8 should not be exceeded when the conductors are at their normal operating temperature. If the conductors are at a different temperature when tested, then the reading should be adjusted accordingly.

Using test equipment (similar to that listed in Appendix 10.1 and sub-section 10.7) completes the tests shown in Figure 10.6.

Ensure that all main equipotential bonding is in place by connecting the test equipment to the phase, neutral and Earth terminals at the remote end of the circuit under test – then press to test and record.

Table 10.8 British Standards for fuse links

Circuit nominal voltage (V)	*Test voltage d.c. (V)*	*Minimum insulation resistance (MΩ)*
SELV and PELV	250	0.25
Up to and including 500 V (with the exception of the above systems)	500	0.5
Above 500 V	1000	1.0

(Courtesy Brian Scaddon.)

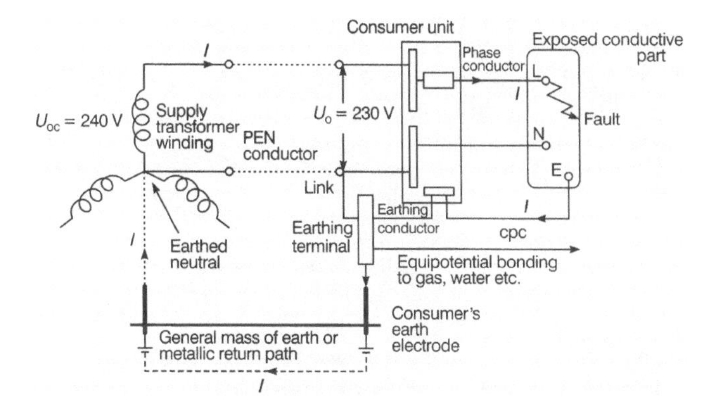

Figure 10.6 Testing Earth fault loop impedance

Further information on measurement of Earth fault loop impedance can be found in Appendix 14 of BS 7671:2018.

10.3.12.3 Verification of voltage drop

When required, for compliance with the Regulations the voltage drop may be evaluated:

- by measuring the circuit impedance;
- by using calculations.

10.3.12.4 Prospective fault current

Verify, test and ensure that:

- prospective fault currents (under both short-circuit and Earth fault conditions):
  - have been assessed for each supply source;
  - are calculated at every relevant point of the complete installation either by enquiry or by measurement;
  - are measured at the origin and at other relevant points in the installation.
- protection of wiring systems against overcurrent takes into account minimum and maximum fault current conditions;

- fault current protective devices are provided:
    - at the supply end of each parallel conductor where two conductors are in parallel;
    - at the supply and load ends of each parallel conductor where more than two conductors are in parallel.
- fault current protective devices are:
    - less than 3 m in length between the point where the value of current-carrying capacity is reduced, and the position of the protective device;
    - installed so as to minimise the risk of fault current;
    - installed so as to minimise the risk of fire or danger to persons.
- fault current protective devices are placed (on the load side) at the point where the current carrying capacity of the installation's conductors is likely to be lessened owing to:
    - the method of installation;
    - the cross-sectional area;
    - the type of cable or conductor used;
    - Inherent environmental conditions.
- conductors are capable to carrying fault current without overheating.

Note: A single protective device may be used to protect conductors in parallel against the effects of fault current occurring.

Ensure that:

- fault current protection devices are capable of breaking; and that
- the breaking capacity rating of each device is not less than the prospective short-circuit current or Earth fault current at the point at which the device is installed.

Note: A lower breaking capacity is permitted if another protective device is installed on the supply side.

- the characteristics of each device used for overload current and/or for fault current protection has been co-ordinated so that the energy let through (i.e. by the fault current protective device) does not exceed the overload current protective device's limiting values;
- devices providing protection against both overload current and fault current are capable of breaking;

- circuit breakers used as fault current protection devices:
    - are capable of making any fault current up to and including the prospective fault current;
    - break any fault current flowing before that current causes danger due to thermal or mechanical effects produced in circuit conductors or associated connections;
    - break and make up any overcurrent up to and including the prospective fault current at the point where the device is installed.
- safety services with sources that are incapable of operating in parallel, are protected against electric shock and fault current.

10.3.13 Insulation tests

When an electrical installation fails an insulation test, the installation must be corrected and the test made again. If the failure influences any previous tests that were made, then those tests must also be repeated.

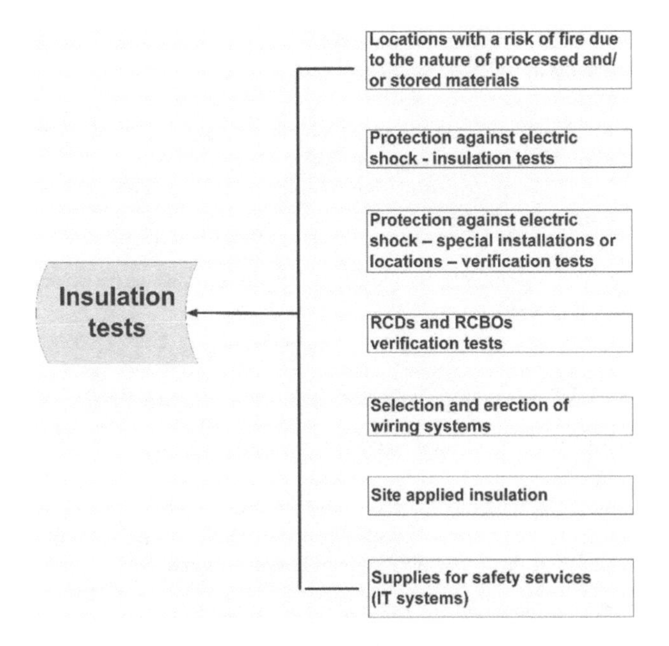

Figure 10.7 Site applied insulation

Other than protection against indirect contact, the following tests are to be completed:

- Locations with a risk of fire due to the nature of processed and/or stored materials;
- Protection against electric shock:
  - insulation tests;
  - special installations or locations – verification tests.
- RCDs and RCBOs – verification tests;
- Selection and erection of wiring systems;
- Site applied insulation;
- Supplies for safety services (IT systems).

10.3.13.1 Locations with a risk of fire due to the nature of processed and/or stored materials

Test that wiring systems (less those using mineral insulated cables and busbar trunking arrangements) have been protected against Earth insulation faults as follows:

- in TN and TT systems, by RDCs having a rated residual operating current (IAn) not exceeding 300 mA;
- in IT systems, by insulation monitoring devices with audible and visible signals.

Note: The disconnection time of the overcurrent protective device, in the event of a second fault, shall not exceed 5 s.

10.3.13.2 Protection against electric shock – Insulation tests

Confirm, measure and test that:

- circuit conductors for each SELV system are physically separated from those of any other system or (i.e. where this proves impracticable) confirm that SELV circuit conductors are:
  - insulated for the highest voltage present;
  - enclosed in an insulating sheath additional to their basic insulation;

- equipment is capable of withstanding all mechanical, chemical, electrical and thermal influences stresses normally encountered during service;
- exposed-conductive-parts that might attain different potentials through failure of the basic insulation of live parts have been arranged so that a person will not come into simultaneous contact with two exposed-conductive-parts, or an exposed-conductive-part and any extraneous-conductive-part;
- if the nominal voltage of a SELV system exceeds 25 V a.c. rms or 60 V ripple-free d.c., then protection against direct contact has been provided by one (or more) of the following:

    - insulation capable of withstanding a type-test voltage of 500 V a.c. rms for 60 seconds; or
    - a barrier (or an enclosure) capable of providing protection to at least IP2X or IPXXB.

- in IT systems, an insulation monitoring device has been provided so as to indicate the occurrence of a first fault from a live part to an exposed-conductive-part or to Earth;
- insulating enclosures:

    - are not pierced by conductive parts (other than circuit conductors) likely to transmit a potential;
    - do not contain any screws of insulating material – the future replacement of which by metallic screws could impair the insulation provided by the enclosure;
    - do not adversely affect the operation of the equipment protected.

- live parts are completely covered with insulation which:

    - can only be removed by destruction;
    - is capable of withstanding electrical, mechanical, thermal and chemical stresses normally encountered during service.

- the basic insulation of operational electrical equipment is at least protected to IP2X or IPXXB;
- where insulation has been applied during the erection of the installation, the quality of the insulation has been verified.

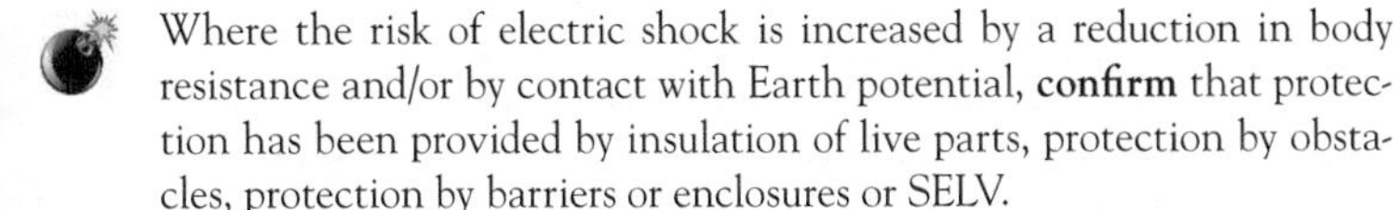

Where the risk of electric shock is increased by a reduction in body resistance and/or by contact with Earth potential, **confirm** that protection has been provided by insulation of live parts, protection by obstacles, protection by barriers or enclosures or SELV.

10.3.13.3 Protection against electric shock – Special installations or locations – Verification tests

Where SELV or PELV is used (whatever the nominal voltage) in locations containing a bath, shower, hot air sauna and/or in a restrictive conductive location, confirm that protection against direct contact has been provided by:

- insulation capable of withstanding a type-test voltage of 500 V a.c. rms for 1 minute; or
- barriers and/or enclosures providing protection to at least IP2X or IPXXB.

The above requirements do not apply to locations where freedom of movement is not physically constrained.

10.3.13.4 RCDs and RCBOs – Verification tests

Test (on the load side of the RCD) that:

- general purpose RCDs:
  - do not open with a leakage current flowing equivalent to 50% of the rated tripping current;
  - open in less than 200 ms with a leakage current flowing equivalent to 100% of the rated RCD tripping.
- general purpose Residual Current Circuit Breakers (RCCD) in compliance with BS EN 61008 or Residual Current Operated Circuit Breakers (RCBO) without integral overcurrent protection to BS EN 61009:
  - do not open with a leakage current flowing equivalency to 50% of the rated tripping current of the RCD;
  - open in less than 300 ms with a leakage current flowing equivalency to 100% of the RCD's tripping current.
- RCD protected socket outlets that comply with BS 7288:
  - do not open with a leakage current flowing equivalent to 50% of the RCD's rated tripping current;
  - open in less than 200 ms with a leakage current flowing equivalency to 100% of the RCD's rated tripping current.

10.3.13.5 Selection and erection of wiring systems

Test to confirm that:

- all electrical joints and connections meet stipulated requirements concerning conductance, insulation, mechanical strength and protection;
- cables that run in thermally insulated spaces are not actually covered by the thermal insulation;
- the current carrying capacity of cables which are installed in thermally insulated walls or above a thermally insulated ceiling conforms with Appendix 4 to the Regulations;
- the current-carrying capacity of cables that are totally surrounded by thermal insulation for less than 0.5 m, has been reduced according to the size of cable, length and thermal properties of the insulation;
- the insulation and/or sheath of cables connected to a bare conductor or busbar is capable of withstanding the maximum operating temperature of the bare conductor or busbar;
- wiring systems are capable of withstanding the highest and lowest local ambient temperature likely to be encountered – or are provided with additional insulation suitable for those temperatures;
- wiring systems have been selected and erected so as to minimise (i.e. during installation, use and maintenance) any damage to the insulation of cables, conductors and their terminations.

10.3.13.6 Site applied insulation

Test to confirm that:

- insulation applied on site to protect against direct contact is capable of withstanding, without breakdown or flashover, an applied test voltage for similar type-tested equipment;
- supplementary insulation applied to equipment during erection (i.e. to protect against indirect contact):
    - protects to at least IP2X or IPXXB; and
    - is capable of withstanding, without breakdown or flashover, an applied test voltage for similar type-tested equipment.

10.3.13.7 Supplies for safety services (IT systems)

In an IT system, confirm that continuous insulation monitoring has been provided to give audible and visible indications of a first fault.

10.3.14 Periodic Inspections and tests

Periodic inspection and testing of all electrical installations must be carried out to confirm that the installation is in a satisfactory condition for continued service.

Note: A generic list of examples of items requiring inspection together with examples of model forms for Certification ad Reporting are given in Appendix 6 to BS 7671:2018;

The aim of periodic inspection and testing is to:

- confirm that the safety of the installation has not deteriorated or been damaged;
- ensure the continued safety of persons and livestock against the effects of electric shock and burns;
- identify installation defects and non-compliance with the requirements of the Regulations, which could give rise to danger;
- protect property being damaged by fire and heat caused by a defective installation.

The frequency of periodic inspection and testing of installations will depend on:

- the type of installation, its use and operation;
- the quality the required maintenance; and
- external influences which the installation is (or might be) subjected to.

10.3.15 Verification tests

All completed installations (including additions and/or alteration to existing installations) shall be inspected and tested for conformance to the requirements of the Wiring and Building Regulations. These tests will include the following:

- accessibility of connections;
- appliances producing hot water or steam;
- cables and conductors for low voltage;
- electric surface heating systems;
- emergency switching – verification tests;
- forced air heating systems;

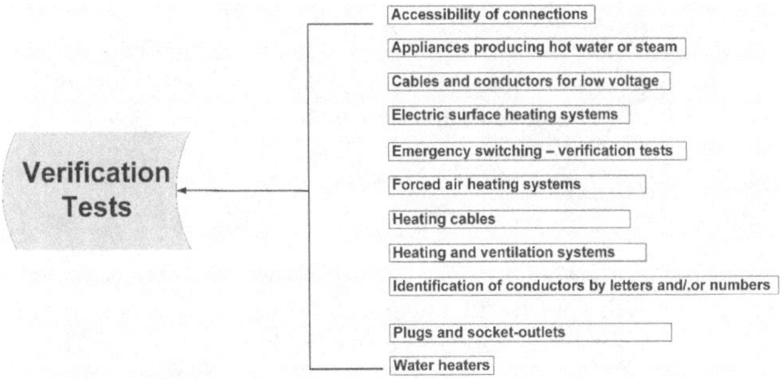

Verification Tests

Accessibility of connections
Appliances producing hot water or steam
Cables and conductors for low voltage
Electric surface heating systems
Emergency switching – verification tests
Forced air heating systems
Heating cables
Heating and ventilation systems
Identification of conductors by letters and/.or numbers
Plugs and socket-outlets
Water heaters

Figure 10.8 Verification tests

- heating cables;
- heating and ventilation systems;
- identification of conductors by letters and/or numbers;
- plugs and socket-outlets;
- water heaters.

10.3.15.1 Accessibility of connections

Confirm that all connections and joints are accessible for inspection, testing and maintenance, unless:

- they are in a compound-filled or encapsulated joint;
- the connection is between a cold tail and a heating element;
- the joint is made by welding, soldering, brazing or compression tool.

10.3.15.2 Appliances producing hot water or steam

Confirm that electric appliances producing hot water or steam have been protected against overheating.

10.3.15.3 Cables and conductors for low voltage

Confirm that flexible and non-flexible cables, flexible cords (and conductors used as an overhead line) operating at low voltage comply with the Relevant British or Harmonized Standard.

10.3.15.4 Electric surface heating systems

Confirm that the equipment, system design, installation and testing of all electric surface heating systems meet the requirements of BS 6351.

10.3.15.5 Emergency switching – Verification tests

For any exterior installation, confirm that the switch is placed outside the building, adjacent to the equipment. Where this is not possible, a notice showing the position of the switch shall be placed adjacent to the equipment and a notice fixed near the switch shall indicate its use.

10.3.15.6 Forced air heating systems

Confirm (by inspection and test) that the electric heating elements of forced air heating systems (other than those of central-storage heaters):

- are incapable of being activated until the prescribed air flow has been established;
- deactivate when the air flow is reduced or stopped;
- do not have two, independent, temperature limiting devices;
- have frames and enclosures, which are constructed out of non-ignitable material.

10.3.15.7 Heating cables

Check that:

- heating conductors and cables that pass through (or are in close proximity to) to a fire hazard:
    - are enclosed in material with an ignitability characteristic 'P' as specified in BS 476;
    - are protected from any mechanical damage.
- heating cables that have been laid (directly) in soil, concrete, cement screed, or other material used for road and building construction are:
    - capable of withstanding mechanical damage;
    - constructed of material that will be resistant to damage from dampness and/or corrosion.

Table 10.9 Maximum conductor operating temperatures for a floor-warming cable

Type of cable	*Maximum conductor operating temperature (C)*
General-purpose pvc over conductor	70
Enamelled conductor, polychiorophene over enamel, pvc overall	70
Enamelled conductor pvc overall	70
Enamelled conductor, pvc over enamel, lead-alloy 'E' sheath overall	70
Heat-resisting pvc over conductor	85
Nylon over conductor, heat-resisting pvc overall	85
Mineral insulation over conductor, copper sheath overall	Temperature dependent on type of seal employed, outer covering, etc.
Silicone-treated woven-glass sleeve over conductor	180
Synthetic rubber or equivalent elastomeric insulation over conductor	85

- heating cables that have been laid (directly) in soil, a road, or the structure of a building are installed so that it:
  - is completely embedded in the substance it is intended to heat;
  - is not damaged by movement (by it or the substance in which it is embedded;
  - complies in all respects with the maker's instructions and recommendations.
- the maximum loading of floor-warming cable under operating conditions is no greater than the temperatures shown in Table 10.9.

10.3.15.8 Heating and ventilation systems

In locations where heating and ventilation systems containing heating elements are installed and where there is a risk of fire due to the nature of processed or stored materials, check to ensure that:

- the dust or fibre content and the temperature of the air does not present a fire hazard;
- temperature limiting devices have a manual reset;

- heating appliances are fixed;
- heating appliances mounted close to combustible materials are protected by barriers to prevent the ignition of such materials;
- heat storage appliances are incapable of igniting combustible dust and/or fibres;
- enclosures of equipment such as heaters and resistors do not attain surface temperatures higher than:
    - 90°C under normal conditions; and
    - 115°C under fault conditions.

10.3.15.9 Identification of conductors by letters and/or numbers

Test to confirm that all individual conductors and groups of conductors:

- have been identified by a label containing either letters or numbers that are clearly legible;
- have numerals that contrast, strongly, with the colour of the insulation;
- numerals **6** and **9** have been underlined to avoid confusion.

10.3.15.10 Plugs and socket-outlets

Inspect and test to ensure that any plug and socket-outlet used in single-phase a.c. or two-wire d.c. circuits that does **not** comply with BS 1363, BS 546, BS 196 or BS EN 60309-2 has either been designed especially for that purpose or:

- is a plug and socket-outlet used for an electric clock and the plug has a fuse not exceeding 3 amperes which complies with BS 646 or BS 1362;
- is a plug and socket-outlet used for an electric shaver that is either part of the shaver supply unit that complies with BS 3535 or, in a room (other than a bathroom) that complies with BS 4573.

10.3.15.11 Water heaters

Confirm that Water heaters (or Boilers) having immersed and uninsulated heating elements are permanently connected to the electricity supply via a double-pole linked switch, which is either:

- separate from and within easy reach of the heater/boiler; or
- part of the boiler/heater;

- verify the effectiveness of RCDs by a test simulating a typical fault condition;
- functionally test assemblies to show that they are properly mounted, adjusted and installed in accordance with the Regulations.

10.3.16 *Electrical connections*

Test and verify that:

- connections between conductors and between a conductor and equipment provide durable electrical continuity and adequate mechanical strength;
- the earthing conductor of main earthing terminals (or bars) is capable of being disconnected to enable the resistance of the earthing arrangements to be measured;
- all joints:
    - are capable of disconnection only by means of a tool;
    - are mechanically strong;
    - ensure the maintenance of electrical continuity.
- the wiring of final and distribution circuits to equipment with a protective conductor current exceeding 10 mA have a protective connection complying with one or more of the following:
    - a single protective conductor with a cross-sectional greater than 10 mm^2;
    - a single (mechanically protected) copper protective conductor with a cross-sectional greater than 4 mm^2;
- two individual protective conductors;
- a BS 4444 Earth monitoring system that will automatically disconnect the supply to the equipment in the event of a continuity fault;
- connection (i.e. of the equipment) to the supply by means of a double wound transformer, which has its secondary winding, connected to the protective conductor of the incoming supply and the exposed conductive-parts of the above.

10.3.17 *Tests for compliance with the Building Regulations*

As shown in Table 10.10, there are four types of installation that have to be inspected and tested for compliance with the Building Regulations.

Table 10.10 Types of installation

Type of inspection	*When is it used?*	*What should it contain?*	*Remarks*
Electrical Installation Certificate	For the initial certification of a new installation or for the alteration or addition to an existing installation where new circuits have been introduced	A schedule of inspections and test results as required by Part 6 (of BS 7671)	The original Certificate shall be given to the person ordering the work and a duplicate retained by the contractor
Minor Electrical Installation Works Certificate	For additions and alterations to an installation such as an extra socket-outlet or lighting point to an existing circuit, the relocation of a light switch, etc.	In accordance with the relevant provisions of Part 6 of BS 7671	This Certificate may also be used for the replacement of equipment such as accessories or luminaires, but **not** for the replacement of distribution boards (or similar items) or the provision of a new circuit
Electrical Installation Report	For the inspection of an existing electrical installation	A schedule of inspections and a schedule of test results as required by Part 6 (of BS 7671)	For safety reasons, the electrical installation will need to be inspected at appropriate intervals by a competent person
Building Regulations Compliance Certificate	Confirmation that the work carried out complies with the Building Regulations	The basic details of the installation, the location, completion date and the name of the installer	A purchaser's solicitor may request this document when you come to sell your property. Looking further ahead, it may be required as one of the documents that will make up your 'Home Information Pack'

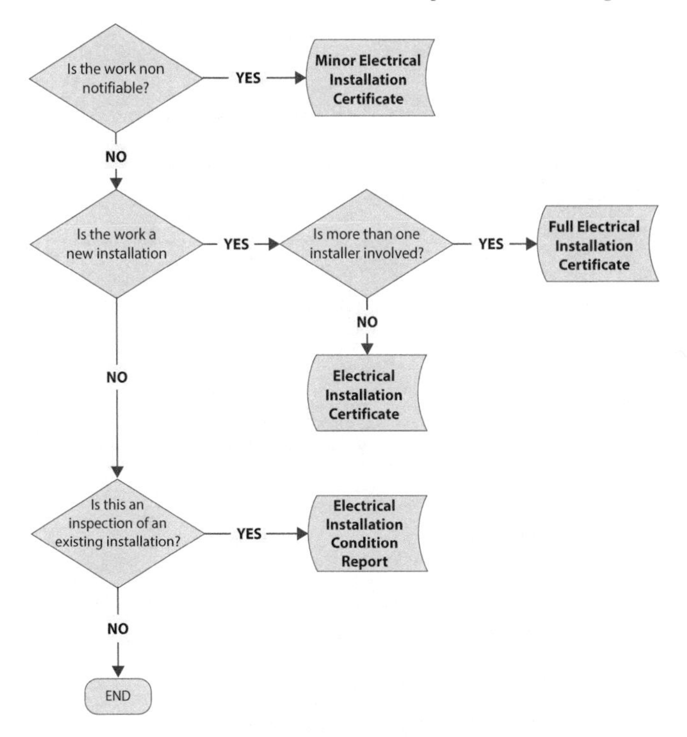

Figure 10.9 Choosing the correct inspection certificate

Part P of the Building Regulations applies only to fixed electrical installations that are intended to operate at low voltage or extra-low voltage which are not controlled by the Electricity Supply Regulations 1988 as amended, or the Electricity at Work Regulations 1989 as amended.

10.3.17.1 Additional tests required for special installations and locations

Part 7 of the Wiring Regulations contains additional requirements in respect of installations where the risk of electric shock is increased by a reduction in body resistance or by contact with Earth potential (e.g. locations containing a bath or shower, swimming pools, hot air sauna, construction installations, agricultural and horticultural premises, caravans, motor caravans and highway power supplies, etc.).

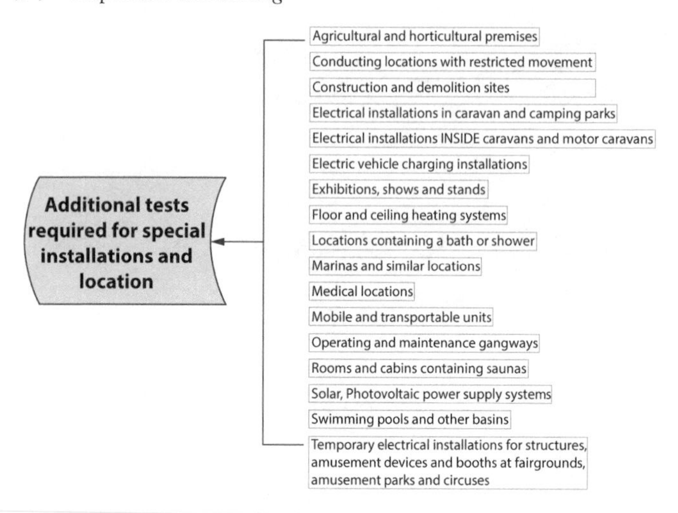

Figure 10.10 Special installations and locations

The prime installations and locations that the Wiring and Building Regulations are concerned with comprise:

- agricultural and horticultural premises;
- conducting locations with restricted movement;
- construction and demolition sites;
- electrical installations in caravan and camping parks;
- electrical installations INSIDE caravans and motor caravans;
- electric vehicle charging installations;
- exhibitions, shows and stands;
- floor and ceiling heating systems;
- locations containing a bath or shower;
- marinas and similar locations;
- medical locations;
- mobile and transportable units;
- operating and maintenance gangways;
- rooms and cabins containing saunas;
- solar, Photovoltaic power supply systems;
- swimming pools and other basins;
- temporary electrical installations for structures, amusement devices and booths at fairgrounds, amusement parks and circuses.

10.3.17.2 Agricultural and horticultural premises

As the possibility of animals unintentionally coming in direct contact with a live installation is greater than for humans all fixed agricultural and horticultural installations (outdoors and indoors) and locations where livestock is kept (such as stables, chicken houses, piggeries, feed-processing locations, lofts and storage areas for hay, straw and fertilisers) shall be inspected to confirm that they comply with Part 705 of the Wiring Regulations.

Also see Section 8.2.1 for further details of inspections and tests that need to be completed.

10.3.17.3 Conducting locations with restricted movement

A restrictive conductive location is one in which the surroundings consist mainly of metallic or conductive parts such as a large metal container or boiler. People employed inside these locations (e.g. a person working inside the boiler whilst using an electric drill or grinder) would have their freedom of movement physically restrained and a large proportion of their body would be in contact with the sides of that location and, therefore, prone to shock hazards.

The supplies to mobile equipment for use in such locations will have to be inspected to confirm that they comply with Section 706 of the Wiring Regulations.

Also see Section 8.2.2 for further details of inspections and tests that need to be completed.

10.3.17.4 Construction and demolition sites

Electrical installations at construction sites are primarily there so as to provide lighting and power for enabling work to proceed. As workman will probably be working ankle deep in wet muddy conditions and using a selection of portable tools such as drills and grinders, they will be particularly susceptible to electric shock and to avoid this happening PME earthing facility shall not be used for earthing an installation all extraneous-conductive parts must be reliably connected to the main earthing terminal.

Installations providing electricity supply for:

- new building construction;
- repairs, alterations, extensions or demolition of existing buildings;
- engineering construction;
- Earthworks;

will have therefore, to be inspected to confirm that they comply with Part 704 of the Wiring Regulations.

Also see Section 8.2.2 further details of inspections and tests that need to be completed.

Note: These requirements do **not** apply to:

- to construction site offices, cloakrooms, meeting rooms, canteens, restaurants, dormitories and toilets;
- installations covered by BS 6907.

10.3.17.5 Electrical installations in caravan and camping parks

For electrical installations in a caravan park or camping park, special consideration needs to be given to the:

- protection of people – due to the fact that the human body may be in contact with Earth potential;
- protection of wiring – due to tent pegs or ground anchors; and
- movement of heavy or high vehicles;

and these installations will have to be inspected to confirm that they comply with:

- Section 708 (for electrical installations within caravan parks, camping parks and similar locations); and or
- Section 721 (for electrical installations inside caravans and motor caravans) of the Wiring Regulations.

Also see Section 8.2.3 for further details of inspections and tests that need to be completed.

10.3.17.6 Electrical installations in caravans and motor caravans

Caravans and motor caravans are designed as leisure accommodation vehicles which are either towed (e.g. by a car) or self-propelled to a caravan site. They will often contain a bath or a shower and special requirements for such installations will apply.

In addition to the normal dangers associated with fixed electrical installations, there is also the potential hazard of totally unskilled people moving the caravan/motor caravan, connecting and disconnecting the mains supply, and not ensuring that it is correctly earthed.

All electrical installations in caravans and motor caravans, therefore, need to be inspected to confirm that they comply with Part 721 of the Wiring Regulations.

Also see Section 8.2.4 for further details of inspections and tests that need to be completed on these particular installations.

Note: It should be remembered that the requirements of this section do **not** apply to:

- electrical circuits and equipment covered by the Road Vehicles Lighting Regulations 1989;
- installations covered by BS EN 1648-1 and BS EN 1648-2;
- internal electrical installations of mobile homes, fixed recreational vehicles, transportable sheds and the like, temporary premises or structures.

10.3.17.7 Electric vehicle charging installations

With the increased number of all-electric or hybrid-electric cars currently being manufactured, the 2018 edition of BS 7671 now contains a new Section (722) to cover the requirements for circuits intended to supply electric vehicles for charging purposes.

Also see Section 8.2.5 for further details of inspections and tests that need to be completed.

10.3.17.8 Exhibitions, shows and stands

Whilst Chapter 51 of BS 7671:2018 still contains the requirements for external influences, the 18th Edition of the Wiring Regulations has now been further developed in accordance with new ISO standards and the BS EN 60721 and BS EN 61000 series on environmental conditions.

Accordingly, the revised Regulations now require that all temporary electrical installations in exhibitions, shows and stands (including mobile and portable displays and equipment), must be inspected to confirm that they comply with Part 711 of the Wiring Regulations.

Also see Section 8.2.6 for further details of inspections and tests that need to be completed.

10.3.17.9 Floor and ceiling heating systems

This particular Section of BS 7671 has been completely revised since it was last published in 2008 and now applies to the installation of embedded electric floor and ceiling heating systems. For example, to enable surface heating these can be erected as either a thermal storage heating system or a direct heating system.

The requirements of this Section also apply to electric heating systems for de-icing, frost prevention and similar applications, and cover both indoor and outdoor systems such as walls, ceilings, floors, roofs, drainpipes, gutters, pipes, stairs, roadways, and non-hardened compacted areas (e.g. football fields, lawns).

In accordance with these revised Regulations, electric floor and ceiling heating systems which are erected as either thermal storage heating systems or direct heating systems, must be inspected to confirm that they comply with Part 753 of the Wiring Regulations.

Note: The regulations do not, however, apply to the installation of wall heating systems nor heating systems for industrial and commercial applications.

Also see Section 8.2.7 for further details of inspections and tests that need to be completed.

10.3.17.10 Locations containing a bath or shower

When people use bathrooms and showers, most of the time they are naturally unclothed and wet and thus very vulnerable to electric shock due to their reduced body resistance Special measures are, therefore, required to ensure that the possibility of direct and/or indirect contact is reduced.

Locations containing a fixed bath (bath tub or birthing pool), showers and their surrounding must, therefore, be inspected to confirm that they comply with Part 701 of the Wiring Regulations.

Also see Section 8.2.8 for further details of inspections and tests that need to be completed.

Locations containing baths or showers for medical treatment, or for disabled persons, may have special requirements.

10.3.17.11 Marinas and similar locations

For marinas, particular attention has to be given to the increased risk of electric shock owing to:

- contact of the body with Earth potential;
- presence of water;
- reduction in body resistance.

Circuits intended to supply pleasure craft or houseboats in marinas and similar locations must therefore be inspected to confirm that they comply with Section 709 of the Wiring Regulations.

Also see Section 8.2.9 for further details of inspections and tests that need to be completed.

Note: The regulations do **not** apply to the supply for houseboats (if they are supplied directly from the public network) or to the internal electrical installations of pleasure craft or houseboats.

10.3.17.12 Medical locations

Section 710 of the Wiring Regulations is aimed at patient healthcare facilities and although the requirements mainly refer to hospitals, private clinics, medical and dental practices, healthcare centres and dedicated medical rooms in the workplace, this Section also applies to electrical installations in locations designed for medical research and (where applicable) to veterinary clinics.

The International Electrotechnical Commission (IEC) and British Standards (BS) manufacturers' standards for medical electrical equipment consist of two types of testing, namely:

'*Type testing*' – which is carried out by an approved test house on a single sample of an equipment (or piece of equipment) which a compliance certification against a particular standard is being sought.

'*Routine testing*', on the other hand, is intended to provide an indication of the basic safety of the equipment without subjecting it to undue stress that would be liable to cause deterioration.

A full list of tests and inspections that need to be completed is contained in Section 7.3.10 of the Wiring Regulations. Further advice is also contained in Section 8.2.10.

Note: These requirements do **not**, however, apply to the actual medical electrical equipment itself, as this is fully covered in ISO 13485 (The Medical Devices Directive) and the BS EN 60601 series on Medical Electrical Equipment and Systems.

10.3.17.13 Mobile and transportable units

A vehicle and a self-propelled or towed or transportable structure (such as a container or cabin) in which all or part of its electrical installation is contained and which is provided with an external temporary supply (e.g. via, a plug and socket-outlet) shall be inspected to confirm that they comply with Part 717 of the Regulations.

Also see Section 8.2.11 for further details of inspections and tests that need to be completed.

10.3.17.14 Operating and maintenance gangways

This (comparatively small) Section of BS 7671:2018 centres on the operation and safe maintenance of switchgear and controlgear within areas that include gangways and where access is restricted to skilled or instructed persons. Access areas and the requirements for operating and maintenance gangways need to comply with Part 729 of the Wiring Regulations.

Also see Section 8.2.13 for further details of inspections and tests that need to be completed.

10.3.17.15 Rooms and cabins containing saunas

Saunas (similar to swimming pools, bathrooms and showers, etc.) are primarily used by people who are unclothed and wet and thus very vulnerable to electric shock due to their reduced body resistance.

Special measures are, therefore, needed to ensure that the possibility of direct and/or indirect contact is reduced and installations supplying electricity for locations in which hot air sauna heating equipment (in accordance with BS EN 60335-2-53) must be inspected to confirm that they comply with Part 703 of the Regulations.

Also see Section 8.2.14 for further details of inspections and tests that need to be completed.

10.3.17.16 Solar, photovoltaic power supply systems

PV converts light directly into electricity by using photons from sunlight to knock electrons into a higher state of energy, thereby creating electricity. Solar cells are packaged in PV modules (often electrically connected in multiples as solar PV arrays) and convert energy from the sun into electricity.

Electrical installations of PV power supply systems (including subsystems with a.c. modules), need to be inspected to confirm that they comply with Part 712 of the Regulations.

Also see Section 8.2.15 for further details of inspections and tests that need to be completed.

10.3.17.17 Swimming pools and other basins

Swimming pools are, by design, 'wet areas' where people are highly venerable to electric shock. Special measures are, therefore, needed to ensure that all possibility of direct and/or indirect contact is reduced.

Requirements applicable to basins of swimming pools, paddling pools and other basins plus their surrounding zones must be inspected to confirm that they comply with Part 702 of the Regulations.

Also see Section 8.2.16 for further details of inspections and tests that need to be completed.

10.3.17.18 Temporary electrical installations for structures, amusement devices and booths at fairgrounds, amusement parks and circuses

Temporarily erected mobile or transportable electrical machines, structures and electrical equipment are typically installed repeatedly and temporarily, at fairgrounds, amusement parks, circuses and/or similar places.

It is essential that an adequate number of socket-outlets are installed to allow the user's requirements to be safely met and their safe design, installation and operation shall be inspected to confirm that they comply with Part 740 of the Regulations.

Also see Section 8.2.17 for further details of inspections and tests that need to be completed.

10.4 Identification and notices

For safety purposes, the Regulations require a number of notices and labels (see Figure 10.11) to be used for electrical installations and these will need to be checked during initial and periodic inspections.

10.4.1 *General*

Whilst Table 51 of BS 7671:2018 provides full details of conductor and conduit colours, Chapter 7 (Section 7.2.3) deals with the main ones to remember.

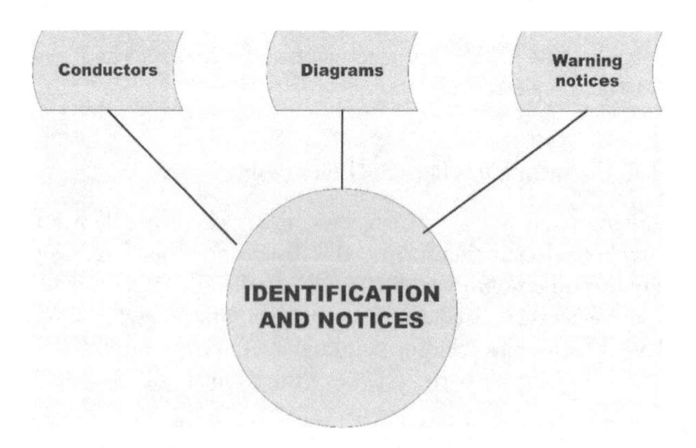

Figure 10.11 Identification and notices

10.4.2 Conductors

Verify that:

- conductor cable cores are identified by colour and /or lettering and/ or numbering;
- binding and sleeves used for identifying protective conductors comply with BS 3858;
- neutral or mid-point conductors are coloured blue;
- protective conductor cable cores are identifiable at all terminations (and preferably throughout their length);
- protective conductors are a bi-colour combination of green and yellow and that neither colour covers more than 70% of the surface being coloured.

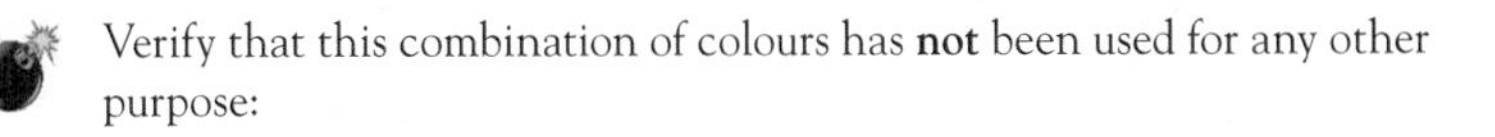

Verify that this combination of colours has **not** been used for any other purpose:

- single-core cables used as protective conductors are coloured green-and-yellow throughout their length;
- bare conductors or busbars used as protective conductors are identified by equal green and yellow stripes that are 15 mm to 100 mm wide.

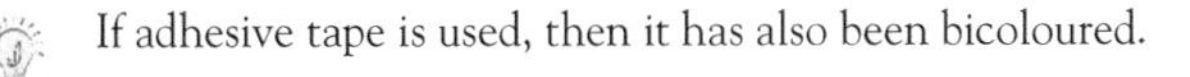

If adhesive tape is used, then it has also been bicoloured.

- PEN conductors (when insulated) shall be coloured either:
  - green-and-yellow throughout their length with, blue markings at the terminations; or
  - blue throughout its length with, green-and-yellow markings at the terminations.
- bare conductors are painted or identified by a coloured tape, sleeve or disc.

ALL other conductors (including those used to identify conductor switchboard busbars and conductors) **shall** be coloured as shown in accordance with Table 10.11.

The colour **green** shall **not** be used on its own.

Table 10.11 Identification of conductors

Function	*Alphanumeric*	*Colour*
a.c. power circuit (Note 1)		
Control circuits, ELV and other applications		
Functional earthing conductor		Cream
Mid-wire of three-wire circuit (Notes 2 and 3)	M	Blue
Negative (of negative earthed) circuit (Note 2)	M	Blue
Negative (of positive earthed) circuit	L−	Grey
Negative of three-wire circuit	L−	Grey
Negative of two-wire circuit	L−	Grey
Neutral of single- or three-phase circuit	N	Blue
Neutral or mid-wire (Note 4)	N or M	Blue
Outer negative of two-wire circuit derived from three-wire system	L−	Grey
Outer positive of two-wire circuit derived from three-wire system	L+	Brown
Phase 2 of three-phase a.c. circuit	L2	Black
Phase 3 of three-phase a.c. circuit	L3	Grey
Phase conductor	L	Brown, Black, Red, Orange, Yellow, Violet, Grey, White, Pink or Turquoise
Phase 1 of three-phase a.c. circuit	L1	Brown
Phase of single-phase circuit	L	Brown
Positive (of negative earthed) circuit	L+	Brown
Positive (of positive earthed) circuit (Note 2)	M	Blue
Positive of three-wire circuit	L+	Brown
Positive of two-wire circuit	L+	Brown
Protective conductors		Green-and-yellow
Three-wire d.c. power circuit		
Two-wire earthed d.c. power circuit		
Two-wire unearthed d.c. power circuit		

Notes:

1 Power circuits include lighting circuits.
2 M identifies either the mid-wire of a three-wire d.c. circuit, or the earthed conductor of a two-wire earthed d.c. circuit.
3 Only the middle wire of three-wire circuits may be earthed.
4 An earthed PELV conductor is **blue**.

Further information, concerning cable identification colours for extra-low voltage and d.c. power circuits, is available from the IET website at www.iet.org.

10.4.2.1 Identification of conductors by letters and/or numbers

Where letters and/or numbers are used to identify conductors, check to confirm that:

- individual conductors and/or groups of conductors are identified by either letters or numbers that are clearly legible;
- all numerals contrast, strongly, with the colour of the insulation;
- numerals 6 and 9 are underlined;
- protective devices are arranged and identified so that the circuit being protected is easily recognisable;
- protective conductors coloured green-and- yellow are not numbered other than for the purpose of circuit identification;
- the alphanumeric numbering system is in accordance with Table 10.11.

10.4.2.2 Identification by colour or marking

Colour or marking is **not** required for:

- concentric conductors of cables;
- metal sheath or the armour of cables (when used as a protective conductor);
- bare conductors (where permanent identification is impracticable);
- extraneous conductive-parts used as a protective conductor;
- exposed-conductive-parts used as a protective conductor.

10.4.3 Diagrams

Verify that all available diagrams, charts, tables or schedules that have been used, indicate:

- the type and composition of each circuit (points of utilisation served, number and size of conductors, type of wiring); and
- the method used for protection against indirect contact;
- the data (where appropriate) and characteristics of each protective device used for automatic disconnection;
- the identification (and location) of all protection, isolation and switching devices;
- the circuits or equipment that are susceptible to a particular test.

Verify that distribution board schedules **have been** provided within (or adjacent to) each distribution board.

All symbols used must comply with BS EN 60617.

10.4.4 Warning notices

Check to confirm that the following warning notices are appropriate to the situation, correctly positioned and contain the right information.

- Earth-free local equipotential bonding;
- earthing and bonding connections;
- electrical installations in caravans, motor caravans and caravan parks;
- emergency switching;
- final circuit distribution boards;
- inspection and testing;

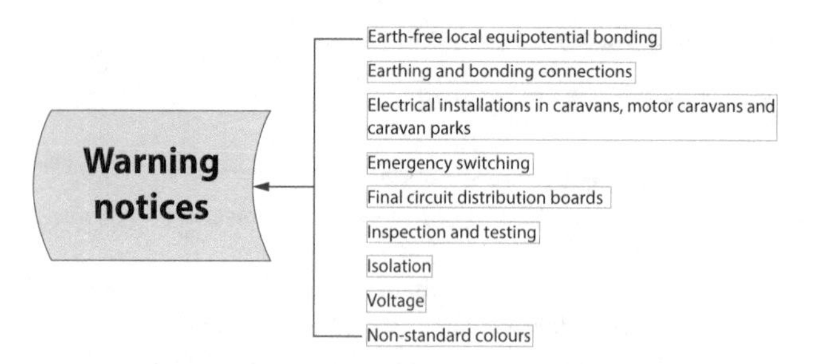

Figure 10.12 Warning notices

- isolation;
- voltage;
- non-standard colours.

10.4.4.1 Earth-free local equipotential bonding

Where Earth-free local equipotential bonding has been used confirm that an appropriate notice has been fixed in a prominent position adjacent to every point of access to the location concerned.

10.4.4.2 Earthing and bonding connections

Confirm that a permanent label with the words shown in Figure 10.13 has been permanently fixed at or near:

- the connection point of every earthing conductor to an Earth electrode;
- the connection point of every bonding conductor to an extraneous-conductive-part;
- the main Earth terminal (when separated from the main switchgear).

Figure 10.13 Safety Notice

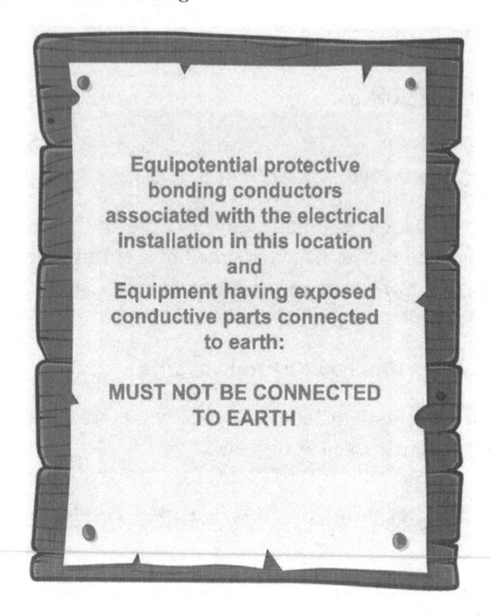

Figure 10.14 Earthing, bonding and electrical separation notice

Where protection is by Earth-free local equipotential bonding or electrical separation, confirm that the warning notice reads as per Figure 10.14.

10.4.4.3 Electrical installations in caravans, motor caravans and caravan parks

Confirm that a notice has been fixed on (or near) the electrical inlet recess of the caravan or motor caravan installation containing details concerning the:

- the supply to the caravan is suitable for its intended purpose;
- the operation of the RCDs (if fitted) are suitably installed;
- a notice, worded (as shown in BS 7671:2018 Figure 721) has been permanently fixed near the main isolating switch.

10.4.4.4 Emergency switching

For any exterior installation making use of an emergency switching device, confirm that the switch is placed outside the building, adjacent to the equipment with a notice fixed near the switch that indicates its use.

10.4.4.5 *Final circuit distribution boards*

Confirm that distribution boards for socket outlet final circuits have a notice that clearly indicates circuits which have a high protective conductor current and that this information is positioned so as to be plainly visible to a person employed in modifying or extending the circuit.

10.4.4.6 *Inspection and testing*

Confirm that:

- a notice has been fixed in a prominent position at or near the origin of every installation upon completion of any work (for example. initial verification, alterations and additions to an installation and periodic inspection and testing);
- the notice has been inscribed in indelible characters and reads as shown in Figure 10.15;
- If an installation includes an RCD then it has a notice (fixed in a prominent position) that reads as shown in Figure 10.16.

Figure 10.15 Inspection and testing notice

Figure 10.16 Earth inspection and testing notice

Confirm that following initial verification, an Electrical Installation Certificate (together with a schedule of inspections and a schedule of test results), has been given to the person ordering the work and that:

- the schedule of test results identified every circuit, its related protective device(s) and a full record of the results of tests and measurements;
- the Certificate took account of the respective responsibilities for the safety of that installation and the relevant schedules;
- defects or omissions revealed during inspection and testing of the installation work covered by the Certificate were rectified before the Certificate is issued.

Confirm that:

- an Electrical Installation Certificate (containing details of the installation together with a record of the inspections made test results and any faults found) or a Minor Electrical Installation

Works Certificate (for all minor electrical installations that do not include the provision of a new circuit) was provided for all alterations or additions to electrical circuits.

10.4.4.7 Isolation

Confirm that a notice is fixed in each position where there are live parts that are not capable of being isolated by a single device.

If an installation is supplied from more than one source confirm that:

- a main switch is provided for each source of supply;
- a notice has been placed warning operators that more than one switch needs to be operated.

Unless an interlocking arrangement is provided, confirm that a notice has been provided warning people that they will need to use the appropriate isolating devices if an equipment or enclosure that contains live parts cannot be isolated by a single device.

10.4.4.8 Voltage

Verify that:

- items of equipment (or enclosures) whose nominal voltage exceeds 230 V (and where the presence of such a voltage would **not** normally be expected) have a warning of the maximum voltage present is clearly visible;
- terminals are housed in separate enclosures a notice showing the maximum voltage that exists between those parts;
- means of access to all live parts of switchgear and other fixed live parts where different nominal voltages exist are marked to indicate the voltages present.

10.4.4.9 Non-standard colours

If an installation that was wired to a previous version of the Regulations is partially altered or rewired according to the current requirements of BS 7671:2018, then a warning notice (see Figure 10.17) has been placed at (or near) the appropriate distribution board.

Figure 10.17 Non-standard colours

10.5 What type of certificates and reports are there?

As previously mentioned, there are three types of Certificates associated with electrical installations. These are:

The Electrical Installation Certificate	For of the design, construction, inspection and testing of major work
Minor Electrical Installation Works Certificate	For the design, construction, inspection and testing of minor work
The Electrical Installation Condition Report	For the regular inspection and testing of electrical installations

The template of these Certificates and information concerning them can be downloaded from https://electrical.theiet.org/bs-7671/model-forms/ (or from a multitude of other websites) and provided your version of the certificate is based on the models shown in Appendix 6 of BS 7671:2018, it can be modified to suit your own particular Company's quality management documentation and include your Company logo, address, etc.

All certificates need to be made out and signed (or otherwise authenticated) by a competent person or persons and may be produced in any durable medium, including written and electronic media.

10.5.1 Electrical installation certificate

Upon completion of the verification of a new installation or changes to an existing installation, an Electrical Installation Certificate (based on the model shown in Figure 10.19, shall be provided) together with a schedule of inspections and a schedule of test results, shall be given to the person ordering the work.

The schedule of test results shall identify every circuit, including its related protective device(s), and shall require that the results of the appropriate tests and measurements are recorded.

The Electrical Installation Certificate contains details of the installation together with a record of the inspections made and the test results. It is **only** used for the initial certification of a new installation or for an alteration or addition to an existing installation where new circuits have been introduced.

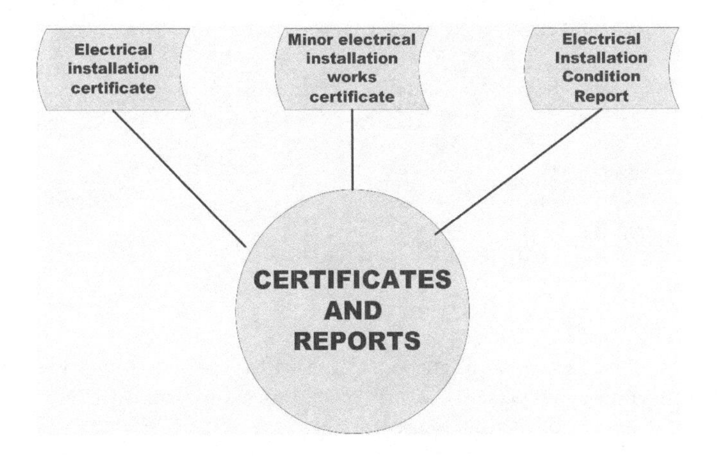

Figure 10.18 What type of certificates and reports are there

ELECTRICAL INSTALLATION CERTIFICATE

STINGRAY

(REQUIREMENTS FOR ELECTRICAL INSTALLATIONS – BS 7671:2018)

DETAILS OF THE CLIENT ☐

....

INSTALLATION ADDRESS ☐

....

....

DESCRIPTION AND EXTENT OF THE INSTALLATION ☐

Description of installation:	New installation
Extent of installation covered by this Certificate:	Addition to an existing installation
(Use continuation sheet if necessary) see continuation sheet No:	Alteration to an existing installation

FOR DESIGN

I/We being the person(s) responsible for the design of the electrical installation (as indicated by my/our signatures below), particulars of which are described above, having exercised reasonable skill and care when carrying out the design and additionally where this certificate applies to an addition or alteration, the safety of the existing installation is not impaired, hereby CERTIFY that the design work for which I/we have been responsible is to the best of my/our knowledge and belief in accordance with BS 7671:2018, amended to .... (date) except for the departures, if any, detailed as follows:

Details of departures from BS 7671 (Regulations 120.3, 133.1.3 and 133.5):

Details of permitted exceptions (Regulation 411.3.3). Where applicable, a suitable risk assessment(s) must be attached to this Certificate.

Risk assessment attached

The extent of liability of the signatory or signatories is limited to the work described above as the subject of this Certificate.

For the DESIGN of the installation: **(Where there is mutual responsibility for the design)

Signature: .... Date: .... Name (IN BLOCK LETTERS): .... Designer No 1

Signature: .... Date: .... Name (IN BLOCK LETTERS): .... Designer No 2**

FOR CONSTRUCTION

I being the person responsible for the construction of the electrical installation (as indicated by my signature below), particulars of which are described above, having exercised reasonable skill and care when carrying out the construction hereby CERTIFY that the construction work for which I have been responsible is to the best of my knowledge and belief in accordance with BS 7671:2018, amended to ....(date) except for the departures, if any, detailed as follows:

Details of departures from BS 7671 (Regulations 120.3 and 133.5):

The extent of liability of the signatory is limited to the work described above as the subject of this Certificate.

For CONSTRUCTION of the installation:

Signature: .... Date: .... Name (IN BLOCK LETTERS): .... Constructor

FOR INSPECTION & TESTING

I being the person responsible for the inspection & testing of the electrical installation (as indicated by my signature below), particulars of which are described above, having exercised reasonable skill and care when carrying out the inspection & testing hereby CERTIFY that the work for which I have been responsible is to the best of my knowledge and belief in accordance with BS 7671:2018, amended to ....(date) except for the departures, if any, detailed as follows:

Details of departures from BS 7671 (Regulations 120.3 and 133.5):

The extent of liability of the signatory is limited to the work described above as the subject of this Certificate.

For INSPECTION AND TESTING of the installation:

Signature: .... Date: .... Name (IN BLOCK LETTERS): .... Inspector

NEXT INSPECTION

I/We the designer(s), recommend that this installation is further inspected and tested after an interval of not more than .... years/months.

Figure 10.19 Electrical installation certificate

Notes:

1 The Electrical Installation Certificate is **not** to be used for a Periodic Inspection (for which an Electrical Installation Condition Report form should be used).

2 The original Certificate shall be given to the person ordering the work and a duplicate retained by the contractor.

3 The Electrical Installation Certificate is **only** valid if accompanied by the Schedule of Inspections and the Schedule(s) of Test Results.
4 For an alteration or addition which does not involve the introduction of a new circuit(s), a Minor Electrical Installation Works Certificate should be used.

The Electrical Installation Certificate **shall** be completed and signed by a competent person(s).

The Electrical Installation Report is 14 pages long and (speaking from experience) most electricians find that having to complete this lengthy form is (to put it mildly) a bit of a bore! However, it is a very important document and one that is repeatedly relied on for completing other inspections (such as the Minor Works Certificate or the Installation Condition Report).

10.5.2 Minor Electrical Installation Works Certificate

A Minor Electrical Installation Works Certificate (also referred to as the Minor Works Certificate) is used for additions and alterations to an installation such as an extra socket-outlet or lighting point to an existing circuit, the relocation of a light switch, etc. This Certificate may also be used for the replacement of equipment such as accessories or luminaires, but **not** for the replacement of distribution boards (or similar items) or the provision of a new circuit.

The object of this certificate is to confirm that a minor electrical installation alteration or improvement has been designed, constructed and tested in accordance with BS 7671:2018.

It is a legal requirement to which everybody must comply and includes:

- full details of al minor works completed (e.g. departures from BS 7671:2018, location and description, etc.);
- details of the modified circuit (e.g. system type, earthing, wiring system used, protective measures, etc.);
- confirmation that all necessary inspections and tests have been completed (e.g. insulation resistance, adequate protective bonding, RCD operation, etc.).

10.5.3 Electrical Installation Condition Report

An Electrical Installation Condition Report (shown in Figure 10.21) is used for reporting on the condition of an existing installation and will include schedules of both the inspection and the test results.

MINOR ELECTRICAL INSTALLATION WORKS CERTIFICATE

STINGRAY

(REQUIREMENTS FOR ELECTRICAL INSTALLATIONS – BS 7671:2018)

To be used only for minor electrical work which does not include the provision of a new circuit

PART 1: Description of the minor works

1. Details of the Client .... Date minor works completed ....
2. Installation location/address ....
3. Description of the minor works ....
4. Details of departures, if any, from BS 7671:2018 for the circuit altered or extended (Regulation 120.3, 133.1.3 and 133.5):
   Where applicable, a suitable risk assessment(s) must be attached to the Certificate
   .... Rrisk assessment attached
5. Comments on (including any defects observed in) the existing installation (Regulation 644.1.2):
   ....

PART 2: Presence and adequacy of installation earthing and bonding arrangements (Regulation 132.16)

1. System earthing arrangement: TN-S TN-C-S TT
2. Earth fault loop impedance at distribution board (Z_{db}) supplying the final circuit .... Ω
3. Presence of adequate main protective conductors:
   Earthing conductor
   Main protective bonding conductor(s) to: Water Gas Oil Structural steel Other..................

PART 3: Circuit details

DB Reference No.: .... DB Location and type: ....

Circuit No.: .... Circuit description: ....

Circuit overcurrent protective device: BS (EN) .... Type .... Rating .... A

Conductor sizes: Live .... mm² cpc .... mm²

PART 4: Test results for the circuit altered or extended (where relevant and practicable)

Protective conductor continuity: $R_1 + R_2$ .... Ω or R_2.............. Ω

Continuity of ring final circuit conductors: L/L.............. Ω N/N .... Ω cpc/cpc .... Ω

Insulation resistance: Live - Live .... MΩ Live - Earth .... MΩ

Polarity satisfactory: Maximum measured earth fault loop impedance: Z_s .... Ω

RCD operation: Rated residual operating current ($I_{\Delta n}$) .... mA

Disconnection time .... ms

Satisfactory test button operation

PART 5: Declaration

I certify that the work covered by this certificate does not impair the safety of the existing installation and the work has been designed, constructed, inspected and tested in accordance with BS 7671:2018 (IET Wiring Regulations) amended to .... (date) and that to the best of my knowledge and belief, at the time of my inspection, complied with BS 7671 except as detailed in Part 1 above.

Name: ....

For and on behalf of: ....

Address: ....

....

....

Signature: ....

Position: ....

Date: ....

Figure 10.20 Minor Electrical Installation Works Certificate

ELECTRICAL INSTALLATION CONDITION REPORT

STINGRAY

SECTION A. DETAILS OF THE PERSON ORDERING THE REPORT
Name ....
Address ....
....

SECTION B. REASON FOR PRODUCING THIS REPORT ....
....
Date(s) on which inspection and testing was carried out ....

SECTION C. DETAILS OF THE INSTALLATION WHICH IS THE SUBJECT OF THIS REPORT
Occupier ....
Address ....
....
Description of premises
Domestic ☐ Commercial ☐ Industrial ☐ Other (include brief description) ☐ ....
Estimated age of wiring system ....years
Evidence of additions / alterations Yes ☐ No ☐ Not apparent ☐ If yes, estimate age ....years
Installation records available? (Regulation 651.1) Yes ☐ No ☐ Date of last inspection .... (date)

SECTION D. EXTENT AND LIMITATIONS OF INSPECTION AND TESTING
Extent of the electrical installation covered by this report
....
....
Agreed limitations including the reasons (see Regulation 653.2) ....
....
Agreed with: ....
Operational limitations including the reasons (see page no..........) ....
....
The inspection and testing detailed in this report and accompanying schedules have been carried out in accordance with BS 7671:2018 (IET Wiring Regulations) as amended to ....
It should be noted that cables concealed within trunking and conduits, under floors, in roof spaces, and generally within the fabric of the building or underground, have **not** been inspected unless specifically agreed between the client and inspector prior to the inspection. An inspection should be made within an accessible roof space housing other electrical equipment.

SECTION E. SUMMARY OF THE CONDITION OF THE INSTALLATION
General condition of the installation (in terms of electrical safety) ....
....
....
Overall assessment of the installation in terms of its suitability for continued use
SATISFACTORY / UNSATISFACTORY* (Delete as appropriate)
*An unsatisfactory assessment indicates that dangerous (code C1) and/or potentially dangerous (code C2) conditions have been identified.

SECTION F. RECOMMENDATIONS
Where the overall assessment of the suitability of the installation for continued use above is stated as UNSATISFACTORY, I / we recommend that any observations classified as *'Danger present'* (code C1) or *'Potentially dangerous'* (code C2) are acted upon as a matter of urgency.
Investigation without delay is recommended for observations identified as *'Further investigation required' (code FI)*.
Observations classified as *'Improvement recommended'* (code C3) should be given due consideration.

Subject to the necessary remedial action being taken, I / we recommend that the installation is further inspected and tested by ....(date)

SECTION G. DECLARATION
I/We, being the person(s) responsible for the inspection and testing of the electrical installation (as indicated by my/our signatures below), particulars of which are described above, having exercised reasonable skill and care when carrying out the inspection and testing, hereby declare that the information in this report, including the observations and the attached schedules, provides an accurate assessment of the condition of the electrical installation taking into account the stated extent and limitations in section D of this report.

Inspected and tested by:	**Report authorised for issue by:**
Name (Capitals)	Name (Capitals) ....
Signature	Signature ....
For/on behalf of	For/on behalf of ....
Position	Position ....
Address	Address ....
Date	Date ....

SECTION H. SCHEDULE(S)
..........schedule(s) of inspection and ....schedule(s) of test results are attached.
The attached schedule(s) are part of this document and this report is valid only when they are attached to it.

Figure 10.21 Electrical Installation Condition Report

An installation which was designed to an earlier edition of the Regulations and which does not fully comply with the current edition is not necessarily unsafe for continued use, or requires upgrading. Only damage, deterioration, defects, dangerous conditions and non-compliance with the requirements of the Regulations – which may give rise to danger – should be recorded.

10.6 Mandatory test requirements specific for compliance to the Building Regulations

10.6.1 Part P – Electrical safety

Confirm that:

- reasonable provision has been made in the design, installation, inspection and testing of electrical installations to protect persons from fire or injury;
- sufficient information has been provided so that persons wishing to operate, maintain or alter an electrical installation can do so with reasonable safety.

10.6.2 Part M – Access and facilities for disabled people

Confirm that (in addition to the requirements of the Disability Discrimination Act 1995) precautions have been taken to ensure that electrical installations in:

- new non-domestic buildings and/or dwellings (e.g. houses and flats used for student living accommodation, etc.);
- extensions to existing non-domestic buildings;
- non-domestic buildings that have been subject to a material change of use (e.g. so that they become a hotel, boarding house, institution, public building or shop);

are capable of allowing people, regardless of their disability, age or gender to:

- gain access to buildings;
- gain access within buildings;

- be able to use the facilities of the building;
- use sanitary conveniences in the principal storey of any new dwelling.

Note: From 1 October 2010, the Equality Act replaced most of the Disability Discrimination Act (DDA). However, the Disability Equality Duty in the DDA continues to apply.

10.6.3 *Part L – Conservation of fuel and power*

- Confirm that energy efficiency measures have been provided which ensure that lighting systems utilise energy-efficient lamps with:
  - manual switching controls; or
  - automatic switching (in the case of external lighting fixed to the building); or
  - both manual and automatic switching controls;

so that the lighting systems can be operated effectively with regards the conservation of fuel and power.

- Confirm that building occupiers have been supplied with sufficient information to show how the heating and hot water services can be operated and maintained.

10.6.4 *Inspection and test*

Verify that all electrical installations have been inspected and tested during, at the end of installation and before they are taken into service.

Confirm that all components that are part of an electrical installation have been inspected (during installation as well as on completion) to verify that the components have been:

- selected and installed in accordance with the requirements of BS 7671:2018;
- made in compliance with appropriate British or Harmonised European Standards;
- evaluated against external influences (such as the presence of moisture);
- checked to see that they have not been visibly damaged or defective so as to be unsafe;

- tested to check satisfactory performance with respect to continuity of conductors, insulation resistance, separation of circuits, polarity, earthing and bonding arrangements, Earth fault loop impedance and functionality of all protective devices including RCDs;
- had their test results recorded;
- had their test results compared with the relevant performance criteria to confirm compliance.

Note: Inspections and testing of DIY work should **also** meet the above requirements.

10.6.5 Consumer units

Ensure that accessible consumer units have been fitted with a child-proof cover or installed in a lockable cupboard.

10.6.6 Design

Confirm that electrical installations have been designed and installed so that they:

- meet the requirements of the Building Regulations;
- do not present an electric shock or fire hazard to people;
- provide adequate protection for persons against the risks of electric shock, burn or fire injuries;
- provide adequate protection against mechanical and thermal damage.

10.6.7 Earthing

Inspect and confirm that:

- electrical installations have been properly earthed;
- lighting circuits include a circuit protective conductor;
- socket-outlets which have a rating of 32 A or less and which may be used to supply portable equipment for use outdoors, are protected by an RCD;
- the Distributor has provided an earthing facility for all new connections;

- new or replacement, non-metallic light fittings, switches or other components do not require earthing unless new circuit protective (earthing) conductors have been provided;
- if there are any socket-outlets that will accept unearthed (e.g. 2-pin) plugs do not use supply equipment that needs to be earthed.

10.6.8 Electricity Distributors responsibilities

Prior to starting works, confirm that the Electricity Distributor has:

- accepted responsibility for ensuring that the supply is mechanically protected and can be safely maintained;
- evaluated and agreed the proposal for a new (or significantly altered) installation;
- installed the cut-out and metre in a safe location.

Confirm that the Distributor has:

- installed and maintained the supply within defined tolerance limits;
- provided certain technical and safety information to the consumer to enable them to design their installation(s);
- ensured that their equipment on consumers' premises:
  - is suitable for its purpose;
  - is safe in its particular environment;
  - clearly shows the polarity of the conductors.

10.6.9 Electrical installations

Verify (by inspection and test) that during installation, at the end of installation and before they are taken into service that all electrical installation work:

- has been carried out professionally;
- complies with the Electricity at Work Regulations 1989 as amended;
- has been carried out by persons who are competent to prevent danger and injury while doing it, or who are appropriately supervised.

10.6.10 Extensions, material alterations and material changes of use

Where any electrical installation work is classified as an extension, a material alteration or a material change of use, confirm that:

- the existing fixed electrical installation in the building is capable of supporting the amount of additions and alterations that will be required;
- the earthing and bonding systems are satisfactory and meet requirements;
- the mains supply equipment is suitable and can carry any additional loads envisaged;
- any additions and alterations to the circuits which feed them comply with the requirements of the regulations;
- all protective measures required, meet the requirements;
- the rating and the condition of existing equipment (belonging to both the consumer and the electricity distributor) is sufficient.

10.6.11 Wiring and wiring systems

Confirm that:

- cables concealed in floors and walls have (if required):
    - an earthed metal covering; or
    - are enclosed in steel conduit; or
    - have some form of additional mechanical protection.
- cables to an outside building (e.g. garage or shed), if run underground, have been routed and positioned so as to give protection against electric shock and fire as a result of mechanical damage to a cable;
- heat-resisting flexible cables (if required) have been supplied for the final connections to certain equipment (see makers instructions).

10.6.11.1 Continuity of conductors

The continuity of conductors and connections to exposed-conductive-parts and extraneous conductive-parts, if any, shall be verified by a measurement of resistance on:

- protective conductors, including protective bonding conductors; and
- in the case of ring final circuits, live conductors.

10.6.11.2 Equipotential bonding conductors

Confirm that:

- main equipotential bonding conductors for water service pipes, gas installation pipes, oil supply pipes plus and certain other 'earthy' metalwork have been provided;
- where there is an increased risk of electric shock (e.g. such as in bathrooms and shower rooms), supplementary equipotential bonding conductors have been installed.

Note: The minimum size of supplementary equipotential bonding conductors (without mechanical protection) is 4 mm^2.

10.6.12 Socket outlets

Confirm by inspection that:

- older types of socket-outlet that have been designed for non-fused plugs, are not connected to a ring circuit;
- RCD protection has been provided for all socket-outlets which have a rating of 32 A or less and which may be used to supply portable equipment for use outdoors;
- switched socket outlets indicate whether they are 'ON';
- socket-outlets that will accept unearthed (2-pin) plugs are not used to supply equipment that needs to be earthed;
- the following requirements (see Table 10.12) for wall sockets have been met.

Confirm (by inspection and testing) that all socket outlets used for lighting:

- are wall-mounted;
- are easily reachable;
- have been installed between 450 mm and 1200 mm from the finished floor level;
- are located no nearer than 350 mm from room corners;
- if a switched variety, indicate whether they are 'ON'.

Table 10.12 Building Regulations requirements for wall sockets

Type of wall	*Requirement*
Cavity masonry	The position of sockets has been staggered on opposite sides of the separating wall Deep sockets and chases have not been used in a separating wall Deep sockets and chases in a separating wall have **not** been placed back to back
Framed walls with absorbent material	Sockets have: • been positioned on opposite sides of a separating wall • not been connected back to back • been staggered a minimum of 150 mm edge to edge
Solid masonry	Deep sockets and chases have **not** been used in separating walls The position of sockets has been staggered on opposite sides of the separating wall
Timber framed	Power points: • that have been set in the linings have a similar thickness of cladding behind the socket box. • have not been placed back to back across the wall.

10.6.13 Switches

Ensure that all controls and switches:

- are easy to operate, visible and free from obstruction;
- have been located between 750 mm and 1200 mm above the floor;
- do not require the simultaneous use of both hands (unless necessary for safety reasons) to operate.

And that:

- mains and circuit isolator switches clearly indicate whether they 'ON' or 'OFF';

- individual switches on panels and on multiple socket outlets have been well separated;
- front plates contrast visually with their backgrounds;
- controls that need close vision (e.g. thermostats) have been located between 1200 mm and 1400 mm above the floor;
- where possible, light switches with large push pads have been used in preference to pull cords;
- the colours red and green have **not** been used in combination as indicators of 'ON' and 'OFF' for switches and controls;
- all switches used for lighting:
  - are easily reachable;
  - have been installed between 450 mm and 1200 mm from the finished floor level.

Confirm that light switches:

- have large push pads (in preference to pull cords);
- align horizontally with door handles;
- are within the 900 to 1100 mm from the entrance door opening;
- are located between 750 mm and 1200 mm above the floor;
- are **not** coloured red and green (i.e. as a combination) as indicators for 'ON' and 'OFF'.

10.6.14 *Telephone points and TV sockets*

Confirm that all telephone points and TV sockets have been located between 400 mm and 1000 mm above the floor (or 400 mm and 1200 mm above the floor for permanently wired appliances).

10.6.15 *Equipment and components*

The most important safety requirements from the point of view of the Building Regulations are:

- emergency alarms;
- fire alarms;
- heat emitters;
- portable equipment for use outdoors;
- power operated entrance doors.

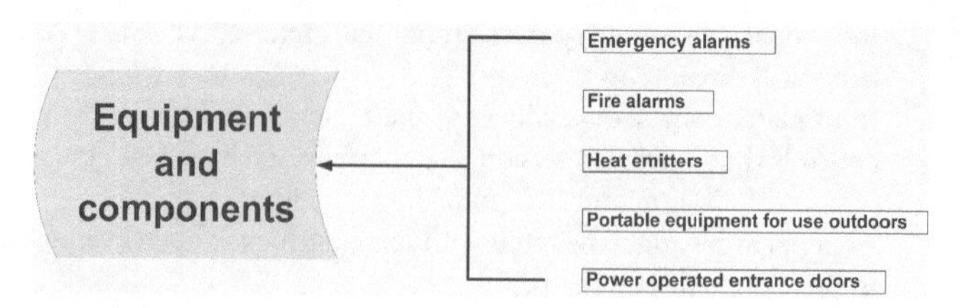

Figure 10.22 Equipment and components for the Building Regulations

10.6.15.1 Emergency alarms

Test and inspect to ensure that:

- emergency assistance alarm systems have:
  - visual and audible indicators to confirm that an emergency call has been received;
  - a reset control that is reachable from a wheelchair, WC, or from a shower/changing seat;
  - a signal that is distinguishable visually and audibly from the fire alarm;
- emergency alarm pull cords are (or should be):
  - coloured red;
  - located as close to a wall as possible;
  - have two red 50 mm diameter bangles.
- front plates contrast visually with their backgrounds;
- the colours red and green have **not** been used (in combination) to indicate 'ON' and 'OFF' for switches and controls.

10.6.15.2 Fire alarms

Verify (by test **and** inspection) that fire detection and fire-warning systems have been properly designed, installed and maintained and that:

- all buildings have arrangements for detecting fire;
- all buildings have been fitted with a suitable (electrically operated) fire warning system (in compliance with BS 5839) or have means of raising an alarm in case of fire (e.g. rotary gongs, handbells or by shouting 'FIRE');

- fire alarms emit an audio and visual signal to warn occupants with hearing or visual impairments;
- the fire warning signal is distinct from other signals which may be in general use;
- in premises that are used by the general public (e.g. large shops and places of assembly) a staff alarm system (complying with BS 5839) has been used.

10.6.15.3 Heat emitters

Check that heat emitters:

- are either screened or have their exposed surfaces kept at a temperature below 43°C;
- that are located in toilets and bathrooms, do not restrict:
  - the minimum clear wheelchair manoeuvring space;
  - there is sufficient space beside a WC used to transfer from the wheelchair to the WC.

10.6.15.3.1 PORTABLE EQUIPMENT FOR USE OUTDOORS

Verify that RCDs have been provided for all socket-outlets which have a rating of 32 A or less and which may be used to supply portable equipment for use outdoors.

10.6.15.4 Power operated entrance doors

- Confirm that all power operated doors have been provided with:
  - safety features to prevent injury to people who are struck or trapped (such as a pressure sensitive door edge which operates the power switch);
  - a readily identifiable (and accessible) stop switch;
  - a manual or automatic opening device in the event of a power failure where and when necessary for health or safety.

Confirm that:

- all doors to accessible entrances have been provided with a power operated door opening and closing system if a force greater than 20 N is required to open or shut a door;

- once open, all doors to accessible entrances are wide enough to allow unrestricted passage for a variety of users, including wheelchair users, people carrying luggage, people with assistance dogs, and parents with pushchairs and small children;
- power operated entrance doors:
  - have a sliding, swinging or folding action controlled manually (by a push pad, card swipe, coded entry, or remote control) or automatically controlled by a motion sensor or proximity sensor such as contact mat;
  - open towards people approaching the doors;
  - provide visual and audible warnings that they are operating (or about to operate);
  - incorporate automatic sensors to ensure that they open early enough (and stay open long enough) to permit safe entry and exit;
  - incorporate a safety stop that is activated if the doors begin to close when a person is passing through;
  - revert to manual control (or fail safe) in the open position in the event of a power failure;
  - when open, do not project into any adjacent access route;
- its manual controls are:
  - located between 750 mm and 1000 mm above floor level;
  - operable with a closed fist;
  - set back 1400 mm from the leading edge of the door when fully open;
  - clearly distinguishable against the background;
  - contrast visually with the background.

10.6.16 *Thermostats*

Check that all controls that need close vision (e.g. thermostats) are located between 1200 mm and 1400 mm above the floor.

10.6.17 *Smoke alarms – Dwellings*

Confirm by test and inspection that smoke alarms have been positioned:

- in the circulation space within 7.5 m of the door to every habitable room;

- in the circulation spaces between sleeping spaces and places where fires are most likely to start (e.g. kitchens and living rooms);
- on every storey of a house (including bungalows).

Confirm by test and inspection that:

- kitchen areas that are not separated from the stairway or circulation space by a door, have been equipped with an additional heat detector in the kitchen, that is interlinked to the other alarms;
- if more than one smoke alarm has been installed in a dwelling then they have been linked so that if a unit detects smoke it will operate the alarm signal of all the smoke detectors.

Verify by inspection that smoke alarms:

- have ideally been mounted, 25–600 mm below the ceiling (25–150 mm in the case of heat detectors) and at least 300 mm from walls and light fittings;
- have not been fixed over a stair shaft or any other opening between floors;
- have not been fitted:
    - in places that get very hot (such as a boiler room);
    - in a very cold area (such as an unheated porch);
    - in bathrooms, showers, cooking areas or garages, or any other place where steam, condensation or fumes could give false alarms;
    - next to or directly above heaters or air conditioning outlets;
    - on surfaces which are normally much warmer or colder than the rest of the space.

Test, inspect and confirm that the power supply for a smoke alarm system:

- has been derived from the dwelling's mains electricity supply via a single independent circuit at the dwelling's main distribution board (consumer unit);
- includes a stand-by power supply that will operate during mains failure;
- is not (preferably) protected by an RCD.

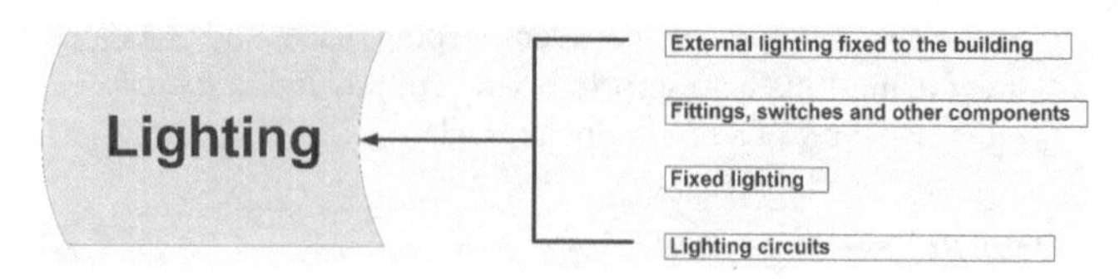

Figure 10.23 Essential lighting checks

10.6.18 *Lighting*

The following circuits and equipment need to be checked:

- External lighting fixed to the building;
- Fittings, switches and other components;
- Fixed lighting;
- Lighting circuits.

10.6.18.1 External lighting fixed to the building

Confirm (by test and inspection) that all external lighting (including lighting in porches, but not lighting in garages and carports):

- automatically extinguishes when there is enough daylight and when not required (e.g. at night);
- have sockets that can only be used with lamps having an efficacy greater than 40 lumens per circuit Watt (such as fluorescent or compact fluorescent lamp types, and not GLS tungsten lamps with bayonet cap or Edison screw bases).

10.6.18.2 Fittings, switches and other components

Confirm that new or replacement, non-metallic light fittings, switches or other components that require earthing (e.g. non-metallic varieties) have been provided with new circuit protective (earthing) conductors.

10.6.18.3 Fixed lighting

Ensure that in locations, where lighting can be expected to have most use, fixed lighting (e.g. fluorescent tubes and compact fluorescent lamps – but not GLS tungsten lamps with bayonet cap or Edison screw bases) with a luminous efficacy greater than 40 lumens per circuit-watt have been made available.

10.6.18.4 *Lighting circuits*

Verify that all lighting circuits include a circuit protective conductor.

10.6.19 *Lecture/conference facilities*

In lecture halls and conference facilities, confirm that artificial lighting has been designed to:

- give good colour rendering of all surfaces;
- be compatible with other electronic and radio frequency installations.

10.6.20 *Cellars or basements*

Ensure that LPG storage vessels and LPG fired appliances that are fitted with automatic ignition devices or pilot lights have not been installed in cellars or basements.

10.7 What about test equipment

Obviously, the actual choice of test equipment that the electrician chooses to use will normally be based on personal preference and experience – and as shown in Figure 10.24, the choice can be enormous. Nevertheless, it is essential that any piece of test equipment (including software) that is used when installing or inspecting electrical installations for compliance with the Regulations, can be relied on to produce accurate results (which is why it is important to erect a notice similar to Figure 10.5 to remind people).

ISO 9001:2015 (i.e. the internationally recognised standard for Quality Management) specifies the requirements for the control of test equipment (although they actually refer to them as '*measuring and monitoring devices*') as follows:

Proof	The controls that an organisation has in place to ensure that equipment (including software) used for conformance to specified requirements is properly maintained.
Likely documentation	Equipment records of maintenance and calibration. Work instructions.

Figure 10.24 A selection of test equipment

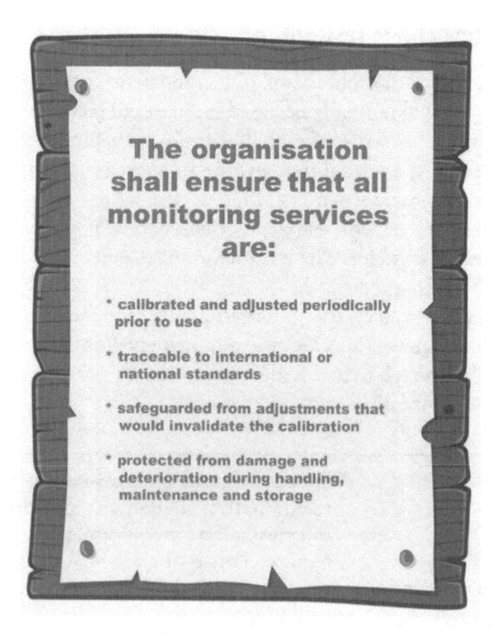

Figure 10.25 Mandatory requirements from the Building Regulations

Although the majority of electricians probably work on an individual basis and the requirement to operate as an accredited and registered ISO 9001:2015 company doesn't really apply, following the recommendations of this standard can only help to improve the quality of any organisation – no matter its size.

In general, therefore:

- all measuring and test equipment that is used by an electrician needs to be well maintained, in good condition and capable of safe and effective operation within a specified tolerance of accuracy;
- all measuring and test equipment should be regularly inspected and/or calibrated to ensure that it is capable of accurate operation (and where necessary by comparison with external sources traceable back to National Standards);
- any electrostatic protection equipment that is utilised when handling sensitive components is regularly checked to ensure that it remains fully functional;
- the control of measuring and test equipment (whether owned by the electrician, on loan, hired or provided by the customer) should always include a check that the equipment is exactly:
    - what is required;
    - has been initially calibrated before use;
    - operates within the required tolerances;
    - is regularly recalibrated; and that
    - facilities exist (either within the organisation or via a third party) to adjust, repair or recalibrate as necessary.

If the measuring and test equipment is used to verify process outputs against a specified requirement, then the equipment needs to be maintained and calibrated against National and international standards and the results of any calibrations carried out **must** be retained and the validity of previous results re-assessed if they are subsequently found to be out of calibration.

10.7.1 *Control of inspection, measuring and test equipment*

Measuring and test equipment should always be stored correctly and satisfactorily protected between use (to ensure their bias and precision) and should be verified and/or recalibrated at appropriate intervals.

10.7.2 *Computers*

Special attention **must** be paid to computers if they are used in controlling processes and particularly to the maintenance and accreditation of any related software.

10.7.3 *Software*

Software used for measuring, monitoring and/or testing specified requirements should be validated prior to use.

10.7.4 *Calibration*

Without exception, all measuring instruments can be subject to damage, deterioration or just general wear and tear when they are in regular use. The electrician should, therefore, take account of this fact and ensure that all of his test equipment is regularly calibrated against a known working standard.

The accuracy of the instrument will depend very much on what items it is going to be used to test, the frequency of use of the test instrument and the electrician will have to decide on the maximum tolerance of accuracy for each item of test equipment.

Of course, calibrating against a 'working standard' is pretty pointless if that particular standard cannot be relied upon and so the workshop standard **must also** be calibrated, on a regular basis, at either a recognised calibration centre or at the UK Physical Laboratory against one of the National Standards.

The electrician will, therefore, have to make allowances for:

- the calibration and adjustment of all measuring and test equipment that can affect product quality of their inspection and/or test;
- the documentation and maintenance of calibration procedures and records;
- the regular inspection of all measuring or test equipment to ensure that they are capable of the accuracy and precision that is required;
- the environmental conditions being suitable for the calibrations, inspections, measurements and tests to be completed.

If the instrument is found to be outside of its tolerance of accuracy, any items previously tested with the instrument must be regarded as suspect.

In these circumstances, it would be wise to review the test results obtained from the individual instrument. This could be achieved by compensating for the extent of inaccuracy to decide if the acceptability of the item would be reversed.

10.7.4.1 Calibration methods

There are various possibilities, such as:

- sending all working equipment to an external calibration laboratory;
- sending one of each item (i.e. a 'workshop standard') to a calibration laboratory, then sub-calibrating each working item against the workshop standard;
- testing by attributes – i.e. take a known 'faulty' product, and a known 'good' product and then test each one to ensure that the test equipment can identify the faulty and good product correctly.

10.7.4.2 Calibration frequency

The calibration frequency depends on how much the instrument is used, its ability to retain its accuracy and how critical the items being tested are.

Infrequently used instruments are often only calibrated prior to their use whilst frequently used items would normally be checked and re-calibrated at regular intervals depending, again, on product criticality, cost, availability, etc.

Normally 12 months is considered as about the maximum calibration interval.

10.7.4.3 Calibration ideals

- Each instrument should be uniquely identified, allowing it to be traced.
- The calibration results should be clearly indicated on the instrument.
- The calibration results should be retained for reference.
- The instrument should be labelled to show the next "calibration due" date to easily avoid its use outside of the period of confidence.
- Any means of adjusting the calibration should be sealed, allowing easy identification if it has been tampered with (e.g. a label across the joint of the casing).

Note: Examples of test equipment normally used by electricians is shown at Appendix 10.1.

Appendix 10.1 – Examples of test equipment used to test electrical installations

The following are examples of instruments that are required to test electrical installations for compliance with the requirements of BS 7671.

Quite a lot of test equipment manufacturers now produce dual or multifunctional instruments and so it is quite common to find an instrument that is capable of measuring a number of different types of tests – for example, continuity and insulation resistance, loop impedance and prospective fault current. It is, therefore, wise to carry out a little research before purchasing!

Continuity tester

All protective and bonding conductors must be tested to ensure that they are electrically safe and correctly connected. Low resistance ohmmeters and simple multimeters are normally used for continuity testing. Ideally, they should have a no-load voltage of between 4 V and 24 V, be capable of producing an a.c. or d.c. short circuit voltage of not less than 200 mA and have a resolution of at least 0.01 mΩ.

Insulation resistance tester

A low resistance between phase and neutral conductors, or from live conductors to Earth, will cause a leakage current which will cause weakening of the insulation, as well as involving a waste of energy which would increase the running costs of the installation. To overcome this problem, the resistance between poles or to Earth needs to be measured (similar to that shown in Figure 10.26) and it must never be less than 0.5 mΩ for the usual supply voltages.

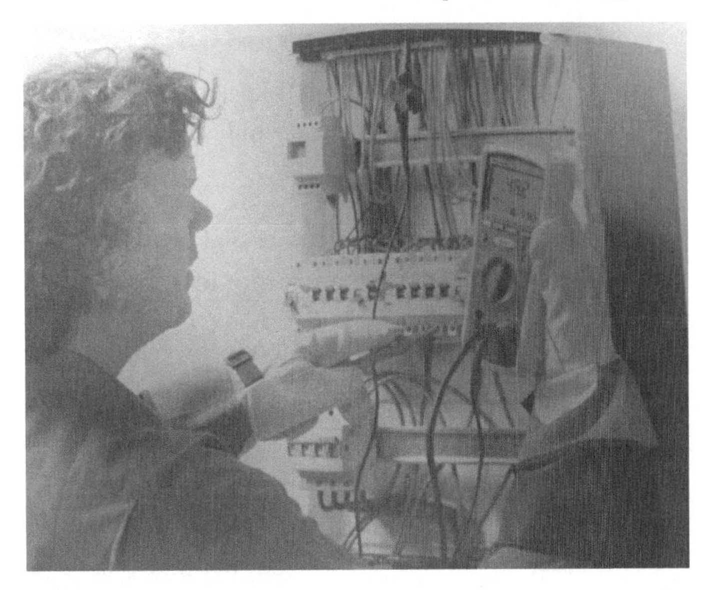

Figure 10.26 Insulation tester

Loop impedance tester

Loop testing is a quick, convenient, and highly specific method of testing an electrical circuit for its ability to engage protective devices (circuit breakers and fuses, etc.) by simulating a fault from live to Earth or from live to neutral (short circuit). The tester first measures the unloaded voltage, then connects a known resistance between the conductors, thereby simulating a fault. The voltage drop is measured across the known resistor, in series with the loop, and the proportion of the supply voltage that appears across the resistor will be dependent on the impedance of the loop.

RCD tester

The standard method for protecting electrical installations is to make sure that an Earth fault results in a fault current that is high enough to operate the protective device quickly so that fatal shock is prevented. However, there are cases where the impedance of the Earth-fault loop, or the impedance of the fault itself, are too high to enable enough fault current to flow. In such a case, either:

- the current will continue to flow to Earth, perhaps generating enough heat to start a fire; or
- metal work which can be touched may be at a high potential relative to Earth, resulting in severe shock danger.

Either or both of these possibilities can be removed by the installation of an RCD.

RCDs are also, sometimes, referred to as:

RCCD	Residual Current Operated Circuit Breaker;
SRCD	Socket outlet incorporating an RCD;
PRCD	portable RCD, usually an RCD incorporated into a plug;
RCBO	an RCCD which includes overcurrent protection;
SRCBO	a Socket outlet incorporating an RCBO and RCD tester allows a selection of out of balance currents to flow through the RCD and cause its operation.

The RCD tester should **not** be operated for longer than 2 s.

Prospective fault current tester

A Prospective Fault Current (PFC) tester is used to measure the prospective phase neutral fault current.

Test lamp or voltage indicator

These types of tester (often referred to as a *tetrascope* or *neon screwdriver*) are frequently used by electricians.

These compact screwdriver multi testers are normally water and impact resistant, with a.c. voltage test, contact test 70–250 VAC, non-contact 100–1000 VAC, Polarity test 1.5–36 VDC, Continuity check 0–5 ohm and Auto power on/off.

Earth electrode resistance

The Earth electrode (when used) is the means of making contact with the general mass of Earth and should be regularly tested to ensure that

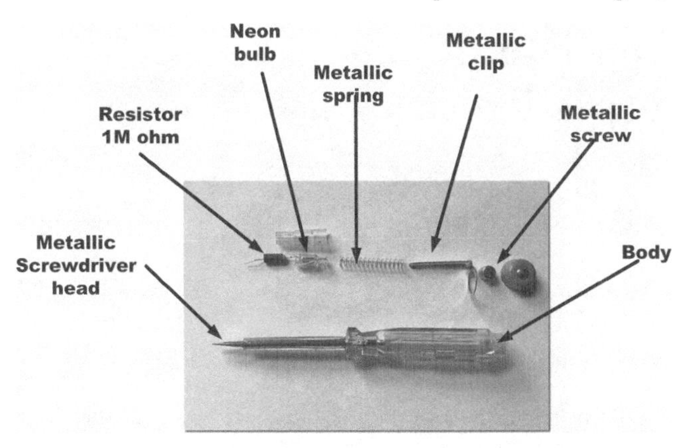

Figure 10.27 Typical neon screwdriver

good contact is made. In all cases, the aim is to ensure that the electrode resistance is not so high that the voltage from earthed metalwork to Earth exceeds 50 V.

Note: Acceptable electrodes are rods, pipes, mats, tapes, wires, plates and structural steelwork buried or driven into the ground. The pipes of other services such as gas and water must **not** be used as Earth electrodes (although they must be bonded to Earth.

Appendix 10.2 – Inspection check list

The Wiring Regulations specifically state that inspections **shall** include the design, construction, inspection and testing of any new electrical installation, or new work associated with an alteration or addition to an existing installation.

Thus, in accordance with the requirements of BS 7671:2018 and for compliance with the Building Regulations 2010, the inspection shall include the following items:

- access to switchgear and equipment;
- cable routing;
- choice and setting of protective and monitoring devices;
- connection of accessories and equipment;
- connection of conductors;
- connection of single-pole devices for protection or switching in phase conductors;
- continuity of protective conductors;
- continuity of ring final circuit conductors;
- Earth electrode resistance;
- Earth fault loop impedance;
- erection methods;
- functional testing;
- identification of conductors;
- insulation of non-conducting floors and walls;
- insulation resistance;
- labelling of protective devices, switches and terminals;
- polarity;
- presence of danger notices and other warning signs;

- presence of diagrams, instructions and similar information;
- presence of fire barriers, suitable seals and protection against thermal effects;
- prevention of mutual (i.e. detrimental) influence;
- presence of undervoltage protective devices;
- prospective fault current;
- protection against electric shock, such as:
  - exposed-conductive-parts;
  - insulating enclosures;
  - insulation of operational electrical equipment;
  - the capability of equipment to withstand mechanical, chemical, electrical and thermal influences and stresses normally encountered during service;
  - verification of the quality of the insulation.
- protection against electric shock and against direct current; using
  - barriers or enclosures;
  - insulation of live parts;
  - obstacles;
  - PELV;
  - placing out of reach.
- protection against external influences;
- protection against indirect contact; such as
  - automatic disconnection of supply;
  - Earth free local equipotential bonding;
  - earthed equipotential bonding;
  - earthing and protective conductors;
  - electrical separation;
  - main equipotential bonding conductors;
  - use of Class II equipment or equivalent insulation;
  - supplementary equipotential bonding conductors.
- selection of conductors for current-carrying capacity and voltage drop;
- selection of equipment appropriate to external influences;
- site applied insulation; particularly
  - protection against direct contact;
  - protection against indirect contact;
  - supplementary insulation.

In addition to the above list of mandatory inspections for compliance with the Wiring and the Building Regulations, the following are some of the additional inspections that electricians usually complete during initial and periodic inspections and tests of electrical installations:

- cables and conductors (current carrying capacity, insulation and/or sheath);
- correct connection of accessories and equipment;
- electrical joints and connections (to ensure that they meet stipulated requirements concerning conductance, insulation, mechanical strength and protection);
- emergency switching;
- insulation;
- insulation on monitoring devices (design, installation and security);
- inspection of associated, electrical installations;
- isolation and switching devices (and their correct location);
- locations with risks of fire due to the nature of processed and/or stored materials;
- plug and socket outlets;
- protection against electric shock – special installations or locations;
- protection against Earth insulation faults;
- protection against mechanical damage;
- protection against overcurrent;
- protection by extra-low voltage systems (other than SELV);
- protection by non-conducting location;
- protection by RCDs
- protection by separation of circuits;
- supplies;
- supplies for safety services;
- wiring systems (selection and erection, temperature variations).

Specific requirements from the Building Regulations

The Building Regulations specifically state that every electrical connection and joint shall be accessible for inspection, except for the following:

- a joint, compound filled or encapsulated;
- a joint designed to be buried in the ground;

- a joint made by welding, soldering, brazing or appropriate compression tool;
- joints or connections made in the equipment by the manufacturer of the product and not intended to be inspected, tested or maintained;
- a connection between a cold tail and the heating element as in ceiling heating, floor heating or a trace heating system;
- equipment complying with BS 7671:2018 for a maintenance free accessory and marked with the symbol:

Inspection shall precede testing and ideally (i.e. from a safety point of view) should be completed with that part of the installation under inspection disconnected from the supply.

The inspection shall include at least the checking of the following items, where relevant to the installation and, where necessary, including any particular requirements for special installations or locations (see Part 7 of BS 7671:2018):

- absence of danger notices and other warning signs;
- absence of diagrams, instructions and similar information;
- adequacy of access to switchgear and equipment;
- connection of single-pole devices for protection;
- correct connection of accessories and equipment;
- identification and connection of conductors;
- erection methods;
- labelling of protective devices, switches and terminals;
- presence of fire barriers, suitable seals and protection against thermal effects;
- prevention of mutual detrimental influences;
- presence of appropriate devices for isolation and switching correctly located;
- presence of undervoltage protective devices;
- protection against electric shock;
- routing of cables in safe zones (or protection against mechanical damage);
- selection of conductors for current-carrying capacity and voltage drop, in accordance with the design;
- selection of equipment and protective measures appropriate to external influences.

Author's End Note

That completes what I consider to be the essential parts of the Wiring Regulations that you should be reminded of. **However**, *for full details of all the official requirements and recommendations for electrical installations, you will need to obtain a copy of BS 7671:2018.*

The IET website (https://electrical.theiet.org/bs-7671) also provides in depth information, on the latest changes made to this standard, its relationship with the Building Regulations and a whole host of other useful information.

'Mr Google' can also provide some very useful on-site hints as well.

This is the final chapter of this book and it is supported by the following Annexes:

Annex A – Symbols used in electrical installations;
Annex B – List of electrical and electromechanical symbols;
Annex C – SI units for existing technology;
Annex D–- IPX coding;
Annex E – Acronyms and abbreviations;
Annex F – Other books associated with the Wiring Regulations

Annex A
Symbols used in electrical installations

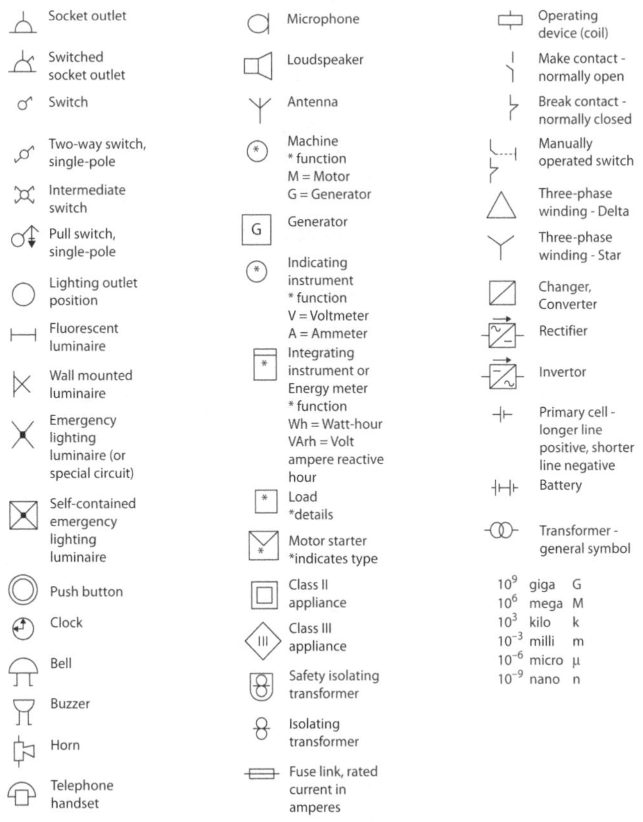

Socket outlet

Switched socket outlet

Switch

Two-way switch, single-pole

Intermediate switch

Pull switch, single-pole

Lighting outlet position

Fluorescent luminaire

Wall mounted luminaire

Emergency lighting luminaire (or special circuit)

Self-contained emergency lighting luminaire

Push button

Clock

Bell

Buzzer

Horn

Telephone handset

Microphone

Loudspeaker

Antenna

Machine
* function
M = Motor
G = Generator

G Generator

Indicating instrument
* function
V = Voltmeter
A = Ammeter

Integrating instrument or Energy meter
* function
Wh = Watt-hour
VArh = Volt ampere reactive hour

Load
*details

Motor starter
*indicates type

Class II appliance

Class III appliance

Safety isolating transformer

Isolating transformer

Fuse link, rated current in amperes

Operating device (coil)

Make contact - normally open

Break contact - normally closed

Manually operated switch

Three-phase winding - Delta

Three-phase winding - Star

Changer, Converter

Rectifier

Invertor

Primary cell - longer line positive, shorter line negative

Battery

Transformer - general symbol

10^9	giga	G
10^6	mega	M
10^3	kilo	k
10^{-3}	milli	m
10^{-6}	micro	µ
10^{-9}	nano	n

Annex B

List of electrical and electromechanical symbols

Symbol	Description
β°	tube oscillating angle
°C	degrees Celsius
Ω	ohm
μg	microgram
$\mu g/m^3$	micrograms per cubic metre
μm	micrometre
μs	microsecond
a	amplitude
A	ampere
A/m	amperes per metre
am	attometre
atm	standard atmosphere
C	coulomb
cd	candela
cd/m^2	candelas per square metre
dB	decibels
dB(A)	decibel amps
dBm	decibel metres
dm^3	cubic decimetre
dm^3/mm	cubic decimetres/millimetre – flow
Em	exametre
eV	electronvolt
f	frequency
F	farad
fm	femtometre

Symbol	Description
ft	foot
g	gram
G	gauss
G	shock
g^2/Hz	accelerated spectral density
GHz	Giga Hertz – frequency
Gm	gigametre
g/m^3	grams per cubic metre
g_n	peak acceleration
G_s	setting value of a characteristic quantity
h	hour
H	henry
ha	hectare
hp	horsepower
hr(s)	hour(s) – alternative to h
Hz	Hertz
I	amps
I^2R	power
in	inch
J	joule
k	constant of the relay
K	kelvin
kA	kiloamps
kA/μs	kiloamps per microsecond
kg	kilogram
kg/m^3	kilograms per cubic metre
kgf	kilogram force
kHz	kilohertz
kPa	kilo Pascal – pressure
ks	kilosecond
kV	kilovolts
kW	kilowatt
kW/m^2	kilowatts per square metre – irradiance
l	litre
lb	pound
lb/in	pounds per square inch
m	metre
m/s	metres per second

Symbol	Description
m/s^2	metres per second per second – amplitude
m^2	square metres
m^3	cubic metres
mbar	millibar – pressure
MHz	Mega Hertz
min	minute
mm	millimetre
Mm	megametre
mm/h	millimetres per hour
mm/m^2	millimetres/square metre – exposure
mol	mole
ms	millisecond
mV	millivolts
MVA	megavolt amps
N	newton
N/m^2	newtons per square metre
NaCl	sodium chloride
nF	nanofarad
nm	nanometre
pH	alkalinity/acidity value
pm	picometre
Pm	petametre
R	intensity of dropfield in mm/h
R	resistance
rad/s	radians per second
s	second
S	siemens
t	tonne
T	time
T	tesla
Tm	terametre
û	amplitude of voltage surge
U_n	nominal voltage
V	volt
V/μs	volts per microsecond
V/km	volts per kilometre
Vm	volts per metre
W	watt

Symbol	Description
Wb	weber
W/m^2	watts per square metre – irradiance
yd	yard
ym	yoctometre
Ym	Yottametre
zm	zeptometre
Zm	zettametre

Annex C
SI units for existing technology

The revised SI (an acronym for the International System of Units, which is informally known as the metric system) rests on a foundation of seven values, known as the constants whose values are the same everywhere in the universe. In the revised SI, these constants completely define the seven base SI units, from the second to the candela.

Table A Basic SI units

SI nomenclature	*Abbreviation*	*Quantity*
metre	m	length
kilogram	kg	mass
second	s	time
ampere	A	electrical current
kelvin	K	temperatures
mole	mol	amount of substance
candela	cd	luminous intensity

C.1 The kilogram (kg)

Of the seven units, only the kilogram (kg) is represented by a physical object, namely a cylinder of platinum-iridium kept at the International Bureau of Weights and Measures at Sèvres, near Paris, with a duplicate at the US Bureau of Standards.

C.2 The metre

The metre (m), on the other hand, 'is the length of the path travelled by light in a vacuum during a time interval of 1/299,792,458 of a second'.

C.3 The second

The second (s) has been defined as '*the duration of 9,192,631,770 periods of radiation corresponding to the energy-level change between the two hyperfine levels of the ground state of caesium-133 atom*'.

C.4 The ampere

The ampere (A) is '*that constant current which, if maintained in two straight parallel conductors of infinite length, of negligible circular cross section and placed 1 m apart in vacuum, would produce between these conductors a force equal to* 2×10^{-7} *newtons per metre length*'.

C.5 The kelvin

The unit of temperature is the kelvin (K), which is a thermodynamic measurement as opposed to one based on the properties of real material. Its origin is at absolute zero and there is a fixed point where the pressure and temperature of water, water vapour and ice are in equilibrium, which is defined as 273.16 K.

C.6 The mole

The mole (mol) is '*that quantity of substance of a system which contains as many elementary entities as there are atoms in 0.012 kg of carbon-12*'. For definition purposes the entities **must** be specified (e.g. atoms, electrons, ions or any other particles or groups of such particles).

C.7 The candela

Finally there is the candela (cd), the unit of light intensity. This is defined as '*the luminous intensity, in the perpendicular direction, of a surface of 1/600,000* m^2 *of a black body at the temperature of freezing platinum under a pressure of 101,325* N/m^2'.

C.8 Small number SI prefixes

Within the SI units there is a distinction between a quantity and a unit. Length is a quantity, but metres (abbreviated to m) is a unit.

Table B Small number SI units

Measurement	*Symbol*	*Equivalent to*
millimetre	mm	0.001 m or 10^{-3} m
micrometre	µm	0.000 001 m or 10^{-6} m
nanometre	Nm	0.000 000 001 m or 10^{-9} m
picometre	pm	0.000 000 000 001 m or 10^{-12} m
femtometre	fm	0.000 000 000 000 001 m or 10^{-15} m
attometre	am	0.000 000 000 000 000 001 m or 10^{-18} m
zeptometre	zm	0.000 000 000 000 000 000 001 m or 10^{-21} m
yoctometre	ym	0.000 000 000 000 000 000 000 001 m or 10^{-24} m

C.9 Large number SI prefixes

Table C Large number SI prefixes

Measurement	*Symbol*	*Equivalent to*
megametre	Mm	1,000,000 m or 10^{6} m
gigametre	Gm	1,000,000 000 m or 10^{9} m
terametre	Tm	1,000,000,000,000 m or 10^{12} m
petametre	Pm	1,000,000,000,000,000 m or 10^{15} m
exametre	Em	1,000,000,000,000,000,000 m or 10^{18} m
zettametre	Zm	1,000,000,000,000,000,000,000 or 10^{21} m
yottametre	Ym	1,000,000,000,000,000,000,000,000 or 10^{24} m

C.10 Deprecated prefixes

Some non-SI fractions and multiples are occasionally used (see below), but they are not encouraged.

Table D Deprecated prefixes

Fractions	*Prefix*	*Abbreviation*	*Multiple*	*Prefix*	*Abbreviation*
10^{-1}	deci	d	10	deka	da
10^{-2}	centi	c	10^{2}	hecto	h

C.11 Derived units

Some units, derived from the basic SI units, have been given special names, many of which originate from a person's name (e.g. Siemens).

Table E Derived units

Quantity	*Name of unit*	*Abbreviation (symbol)*	*Expression in terms of other SI units*
energy	joule	J	Nm
force	newton	N	–
power	watt	W	J/s
electric charge	coulomb	C	As
potential difference (voltage)	volt	V	W/A
electrical resistance (or reactance or impedance)	ohm	Ω	V/A
electrical capacitance	farad	F	C/V
magnetic flux	weber	Wb	Vs
Inductance (note that the plural of henry is henrys)	henry	H	Wb/A
magnetic flux density	tesla	T	Wb/m^2
admittance (electrical conductance)	siemens	S	$A/V(=\Omega^{-1})$
frequency	hertz	Hz	cycles per second (or events per second)

C.12 Units without special names

Other derived units, without special names, are listed below.

Table F Units without special names

Quantity	*Unit*	*Abbreviation*
area	square metres	m^2
volume	cubic metres	m^3
density	kilograms per cubic metre	kg/m^3
velocity	metres per second	m/s
angular velocity (angular frequency)	radians per second	rad/s
acceleration	metres per second per second	m/s^2
pressure	newtons per square metre	N/m^2
electric field strength	volts per metre	Vm
magnetic field strength	amperes per metre	A/m
luminance	candelas per square metre	cd/m^2

C.13 Tolerated units

Some non-SI units are tolerated in conjunction with SI units.

Table G Tolerated units

Quantity	*Unit*	*Abbreviation (symbol)*	*Definition*
area	hectare	ha	10^4 m^2
volume	litre	l	10^{-3} m^3
pressure	standard atmosphere	atm	101,325 Pa
mass	tonne	t	10^3 kg(Mg)
energy	electronvolt	eV	1.6021×10^{19} J
magnetic	gauss	G	10^{-4} T

C.14 Obsolete units

For historical interest (as well as for completeness), the following table gives a list of obsolete units.

Table H Obsolete units

Quantity	*Unit*	*Abbreviation (symbol)*	*Definition*
length	inch	in	0.0254 m
	foot	ft	0.3048 m
	yard	yd	0.9144 m
	mile	mi	1.60394 km
mass	pound	lb	0.4539237 kg
force	dyne	dyn	10^{-5} N
	poundal	pdl	0.138255 N
	pound force	lbf	4.44822 N
	kilogram force	kgf	9.80665 N
pressure	atmosphere	atm	101.325 kN/m^2
	torr	torr	133.322 N/m^2
	pounds per square inch	lb/in	6894.76 N/m^2
energy	erg	erg	10^{-7} J
power	horsepower	hp	745.700 W

C.15 Table of the SI units, symbols, and abbreviations

The table below gives some of the most commonly used SI symbols, units and abbreviations which are seen in scientific and electromechanical engineering application.

Table I SI Units and symbols

SI units and SI unit symbols		
SI unit name	*SI unit symbol*	*Quantity measured*
ampere	A	Electric current
ampere per meter	A/m	Magnetic field strength
ampere per square meter	A/m^2	Current density
becquerel	Bq s^{-1}	Activity – of radionuclide
candela	cd	Luminous intensity
candela per square metre	cd/m^2	Luminance
coulomb	C s·A	Electric charge, quantity of electricity
coulomb per cubic metre	C/m^3	Electric charge density
coulomb per kilogram	C/kg	Exposure (x rays and gamma rays)
coulomb per square metre	C/m^2	Electric flux density
cubic metre	m^3	Volume
cubic metre per kilogram	m^3/kg	Specific volume
degree Celsius	°C	Celsius temperature
farad	F C/V	Capacitance
farad per metre	F/m	Permittivity
grey	Gy	Absorbed dose, specific energy imparted, absorbed dose index
grey per second	Gy/s	Absorbed dose rate
henry	H Wb/A	Inductance
henry per metre	H/m	Permeability
hertz	Hz s^{-1}	Frequency
joule	J N·m	Energy, work, quantity of heat
joule per cubic metre	J/m^3	Energy density
joule per kelvin	J/K	Heat capacity, entropy
joule per kilogram	J/kg	Specific energy

(*Continued*)

Table I (Continued)

SI units and SI unit symbols		
SI unit name	*SI unit symbol*	*Quantity measured*
joule per kilogram kelvin	J/(kg·K)	Specific heat capacity
joule per mole	J/mol	Molar energy
joule per mole kelvin	J/(mol·K)	Molar heat capacity, molar entropy
kelvin	K	Absolute temperature, sometimes referred to as thermodynamic temperature
kilogram	kg	Mass
kilogram per cubic metre	kg/m^3	Density, mass density
lumen	lm	Luminous flux
lux	lx lm/m^2	Illuminance
metre	m	Length
metre per second	m/s	Speed, velocity
metre per second squared	m/s^2	Acceleration
mole	mol	Amount of substance
mole per cubic metre	mol/m^3	Concentration
newton	N	Force
newton metre	N·m	Moment of force
newton per metre	N/m	Surface tension
ohm	Ω V/A	Electric resistance
pascal	Pa N/m^2	Pressure, stress
pascal second	PA·s	Dynamic viscosity
radian	rad	Plane angle
radian per second	rad/s	Angular velocity
radian per second squared	rad/s^2	Angular acceleration
second	s	Time or time interval
siemens	S A/V	Electric conductance (1/electric resistance)
sievert	Sv	Dose equivalent (index)
square metre	m^2	Area
steradian	sr	Solid angle
tesla	T Wb/m^2	Magnetic flux density
volt	V W/A	Electrical potential or potential difference, electromotive force
volt per metre	V/m	Electric field strength

(Continued)

Table I (Continued)

SI units and SI unit symbols		
SI unit name	*SI unit symbol*	*Quantity measured*
watt	W J/s	Power
watt per metre kelvin	W/(m·K)	Thermal conductivity
watt per square metre	W/m^2	Power density, heat flux density, irradiance
watt per square metre steradian	$W \cdot m^{-2} \cdot sr^{-1}$	Radiance
watt per steradian	W/sr	Radiant intensity
weber	Wb V·s	Magnetic flux

Annex D IPX coding

The IP code (or 'Ingress Protection') ratings are an internationally recognised, protection classification system which was created by the International Electrotechnical Commission (IEC), and published as a British, European and International Standard, namely, BS EN 60529:1992.

This particular standard describes, classifies and rates the degree of protection provided by mechanical casings and electrical enclosures against intrusion, dust, moisture and accidental contact by electrical enclosures and mechanical casings.

It is particularly useful in the electrical engineering and manufacturing world, because it gives us a clue as to the types of equipment, we can use in difficult environments.

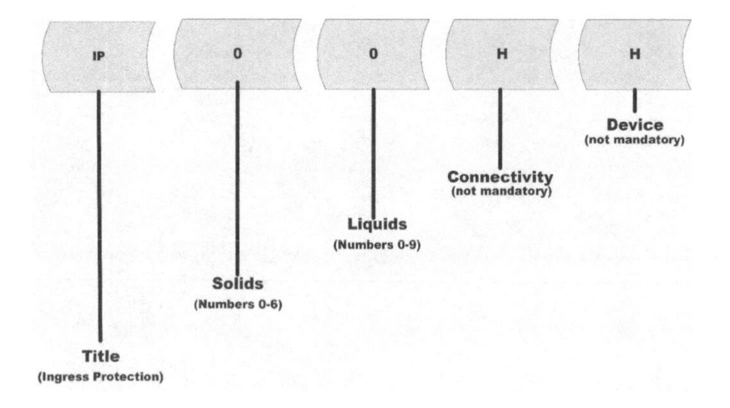

D.1 First digit: Solids

The first digit shows the level of protection against solid foreign objects.

Level	*Object size protected against*	*Effective against*
0	Not protected	No protection against contact and ingress of objects
1	>50 mm	Any large surface of the body, such as the back of the hand, but no protection against deliberate contact with a body part.
2	>12.5 mm	Fingers or similar objects.
3	>2.5 mm	Tools, thick wires, etc.
4	>1 mm	Most wires, screws, etc.
5	Dust protected	Ingress of dust is not entirely prevented, but it must not enter in sufficient quantity to interfere with the satisfactory operation of the equipment; complete protection against contact.
6	Dust tight	No ingress of dust; complete protection against contact.

D.2 Second digit: Liquids

The second digit shows the level of protection against liquids

Level	*Effective against*
0	Non protected
1	Protected against dripping water
2	Protected against dripping water when sited up to 15″
3	Protected against spraying water
4	Protected against splashing water
5	Protected against water jets
6	Protected against heavy seas
7	Protected against the effects of immersion
8	Protected against immersion I water (under pressure)
9	Protected against high temperature water jets

D.3 Optional 3rd and 4th digits

In addition to the tables above, additional letters are also available to provide further related information concerning the protection of the device.

D.3.1 3rd digit

The 3rd digit is available to describe what particular part of the device or circuit needs to be protected.

Level	*Effectivity*
A	Back of hand
B	Finger
C	Tool
D	Wire

D.3.2 4th digit

The 4th digit is available to describe the strength and mobility of the device and the weather.

Level	*Effectivity*
F	Oil resistance
H	High voltage device
M	Device moving during a water test
S	Device standing still during a water test
W	Weather conditions

D.4 IP rating reference chart

Below is an easy to follow reference chart to help you decide which IP rating you require or already have.

IP number	*First digit – solids*	*Second digit – liquids*
IP00	Not protected from solids	Not protected from liquids
IP01	Not protected from solids	Protected from condensation
IP02	Not protected from solids	Protected from water spray less than 15 degrees from vertical
IP03	Not protected from solids	Protected from water spray less than 60 degrees from vertical
IP04	Not protected from solids	Protected from water spray from any direction
IP05	Not protected from solids	Protected from low pressure water jets from any direction
IP06	Not protected from solids	Protected from high pressure water jets from any direction
IP07	Not protected from solids	Protected from immersion between 15 centimetres and 1 metre in depth
IP08	Not protected from solids	Protected from long-term immersion up to a specified pressure
IP10	Protected from touch by hands greater than 50 millimetres	Not protected from liquids
IP11	Protected from touch by hands greater than 50 millimetres	Protected from condensation
IP12	Protected from touch by hands greater than 50 millimetres	Protected from water spray less than 15 degrees from vertical
IP13	Protected from touch by hands greater than 50 millimetres	Protected from water spray less than 60 degrees from vertical
IP14	Protected from touch by hands greater than 50 millimetres	Protected from water spray from any direction
IP15	Protected from touch by hands greater than 50 millimetres	Protected from low pressure water jets from any direction

(Continued)

IP number	*First digit – solids*	*Second digit – liquids*
IP16	Protected from touch by hands greater than 50 millimetres	Protected from high pressure water jets from any direction
IP17	Protected from touch by hands greater than 50 millimetres	Protected from immersion between 15 centimetres and 1 metre in depth
IP18	*Protected from touch by hands greater than 50 millimetres*	*Protected from long-term immersion up to a specified pressure*
IP20	Protected from touch by fingers and objects greater than 12 millimetres	Not protected from liquids
IP21	Protected from touch by fingers and objects greater than 12 millimetres	Protected from condensation
IP22	Protected from touch by fingers and objects greater than 12 millimetres	Protected from water spray less than 15 degrees from vertical
IP23	Protected from touch by fingers and objects greater than 12 millimetres	Protected from water spray less than 60 degrees from vertical
IP24	Protected from touch by fingers and objects greater than 12 millimetres	Protected from water spray from any direction
IP25	Protected from touch by fingers and objects greater than 12 millimetres	Protected from low pressure water jets from any direction
IP26	Protected from touch by fingers and objects greater than 12 millimetres	Protected from high pressure water jets from any direction
IP27	Protected from touch by fingers and objects greater than 12 millimetres	Protected from immersion between 15 centimetres and 1 metre in depth
IP28	Protected from touch by fingers and objects greater than 12 millimetres	Protected from long-term immersion up to a specified pressure
IP30	Protected from tools and wires greater than 2.5 millimetres	Not protected from liquids
IP31	Protected from tools and wires greater than 2.5 millimetres	Protected from condensation

(Continued)

IP number	*First digit – solids*	*Second digit – liquids*
IP32	Protected from tools and wires greater than 2.5 millimetres	Protected from water spray less than 15 degrees from vertical
IP33	Protected from tools and wires greater than 2.5 millimetres	Protected from water spray less than 60 degrees from vertical
IP34	Protected from tools and wires greater than 2.5 millimetres	Protected from water spray from any direction
IP35	Protected from tools and wires greater than 2.5 millimetres	Protected from low pressure water jets from any direction
IP36	Protected from tools and wires greater than 2.5 millimetres	Protected from high pressure water jets from any direction
IP37	Protected from tools and wires greater than 2.5 millimetres	Protected from immersion between 15 centimetres and 1 metre in depth
IP38	Protected from tools and wires greater than 2.5 millimetres	Protected from long-term immersion up to a specified pressure
IP40	Protected from tools and small wires greater than 1 millimetre	Not protected from liquids
IP41	Protected from tools and small wires greater than 1 millimetre	Protected from condensation
IP42	Protected from tools and small wires greater than 1 millimetre	Protected from water spray less than 15 degrees from vertical
IP43	Protected from tools and small wires greater than 1 millimetre	Protected from water spray less than 60 degrees from vertical
IP44	Protected from tools and small wires greater than 1 millimetre	Protected from water spray from any direction
IP45	Protected from tools and small wires greater than 1 millimetre	Protected from low pressure water jets from any direction
IP46	Protected from tools and small wires greater than 1 millimetre	Protected from high pressure water jets from any direction

(Continued)

IP number	*First digit – solids*	*Second digit – liquids*
IP47	Protected from tools and small wires greater than 1 millimetre	Protected from immersion between 15 centimetres and 1 metre in depth
IP48	Protected from tools and small wires greater than 1 millimetre	Protected from long-term immersion up to a specified pressure
IP50	Protected from limited dust ingress	Not protected from liquids
IP51	Protected from limited dust ingress	Protected from condensation
IP52	Protected from limited dust ingress	Protected from water spray less than 15 degrees from vertical
IP53	Protected from limited dust ingress	Protected from water spray less than 60 degrees from vertical
IP54	Protected from limited dust ingress	Protected from water spray from any direction
IP55	Protected from limited dust ingress	Protected from low pressure water jets from any direction
IP56	Protected from limited dust ingress	Protected from high pressure water jets from any direction
IP57	Protected from limited dust ingress	Protected from immersion between 15 centimetres and 1 metre in depth
IP58	Protected from limited dust ingress	Protected from long term immersion up to a specified pressure
IP60	Protected from total dust ingress	Not protected from liquids
IP61	Protected from total dust ingress	Protected from condensation
IP62	Protected from total dust ingress	Protected from water spray less than 15 degrees from vertical
IP63	Protected from total dust ingress	Protected from water spray less than 60 degrees from vertical

(Continued)

IP number	*First digit – solids*	*Second digit – liquids*
IP64	Protected from total dust ingress	Protected from water spray from any direction
IP65	Protected from total dust ingress	Protected from low pressure water jets from any direction
IP66	Protected from total dust ingress	Protected from high pressure water jets from any direction
IP67	Protected from total dust ingress	Protected from immersion between 15 centimetres and 1 metre in depth
IP68	Protected from total dust ingress	Protected from long term immersion up to a specified pressure
IP69	Protected from total dust ingress	Protected from steam-jet cleaning

Annex E
Acronyms and abbreviations

a.c.	Alternating Current
ACS	Assembly for Construction Sites
ADS	Automatic Disconnection of Supply
AFDD	Arc Fault Detection Devices
Band I	Extra-low voltage
Band II	Low voltage
BEC	British Electrotechnical Committee
BS	British Standard
BSI	British Standards Institution
CAD	Computer Aided Design
CEN	Comité Européen de Normalisation
CENELEC	Comité Européen de Normalisation Electrotechnique
CHP	Combined Heat & Power Generation
CPC	Circuit Protective Conductor
CRT-D	Cardiac Resynchronization Therapy Defibrillator
CPS	Control and Protective Switching Device
d.c.	Direct Current
DCL	Device for Connecting a Luminaire
DDA	Disability Discrimination Act
DED	Disability Equality Duty
DIY	Do It Yourself
EAS	Electrotechnical Assessment Scheme
ECA	Electrical Contractors Association
ECG	Electrocardiogram
EEBAD or EEBADS	Earthed Equipotential Bonding and Automatic Disconnection of Supply

EIC	Electrical Installation Certificate
EICR	Electrical Installation Condition Report
ELV	Extra Low Voltage
EMC	Electromagnetic Compatibility
emf	Electromotive force
EMI	Electromagnetic Interference
EN	European Normalisation
ESQCR	Electricity Safety, Quality and Continuity Regulations
EU	European Union
EV	Electric Vehicle
EVSE	Electric Vehicle Supply Equipment
EWR	Electricity at Work Regulations
FE	Functional Earth
FELV	Functional Extra Low Voltage
HSE	Health & Safety Executive
HTM	Health Technical Memorandum
IEC	International Electrotechnical Commission
IET	Institution of Engineering and Technology
IFLS	Insulation fault location system
ILU	Integrated Logistic Unit
IMD	Insulation Monitoring Device
ISO	International Standards Organisation
IT	Information Technology
ITCZ	International Conveyance Zone
LPG	Liquefied Petroleum Gas
LPS	Lightening Protection System
LUR	Logical User Requirement
MDD	Medical Devices Directive
ME	Medical Electrical
MEIC	Minor Electrical Installation Works Certificate
MET	Main Earthing Terminal
MMI	Man Machine Interface
MTBF	Mean Time Between Failures
NICEIC	National Inspection Council for Electrical Installation Counselling
OJT	On-the-Job-Training
OPSI	Office of Public Sector Information
PAT	Portable Appliance Testing
PE	Protective Earth

PELV	Protective Extra Low Voltage
PEN	Combined Protective and Neutral conductors
PHEV	Plug-in Hybrid
PME	Protective Multiple Earthing
PRCD	Portable RCD, usually an RCD incorporated into a plug
PV	Photovoltaic
PVC	PolyVinyl Chloride
RAM	Reliability, Availability and Maintainability
RCBO	An RCCD which includes overcurrent protection
RCCB	Residual Current Operated Circuit Breaker with integral overcurrent protection
RCCD	Residual Current Operated Circuit Breaker
RCD	Residual Current Device
RCM	Residual Current Monitor
SELV	Separated Extra Low Voltage
SPD	Surge Protective Device
SRCBO	Socket outlet incorporating an RCBO and RCD tester
SRCD	Socket outlet incorporating an RCD
TOM	Temporary Overvoltages
TQM	Total Quality Management
UPS	Uninterruptible Power Supply
VSD	Variable Speed Drive
WAUILF	Workplace Applied Uniform Indicated Low Frequency (application)
YFR	Yearly Forecast Rationale

Annex F

Other books associated with the Wiring Regulations

Title	*Extracts from book reviews*	*ISBN*
Wiring Regulations in Brief (Fourth Edition) 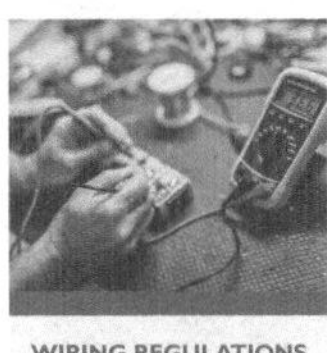 	This newly updated edition of Wiring Regulations in Brief provides a user-friendly guide to the newest amendments to BS 7671 and the IET Wiring Regulations. Topic-based chapters link areas of working practice – such as earthing, cables, installations, testing and inspection, and special locations – with the specifics of the Regulations themselves. This allows quick and easy identification of the official requirements relating to the situation in front of you. The requirements of the regulations, and of related standards, are presented in an informal, easy-to-read style to remove confusion. Packed with useful hints and tips, and highlighting the most important or mandatory requirements, this book is a concise reference on all aspects of the eighteenth edition of the IET Wiring Regulations. This handy guide provides an on-the-job reference source for electricians, designers, service engineers, inspectors, builders, and students.	 Paperback – ISBN 9780367431983 Hardback – ISBN 9780367432010 e-Book – ISBN 9781003001829

(Continued)

Title	Extracts from book reviews	ISBN
Building Regulations in Brief (Tenth Edition)	This Tenth edition of the most popular and trusted guide reflects all the latest amendments to the Building Regulations, planning permission and the Approved Documents in England and Wales. This includes coverage of the new Approved Document P on security, and a second part to Approved Document M which divides the regulations for 'dwellings' and 'buildings other than dwellings'. A new chapter has been added to incorporate these changes and to make the book more user-friendly. Giving practical information throughout on how to work with (and within) the Regulations, this book enables compliance in the simplest and most cost-effective manner possible. The no-nonsense approach of Building Regulations in Brief cuts through any confusion and explains the meaning of the Regulations. Consequently, it has become a favourite for anyone in the building industry or studying, as well as those planning to have work carried out on their home.	Routledge Taylor & Francis Group Paperback – ISBN 9781032007618 Hardback – ISBN 9780367774233 e-Book – ISBN 9781003175483
Building Regulations Pocket Book (Second Edition)	This handy guide provides you with all the information you need to comply with the UK Building Regulations and Approved Documents. On site, in the van, in the office, wherever you are, this is the book you'll refer to time and time again to double check the regulations on your current job. This book is essential reading for all building contractors and sub-contractors, site engineers, building engineers, building control officers, building surveyors, architects, construction site managers and DIYers. Homeowners will also find it useful to understand what they are responsible for when they have work done on their home (ignorance of the regulations is no defence when it comes to compliance!)	Routledge Taylor & Francis Group Paperback – ISBN 9780367774172 Hardback – ISBN 9781032003566 e-Book – ISBN 9781003173786

(Continued)

Title	*Extracts from book reviews*	*ISBN*
Scottish Building Standards in Brief (First Edition) 	Scottish Building Standards in Brief takes the highly successful formula of Ray Tricker's Building Regulations in Brief and applies it to the requirements of the Building (Scotland) Regulations 2004. With the same no-nonsense and simple to follow guidance but written specifically for the Scottish Building Standards it's the ideal book for builders, architects, designers and DIY enthusiasts working in Scotland. The book explains the meaning of the regulations, their history, current status, requirements, associated documentation and how local authorities view their importance, and emphasises the benefits and requirements of each one. There is no easier or clearer guide to help you to comply with the Scottish Building Standards in the simplest and most cost-effective manner possible.	Routledge Taylor & Francis Group Paperback – ISBN 9780750685580 Hardback – ISBN 9781138162365 e-Book – ISBN 9780080942513
Water Regulations In Brief (First Edition) 	Water Regulations in Brief is a unique reference book, providing all the information needed to comply with the regulations, in an easy to use, full colour format. Crucially, unlike other titles on this subject, this book doesn't just cover the Water Regulations, it also clearly shows how they link in with the Building Regulations, Water Bylaws **AND** the Wiring Regulations, providing the only available complete reference to the requirements for water fittings and water systems. Structured in the same logical, time saving way as the author's other bestselling ' in Brief' books, Water Regulations in Brief will be a welcome change to anyone tired of wading through complex, jargon heavy publications in search of the information they need to get the job done.	Routledge Taylor & Francis Group Paperback – ISBN 9781856176286 Hardback – ISBN 9781138408661 e-Book – ISBN 9780080950945

(Continued)

Title	*Extracts from book reviews*	*ISBN*
Quality Management Systems A Practical Guide to Standards Implementation (First Edition) 	This book provides a clear, easy to digest overview of Quality Management Systems (QMS). Critically, it offers the reader an explanation of the International Standards Organization's (ISO) requirement that in future all new and existing Management Systems Standards will need to have the same high-level structure, commonly referred to as Annex SL, with identical core text, as well as common terms and definitions. In addition to explaining what Annex SL entails, this book provides the reader with a guide to the principles, requirements and interoperability of Quality Management System standards, how to complete internal and external management reviews, third-party audits and evaluations, as well as how to become an ISO Certified Organisation once your QMS is fully established. As a simple and straightforward explanation of QMS Standards and their current requirements, this is a perfect guide for practitioners who need a comprehensive overview to put theory into practice, as well as for undergraduate and postgraduate students studying quality management as part of broader Operations and Management courses.	Routledge Taylor & Francis Group Paperback – ISBN 9780367223533 Hardback – ISBN 9780367223519 e-Book – ISBN 9780429274473
ISO 9001:2015 In Brief (Fourth Edition) 	ISO 9001: 2015 In Brief provides an introduction to quality management systems for students, newcomers and busy executives, with a user friendly, simplified explanation of the history, the requirements and benefits of the new standard. This short, easy-to-understand reference tool also helps organisations to quickly set up an ISO 9001:2015 compliant Quality Management System for themselves at minimal expense and without high consultancy fees.	Routledge Taylor & Francis Group Paperback – ISBN 9781138025868 Hardback – ISBN 9781138025851 e-Book – ISBN 9781315774831

(Continued)

Title	*Extracts from book reviews*	*ISBN*
ISO 9001:2015 for Small Businesses (Sixth Edition) 	Small businesses face many challenges today, including the increasing demand by larger companies for ISO 9001compliance, a challenging task for any organisation and in particular for a small business without quality assurance experts on its payroll. Ray Tricker has already guided hundreds of businesses through to ISO accreditation, and this sixth edition of his life-saving ISO guide provides all you need to meet the new 2015 standards. This edition includes an example of a complete, generic Quality Management System consisting of a Quality Manual plus a whole host of Quality Processes, Quality Procedures and Word Instructions; **AND** access to a **FREE**, software copy of these generic QMS files to give you a starting point from which to develop your own documentation.	Routledge Taylor & Francis Group Paperback – ISBN 9781138025868 Hardback – ISBN 9781138025820 e-Book – ISBN 9781315774855
ISO 9001:2015 Audit Procedures (Fourth Edition) 	Revised and fully updated, ISO 9001:2015 Audit Procedures describes the methods for completing management reviews and quality audits and describes the changes made to the standards for 2015 and how they are likely to impact on your own audit procedures. ISO 9001:2015 Audit Procedures is for auditors of small businesses looking to complete a quality audit review for the 2015 standards. This book will also prove invaluable to all professional auditors completing internal, external and third-party audits. The book also includes access to a **FREE**, software copy of these generic ISO 9001:2015 audit files to give you a starting point from which to develop your own Audit Procedures.	Routledge Taylor & Francis Group Paperback – ISBN 9781138025899 Hardback – ISBN 9781138025882 e-Book – ISBN 9781315774817

(Continued)

Title	*Extracts from book reviews*	*ISBN*
"Environmental Requirements for Electromechanical and Electronic Equipment" (First Edition) 	This book contains background guidance, typical ranges, details of recommended test specifications, case studies and regulations covering the environmental requirements required by designers and manufacturers of electrical and electromechanical equipment worldwide. The implementation of the EMC directive is just one aspect of the requirements placed upon manufacturers and designers of electrical equipment. Factors that must be taken into account include temperature, solar radiation, humidity, pressure, weather and the effects of water and salt, pollutants and contaminants, mechanical stresses and vibration, ergonomic considerations, electrical safety including EMC, reliability and performance.	 Hardback – ISBN: 9780750639026 e-Book – ISBN: 9780080505817

Index

Note: Locators in **bold** and *italics* indicate tables and figures in the text